AF622000

TURBULENCE: NEW APPROACHES

TURBULENCE: NEW APPROACHES

O.M. Belotserkovskii, A.M. Oparin
and V.M. Chechetkin

Viva Books

New Delhi | Mumbai | Chennai | Kolkata | Bangalore | Hyderabad | Kochi

Copyright © O.M. Belotserkovskii, A.M. Oparin, V.M. Chechetkin

First Indian Edition 2008

VIVA BOOKS PRIVATE LIMITED

- 4737/23 Ansari Road, Daryaganj, New Delhi 110 002
  E-mail: vivadelhi@vivagroupindia.net, Tel. 42242200
- Plot No. 76, Service Industries, Shirvane, Sector 1, Nerul, Navi Mumbai 400 706
  E-mail: vivamumbai@vivagroupindia.net, Tel. 27721273, 27721274
- Jamals Fazal Chambers, 26 Greams Road, Chennai 600 006
  E-mail: vivachennai@vivagroupindia.net, Tel. 28290304, 28294241
- B-103, Jindal Towers, 21/1A/3 Darga Road, Kolkata 700 017
  E-mail: vivakolkata@vivagroupindia.net, Tel. 22836381, 22816713
- 7, Sovereign Park Apartments, 56-58, K. R. Road, Basavanagudi, Bangalore 560 004
  E-mail: vivabangalore@vivagroupindia.net, Tel. 26607409, 26607410
- 101-102 Mughal Marc Apartments, 3-4-637 to 641, Narayanguda, Hyderabad 500 029
  E-mail: vivahyderabad@vivagroupindia.net, Tel. 27564481, 27564482
- First Floor, Beevi Towers, SRM Road, Kaloor, Kochi 682 018, Kerala
  E-mail: vivakochi@vivagroupindia.net, Tel: 0484-2403055, 2403056

www.vivagroupindia.com

Published by arrangement with

Cambridge International Science Publishing
7, Meadow Walk, Great Abington
Cambridge CB1 6AZ, UK

All rights reserved. No part of this book may be reproduced or utilized in any form or by any means, electronic or mechanical, including photocopying, recording or by any information storage or retrieval systems, without permission in writing from the publisher.

This edition is for sale in the Indian subcontinent only. Not for export elsewhere.

ISBN-10: 81-309-0942-1
ISBN-13: 978-81-309-0942-4

Published by Vinod Vasishtha for Viva Books Private Limited, 4737/23, Ansari Road, Daryaganj, New Delhi-110 002. Printed and bound by Anand Sons, Delhi - 110 092.

TURBULENCE: NEW APPROACHES

O.M. Belotserkovskii, A.M. Oparin
and V.M. Chechetkin

Viva Books

New Delhi | Mumbai | Chennai | Kolkata | Bangalore | Hyderabad | Kochi

Copyright © O.M. Belotserkovskii, A.M. Oparin, V.M. Chechetkin

First Indian Edition 2008

VIVA BOOKS PRIVATE LIMITED

- 4737/23 Ansari Road, Daryaganj, New Delhi 110 002
  E-mail: vivadelhi@vivagroupindia.net, Tel. 42242200
- Plot No. 76, Service Industries, Shirvane, Sector 1, Nerul, Navi Mumbai 400 706
  E-mail: vivamumbai@vivagroupindia.net, Tel. 27721273, 27721274
- Jamals Fazal Chambers, 26 Greams Road, Chennai 600 006
  E-mail: vivachennai@vivagroupindia.net, Tel. 28290304, 28294241
- B-103, Jindal Towers, 21/1A/3 Darga Road, Kolkata 700 017
  E-mail: vivakolkata@vivagroupindia.net, Tel. 22836381, 22816713
- 7, Sovereign Park Apartments, 56-58, K. R. Road, Basavanagudi, Bangalore 560 004
  E-mail: vivabangalore@vivagroupindia.net, Tel. 26607409, 26607410
- 101-102 Mughal Marc Apartments, 3-4-637 to 641, Narayanguda, Hyderabad 500 029
  E-mail: vivahyderabad@vivagroupindia.net, Tel. 27564481, 27564482
- First Floor, Beevi Towers, SRM Road, Kaloor, Kochi 682 018, Kerala
  E-mail: vivakochi@vivagroupindia.net, Tel: 0484-2403055, 2403056

www.vivagroupindia.com

Published by arrangement with

Cambridge International Science Publishing
7, Meadow Walk, Great Abington
Cambridge CB1 6AZ, UK

All rights reserved. No part of this book may be reproduced or utilized in any form or by any means, electronic or mechanical, including photocopying, recording or by any information storage or retrieval systems, without permission in writing from the publisher.

This edition is for sale in the Indian subcontinent only. Not for export elsewhere.

ISBN-10: 81-309-0942-1
ISBN-13: 978-81-309-0942-4

Published by Vinod Vasishtha for Viva Books Private Limited, 4737/23, Ansari Road, Daryaganj, New Delhi-110 002. Printed and bound by Anand Sons, Delhi - 110 092.

Contents

Chapter 1

Fundamental aspects of the direct numerical modelling of free turbulence and hydrodynamic instabilities

1.1. INTRODUCTION

During the last century, starting with Reynolds, the theory of turbulence has been developed as a superposition of flows around the some mean distribution of the structure, in particular the statistical theory of turbulence – on the calculation of the evolution of correlation integrals (see: A.S. Monin. A.M. Yaglom: 'Statistical mechanics of liquid'). The classic conclusion of these assumptions is the energy spectrum in respect of Kolmogorov–Obuchov frequencies.

Recently, special attention has been paid to large scales. It has become clear that, possibly, there is not universal spectrum of turbulence, general for all flow conditions. It is necessary to de-examine the process of turbulisation of flows, paying special attention to the large scales. It has become clear (se: O.M. Belo tserkovskii, A.M. Oparin: 'Numerical experiments in turbulence') that the controlling role in turbulence is played by large structures. They contain a large part of the kinetic energy of turbulence, although the energy itself may represent some percent of the total energy of the flow.

One of the problems of developing the turbulence theory is the relationship to the statistical theory of turbulence which was constructed on the basis of averaging solutions of linearised hydrodynamic equations and subsequent derivation of equations for the average characteristics of the flow. In [148] it was shown that the nonlinear interactions are important for both the regimes of turbulent flow and laminar flow.

In this sense, doubts are cast on the very approach based on the calculation of the average-out characteristics of the flow from linearised equations (including correlation integrals from identical equations). The effect of viscosity on the structure of the flow has not been investigated. No attention has been given to the question of the extent by which the viscosity may be regarded as a universal quantity for different flow scales (Canuto [148]). This also relates to the Reynolds number Re, although in the statistical theory of turbulence this number is used as a parameter of the problem. In the conventional approach, examination of the process of the growth of instabilities is always regarded as a local process with expansion in respect of harmonics. In fact, the process of formation of large scales is of the non-local type. This can be indicated by the physical meaning of the Reynolds number. On small scales, the dissipative forces prevail over the dynamic forces, and the structure of the flow depends on these forces and is described by the Navier–Stokes equations. On large scales of the flow, in the case of the structures with $L > L_{cr}$, the dynamic stresses in the shear flow result in the formation of a couple of forces generating large scales in the turbulent flow.

The problem of formation of the turbulence spectrum is the main problem. Since all interactions are important, the examination of the development of turbulence may be carried out using the numerical modelling of the process of development of turbulence. However, it is necessary to construct a physical model of the development of turbulence within which we shall carry out numerical modelling.

We propose the following model of development of turbulence. When the viscosity is not capable of ensuring the profile of the shear flow, because of the development of the instabilities of the Rayleigh–Taylor type or Rossby waves, and some other processes, large scale vortices appear in the flow. The further development of the pattern of turbulence is associated with the evolution of large vortices and generation, by the vortices, of the high-frequency part of the spectrum. It is important to mentione that the energy of the large vortices is obtained from the shear flow. The vortices themselves do not break up but sometimes merge, reaching the maximum scale possible in the given problem. The generation of small vortices is associated with the interaction of large vortices with the flow or flow boundary. Within the framework of this physical model, we have also carried out the modelling of turbulent flows.

It appears to us that the principal problem of the theory of turbulence is the generation of large scales. This generation may be associated with different hydrodynamic instabilities and inclined shock waves. For example, in [152], the formation of large scales is linked with the Rank–Hilsch effect, leading to the development of Rossby vortices which in turn result in the development of large turbulent scales (Lovelace, Colgate, Nelson, 1998). This group has also investigated the development of large scales of turbulence because of the presence of a pressure gradient or entropy in the gravitation field.

The free shear flow also contains a pressure gradient. Consequently, the physical meaning of the Reynolds number Re = $VL/\nu = \Delta VL/\nu$ may be explained in the following context: turbulence develops if the pressure gradient, which depends on the speed gradient on the scale L, results in the formation of a couple of forces that is larger than the tensor of viscous stresses. Otherwise, the speed distribution of the type of Kutta flow appears. The Kelvin–Helmholtz instability operates in the same manner, generating a pair of forces.

Another possible process is the generation of large-scale turbulent vortices in the vicinity of inclined shock waves, leading to the formation of the tangential component of the speed. However, none of the discussed problems have as yet been solved, although there are exceptions: solutions are proposed in individual studies which we shall discuss in greater detail in the following sections of the book.

It was interesting to note that the Nature itself provides examples for such a physical model. Here, it is important to mention the structure of the Gulfstream flow, the red spot on the Jupiter and so on, when the large structures live for a very long time, without breaking out into smaller structures. In turn, fine formations are detected around the large structures.

The above considerations represent the problem of turbulence as a problem of the generation of large scales, with the subsequent generation of the high-frequency part of the turbulence spectrum. This formulation of the problem differs greatly from the paradigm of the pulsations of the flow around the mean distribution. In our model, the structure of the flow is controlled by the large scales.

At present, one of the important problems of the mechanics of solids and plasma physics is the modelling of complex transition and turbulent movements (including for multidimensional problems, taking into account the compressibility of the medium, in different regimes of movement in a wide range of the variation the parameters of the flow, and so on) using currently available computing methods, algorithms and applied mathematics approaches. The so-called transition flows, often encountered in practice, are characterised by the highly nonstationary nature and nonlinearity of the processes taking place, by large displacements of the medium, different types and complicated mechanisms of interaction, energy dissipation. For example, this includes the gas-dynamic problems such as trans-sonic and diffraction problems of the flow in the wake behind a moving solid, the interaction of the blown-in stream with the main flow, etc. In the mechanics of deformed solids, transition phenomena take place in, for example, examination of the elastic–viscous–plastic states. In plasma physics, these processes are observed in the numerical modelling of problems of the interaction of powerful laser radiation with matter, etc. In a number of cases (flows in a wake in the presence of a 'blown in', the processes taking place during laser compression of shells, etc), the phenomena are of turbulent nature.

The main problem of the theory of turbulence – the examination of general dynamics and the nature of development of turbulence, i.e. examination of the evolution of large-scale formations and statistical representation of the turbulent motion with time. It is usually assumed that the variation the distribution of the speeds in space with time is determined by the Navier–Stokes equations. The development and degeneration of turbulence may be analysed in principle by two methods:*

– obtaining a general solution of the Navier–Stokes equations for an arbitrary initial distribution of the speeds in space;

– to derive a unique dynamic equation, describing the variation of the total distribution of probabilities with time (the problem of representation of stochastic distribution in the functional space).

Naturally, the main characteristics of the turbulent movements – 'structural' nature of turbulence, nonstationary nature and nonlinearity of the processes taking place, the possibility of transfer of 'groups' the molecules (molar transfer), the presence of a continuous flux of energy in a cascade of vortices, different mechanisms of interaction for different scales of movement, the phenomenon of viscous dissipation, etc, – should be reflected accurately in the numerical modelling of turbulence. Generally speaking, the following three main problems can be formulated:

I – the determination of the main characteristics of the large-

---

*See, for example, [34, 96, 119]

scale turbulence without detailed examination of the structure in the limiting conditions of movement (when Re $\rightarrow \infty$);

II – the calculation of the region of transition of the laminar to turbulent flow ('breakdown' of the stable laminar flow at Re_{cr});

III – detailed examination of the the generation of large-scale turbulent structures and further appearance of chaos in the high-frequency part of the spectrum.

The development of high-productivity computers (including parallel computers) and of efficient numerical methods of solving nonlinear problems of mathematical physics and mechanics has created suitable conditions for direct numerical modelling of complicated flows of the liquid and the gas, including the turbulence phenomena. The numerical experiment, combined with the physical experiment, opens new possibilities in understanding the phenomena taking place in the nature, determining the role of various factors in these phenomena, and also makes it possible to determine more accurately and completely the boundaries of applicability of the schemes and mathematical models. Within the framework of the numerical experiment in a single complex, it is possible to examine the mathematical formulation of the problem, the optimum method of solving the problem, and the realisation process. It should be mentioned that the modelling method must be based on the physically substantiated processes. In fact, we examine here some simulation system describing adequately (to a certain degree) the main characteristics of the investigated the phenomena. In the process of calculations, it is possible to improve the accuracy of the initial formulation of the problem, and the presence of such a 'feedback' provides reliable results for comparatively short computing times.

We shall present the results of direct numerical modelling of a wide spectrum of turbulence problems, carried out on the basis of complete models without the application of semi-empirical theories. Special attention is paid here to the examination of separation (in the general case, turbulent) flows in a wake behind a solid for the limiting conditions of movement for very high Reynolds Re [1], attention is given to laminar–turbulent regimes, and also to the phenomena of transition to chaos for a number of real formulations. The main aim is to construct discrete non-equilibrium disregarding models (corresponding to the given class of flow) and the development of 'rational' numerical algorithms available for application of the current level of development of computing methods. This chapter is concerned mainly with the explanation of the general concept of direct numerical modelling of turbulence, developed by O.M. Belotserkovskii over a period of many years (see, for example [15, 16, 19], etc), realised by students and colleagues at the Institute of Automation of Design, The Moscow Physical–Technical Institute, The Computing Centre of the Russian Academy of Sciences, etc. A number of original approaches is included in this monograph (see also [17, 90]).

2. The problem of examination of the properties of separation flows in a wake behind a moving solid is associated with the examination of the free developed shear turbulence. In the examination of the given type of movement, the investigation of the average large-scale microstructures and statistical characteristics of 'intrinsic' turbulence is theoretically possible and physically substantiated. The central question, formed here, is in our view as follows: which models – the model of the ideal medium, the Navier–Stokes equations or on the kinetic level – should be used in the construction of schemes for investigating separation and turbulent flows in different motion conditions. The main concept of our approach is determined by the following considerations.

For a wide range of phenomena of this type with a high Reynolds number in the low frequency and inertia ranges of turbulent motion, the effect of molecular viscosity and small elements of the flow in a large part of the region of perturbations on the general characteristics of the macrostructures of the developed flow and on the streamline pattern is practically very small. Consequently, it is possible to ignore the effects of molecular viscosity in the examination of dynamics of large vortices and examine them on the basis of models of the ideal medium, for example, the Euler equations (with the possible application of the methods of 'rational' averaging, but without using semi-empirical models of turbulence). This includes stream problems of the flows in an accompanying wake behind the solid, the movement of ship hulls with stern shear, the formation of front stalling zones in the flow around blunt solids with streams or needles, directed against the flow, etc. At the same time, the properties of the flows in the boundary layers, narrow mixing layers, in the viscous range of turbulence, and also in the case of moderate Reynolds numbers and in the range of the laminar–turbulent transition are determined basically by molecular diffusion and, in this case, it is necessary to examine the Navier–Stokes models. Pulsation movements in turbulence (for example, in the centre of the wake) are of the unstable, irregular nature and represent a stochastic process. Here, we can talk only about the determination of the average characteristics of such a movement (moments of different order) by means of the appropriate statistical processing of the results, using, for example, kinetic approaches.

The main difficulty in the examination of this class of problems is the development of a general concept of the construction of structural numerical models of turbulence.

Special attention will be given to discussing the approaches in which the free developed turbulent motion for long time periods is described using a complete (and closed) system of dynamic equations for the true values of speed and pressure, and also statistical methods. The combined application of these approaches (based on the examination of the equations of hydrodynamics and Monte-Carlo statistical methods) makes it possible to understand more efficiently the structure of turbulence and determine the rational methods of constructing appropriate mathematical models, greatly reducing the requirements on the computer resources.

3. The numerical realisation of the general concept the turbulence theory is carried out in different directions (see, for example, [34, 121], etc):

– integration of the complete nonstationary Navies–Stokes equations without additional assumptions on the nature of transfer (finite-difference approaches, spectral Fourier methods*, etc, used for the comparatively simple problems of convection, diffusion, in the modelling of the breakdown of the laminar regime, etc);

– the calculation of these models on a coarser net using semi-empirical variable coefficients of transfer (effective viscosity, etc);

– the solution of the Reynolds or Bussinesq equations for the mean elements of motion and the Reynolds stresses together with the approximate transfer equations;

– the application of differential equations for the moments of different order and with different types of 'closing', etc.

The main difficulties, formed in the direct examination of the free developed turbulent flows in the accompanying wake, streams,

---

*The spectral method is based on the possibility of approximating any random function by the sum of non-correlated harmonic functions. By averaging the Navier–Stokes equation, we obtain a non-closed system of equations for the spectral plane which describes, in particular, the time dependence of the energy content of the vortices with different wave numbers. Because the mechanism of energy exchange between the vortices has not as yet been completely explained, different hypotheses are used to form a closed system.

are associated with the fact that these flows are characterised by the scales of excited motion in the range of several orders of magnitude. The spectrum of the variation of the wave numbers in the real turbulent break-away flows reaches 425 orders of magnitude so that in foreseable future it will not be possible to construct a general model of turbulence for large Reynolds numbers without introducing some physical concept of the model.

One of the approaches proposed for the 'softening' the problem is that we should restrict ourselves to the direct examination of the three-dimensional nonstationary turbulent flow only on scales exceeding some given dimension h^* (where, for example, $h^*$is the step of the calculation net). This approach has also physical substantiation. For example, the effect of the high-frequency part on the large scales of the turbulent spectrum is almost negligible. The problem of the annihilation of the kinetic energy of turbulent motion for the high-frequency part of the spectrum may be modelled by the energy sink or by the development of subgrid procedures. The scales of the vortices ($h < h^*$) for which the direct solution is not possible, are modelled as subgrid turbulence with the application of the vortex viscosity or other 'rational' approximation of the transfer processes (Fig. 1.1). This approach is associated with the hope that the fine-scale structure of turbulence has no effect on the structure of turbulence. In turn, fine-scale structures may be almost universal for different problems (in particular, for high wave numbers k: $k/k_K > 10^{-2}$, where k_K is the wave number, corresponding to the Kolmogorov scale, and, consequently, the accurate sultion is not required here). This shows, in particular, that the effect of the high-frequency part of the spectrum on large structures is insignificant and this effect may be ignored or taken into account only very approximately. (As an example, we present Fig. 1.1 from [121] showing the spectrum of energy density $E_1(k)$ of pulsations of the longitudinal components of the speed for different turbulent flows; $\int_0^\infty E_1(k)dk = \langle u'^2 \rangle$, $\Phi = E_1\left(\Theta\nu^3\right)^{-1/4}$ is the dimensionless spectral density, Θ is the local rate of dissipation of energy per mass unit; ν is kinematic viscosity, $k_K = (\Theta\nu^3)^{-1/4}$ is the Kolmogorov wave number). The structure of the large vortices containing the part of turbulence which greatly changes in transition from one flow to another or from one group of the conditions to another, is investigated directly in this case (in principle, this concept is also developed below). Another method of direct

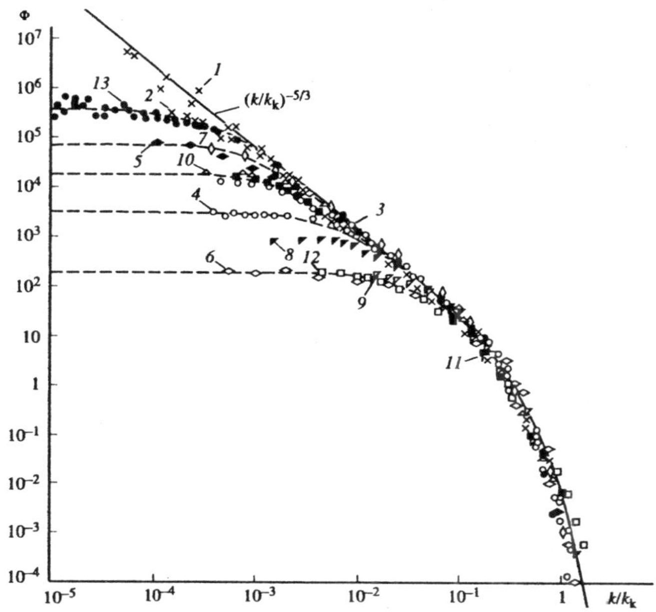

Fig.1.1. Spectra of energy density $E(k)$ of longitudinal component fluctuations for various flows; $\Phi = E_1(\Theta/\eta^3)^{-1/4}$, k is the wave number. (1) $Re_\lambda = 2000$, a tide tank, $Re_\delta \sim 10$; (2) $Re_\lambda = 780$, a round jet; (3) $Re_\lambda = 170$, a tube flow, $Re_\lambda = 5 \cdot 10^5$; (4) $Re_\lambda = 130$, constant shear flow; (5) $Re_\lambda = 380$, the wake behind a cylinder; (6) $Re_\lambda = 23$, the wake behind a cylinder; (7) $Re_\lambda = 540$, turbulence downstream of a honeycomb; (8) $Re_\lambda = 72$, turbulence downstream of a honeycomb; (9) $Re_\lambda = 37$, turbulence downstream of a honeycomb; (10) $Re_\lambda = 23$, boundary layer; $y_\delta = 0.5$, $Re_\delta = 3.1 \cdot 10^5$; (11) $Re_\lambda = 28$, boundary layer, $y_\delta = 0.22$, $Re_\delta = 5.6 \cdot 10^5$; (12) $Re_\lambda = 23$, $y_\delta = 1.2$, $Re_\delta = 3.1 \cdot 10^5$; (13) $Re_\lambda = 850$, the same, above water, $y_\delta = 0.6$, $Re_\delta = 4.0 \cdot 10^5$.

examination is the approach based on the 'vortex dynamics' (or 'discrete vortices') in which the turbulent zone is modelled by a set of many discrete non-viscous vortices with the evolution tracked in respect of time [31,125]. This approach is very promising because the perturbation has always a finite amplitude and a finite wavelength.

4. We are now facing the problem of constructing structural and appropriate numerical algorithms for the solution of the three problems described previously. We shall start with the problem of examination of the large-scale formations at high values of Re. The procedures of examination of freely developed turbulence, based on

the Navier–Stokes equations, are closely linked, as is well-known, with the examination of the solutions of these equations (in particular, with the appearance of stochasticity) at low molecular viscosity (we shall not discuss the justification and possibility of this formulation, see also [89]). Here, we shall use the following hypotheses proposed by Reynolds:

1. At $Re > Re_{cr}$, the laminar motion is unstable;
2. At $Re > Re_{cr}$, the random movements should be described by the Navier–Stokes equations;
3. This system, or its solutions, should be averaged-out in respect of some unknown parameters by means of some (also unknown) function;
4. The averaged-out system of equations is not closed (some model of closure is required).

In this formulation of the problem, the main difficulty in the modelling of large-scale turbulence is the construction, for the high supercritical Reynolds numbers, of a nonstationary stable (for averaged-out characteristics) numerical solution, adequate to the Navier–Stokes equations. In this case, to ensure the conditions of approximation and stability of the solution in the calculations, the step of the grid should be such as to ensure that the error of approximation of the convective members in the Navier–Stokes equations is considerably smaller than the difference representations of the viscosity numbers. The estimates for the modelling equations show [155] that the approximation to the real solution should be achieved under the condition $\mathrm{Re}\, h < \alpha \ll 1$, i.e. the calculation of flows with the molecular mechanism of dissipation for high ('turbulent') values of the Reynolds numbers, and if this can be carried out, then only on very fine difference grids, comparable, generally speaking, with the minimal size of pulses.

It should be added that the fraction of molecular viscosity (ν) in the 'effective' turbulent viscosity (ν_t), formed in turbulent exchange, is very small (usually $\nu_t/\nu \approx 10^{4\div 6}$). In addition to this, in examination of the problems of turbulence, it is essential to investigate the three-dimensional (in respect of spatial variables) nonstationary Navier–Stokes equations. In this case, the solution of these problems at high Reynolds numbers may be non-unique (although it has been confirmed that the nonstationary one-dimensional and two-dimensional problems for the Navier–Stokes equations have always a unique solution in the entire time period $t > 0$). Here, the conditions of formation of turbulence should evidently correspond to the conditions of non-uniqueness of the solutions of the three-dimensional nonstationary Navier–Stokes equations and, generally speaking, the problems of the relationship between the Navier–Stokes equations and turbulence remain open (see, for example, [89], etc). It should be mentioned that the structure of the Navier–Stokes equations leads to large numerical effects in the numerical modelling of turbulence at high Reynolds numbers. In particular, this shows that in direct formulation, the problem of the detailed examination on the basis of the Navier–Stokes model of developed large-scale turbulent structures in the case of very high Reynolds numbers is evidently difficult to carry out, even when using the most powerful computers.[1]

1.2. PROBLEM OF 'RATIONAL' AVERAGING IN NUMERICAL EXAMINATION OF TURBULENCE

1. In recent years, special attention has been paid to two directions in the numerical approaches to examination of turbulence. According to Orseg [104], these are the models of closure of small-scale movements, and direct numerical modelling.

In the first of these approaches, proposed by Deardorf [154, 160], the approximation of the processes of transfer of turbulence is carried out only on the scales of motion which are not resolved in the explicit form in the numerical approximation of the Navier–Stokes equations.[2] In this case, the small scales are interpreted by means of statistical approximation in the process of detailed examination of large scales which on its own is a strong assumption. The effect of non-resolvable small scales on the resolvable large-scales is characterised by means of the coefficient of turbulent viscosity using semi-empirical constants. In fact, in all likelihood, the transfer of the energy of large-scale structures of turbulence to heat is associated with the direct degeneration of the high-

---

[1]It is natural to assume, as correctly noted by V.V. Struminskii, that the 'main form of motion of matter – turbulent – should be described by the laws of mechanics and physics without using additional hypotheses and assumptions... The equations of mechanics and statistics yielded the Boltzmann equation, and the Boltzmann equation gave, as shown, the Navier–Stokes equations describing directly only laminar flows. The main class of turbulent motions has evidently been lost somewhere, so that it is necessary to carry out more detailed investigations' [79,114].

[2] As in direct modelling, the large scales are calculated directly.

frequency part of the spectrum by large structures with subsequent annihilation of the high-frequency part of the spectrum. The approach proposed by Smagorinskii [228] has proved to be most popular. The coefficient of turbulent viscosity ν_t, replaçing the coefficient of molecular viscosity, is selected here in the following form:

For the three-dimensional flow

$$\nu_t = (c\Delta x)^2 \left| \left(\frac{\partial v_i}{\partial x_j} + \frac{\partial v_j}{\partial x_i} \right)^2 \right|^{1/2}$$

and for the two-dimensional flow

$$\nu_t = (c'\Delta x)^3 \left| \left(\frac{\partial}{\partial x_j} (\nabla \times \mathrm{v}) \right)^2 \right|^{1/2}$$

Here Δx is the spatial resolution (step) of the grid, and c and c' are constants equal to approximately 0.1–0.2. These expressions for ν_t are suitable only in the case in which difference approximations of the first or second order of accuracy are used for spatial derivatives [104,228]. It would appear that in this approach we do not impose any direct restriction on the value of the Reynolds number, i.e. we take into account the effect of small-scale turbulence (this is also the advantage of this approach in comparison with direct numerical modelling). Actually, this is not quite the case. The satisfactory accuracy of the closing systems for the small-scale component is obtained only when the separation of the flow into small-scale and large-scale components does not have any significant influence on the evolution of the large-scale structures.

Thus, it is important the problem of the efficiency of achieving invariance of the relative distribution of the scales remains unsolved.[1]

In addition to this, the methods of taking into account the small-scale effects have a number of shortcomings: the application of arbitrary models of transfer for small-scale structures (with semi-empirical coefficients), and also neglecting all stochastic effects of

---

[1]This situation will be referred to as the 'hypothesis of invariance of distribution of scales'.

the fluctuation of the characteristics of this component on the generation of fluctuation in large-scale formations.

In the case of direct numerical modelling of turbulence on the basis of the Navier–Stokes equations with a very high Reynolds number, it is proposed to reduce artificially the Reynolds number to the value at which this flow can be modelled with sufficient accuracy in the currently available computers. In fact, it is assumed that relatively large-scale motions remain unchanged in this case, although it is clear that in this modelling flow the motions on all scales cannot change. In fact, the large-scale characteristics of turbulent flows evidently do not depend on the Reynolds number, if the boundary and initial conditions are also independent of this number [82]. There is a strong tendency for large-scale motions in turbulence to self-regulation resulting in independence of the large-scale structures on the details of the dissipation mechanism [104].

As confirmed in [104], according to the hypothesis on the independence of the macrostructure on the Reynolds number, for the motions with wave numbers $k \ll l/\lambda_0$ not only the static characteristics but also the detailed structure of such a flow are independent of the Reynolds number[2] (here $\lambda_0 = (\nu^3/\Theta)^{1/4} \approx l/\mathrm{Re}^{3/4}$ at $\mathrm{Re} \to \infty$; λ_0, l are the scales of the small and large vortices, respectively). Actually, with increase of the Reynolds number, the small-scale three-dimensional turbulence is set up in such a manner as to ensure the required rate of dissipation. This situation is characteristic of the stationary structures of turbulence, if we can talk generally about stationarity in turbulence.

In this case, the spectrum $k^2E(k)$ of the mean quadratic vorticity spreads into the region of higher and higher values k, and its

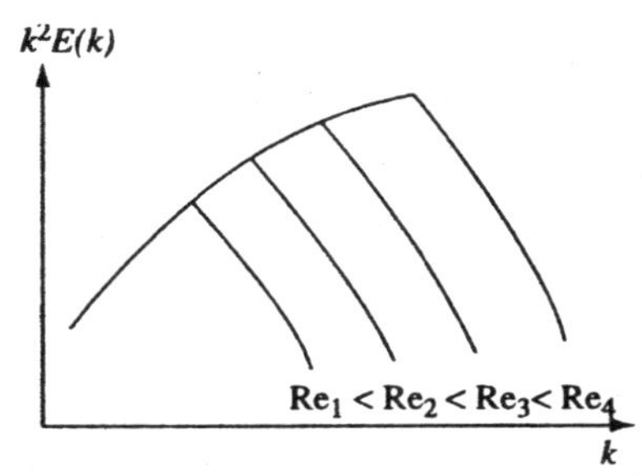

Fig. 1.2. Schematic illustration of the variation of the spectrum of energy dissipation with increase of Re: according to the hypothesis on independence on Re, the scale of motion with the wave number smaller than the maximum value on the spectrum $k^2E(k)$ (i.e. at l/λ_0), are almost completely independent of the Reynolds number [104].

[2] Here, we can consider only the statistical characteristics (see section 6 on the correctness of formulation of the problem.

behaviour in the zone of low k (macrostructure) remains almost constant (Fig. 1.2).

Thus, it may be assumed [34,82] that, regardless of the multiscale nature of turbulent transfer processes, the large-scale motion is almost independent of the Reynolds number for the scales larger than critical (Fig. 1.2). At the same time, for the small scales self-modelling of the turbulent flow is observed with respect to the Reynolds number. This means that the mean values of all quantities, determined by large-scale oscillations of velocity, are independent of the Reynolds number if $Re \to \infty$ (the hypothesis of the statistical independence of the large- and small-scale motions). This principle, generally speaking, cannot be applied to the description of the gradients of gas-dynamic parameters, because the latter are determined by small-scale fluctuations. This principle has been justified by experiments [82].

The hypothesis of the statistical independence is used in a large number of applications because it makes it possible to 'split' a large number of correlations between the macro- and microcharacteristics of turbulence at $Re \to \infty$.

It should be mentioned that the description of only large-scale oscillations of velocity cannot be close (evolution of these oscillations is also determined by the rate of generation of the small-scale component and subsequent viscous dissipation). However, if we examine the characteristics of oscillations of all scales (in this case, it is necessary to take into account the effect of molecular viscosity), then we shall be concerned, generally speaking, with the analysis of excess information (because the main features of turbulence are not independent of Re). Another approach, based on the search for universal relationships between the characteristics of small-scale and large-scale pulsations, is also possible here. As indicated by the theory proposed by Kolmogorov and Obukhov [72, 85], these relationships indeed exist if the characteristic scales of the oscillations, determining the energy of turbulence and its dissipation, greatly differ. These relationships are also a consequence of the principle of self-modelling of turbulence in respect of the Reynolds number for high values of Re [82].

2. The previously mentioned and, in our view, important considerations regarding the separation of the scales, may also be interpreted as follows. On an example of an isotropic turbulent flow, we shall examine the features of behaviour of the spectral distribution for the energy of pulsating motion $E(k)$ of the velocity field. According to the proposal by Kolmogorov ([172], etc), at

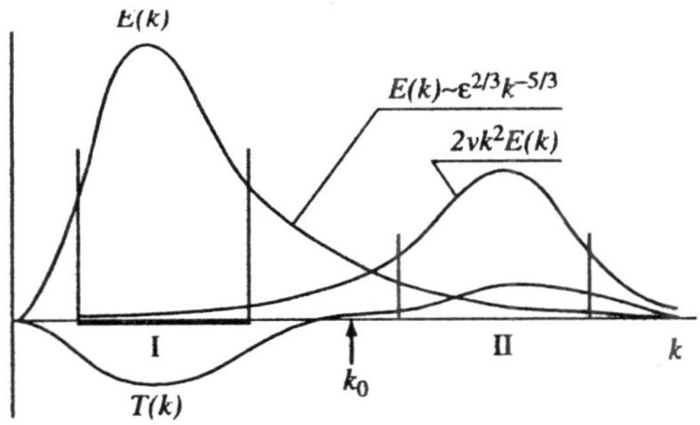

Fig.1.3. Schematic representation at high Re of the energy spectrum $E(k)$, the energy dissipation spectrum $2\nu k^2(E)$ and function $T(k)$ (I is the energy range, II is the dissipation range, the inertia range is between them) [96].

relatively large Reynolds number there should be an intermediate range of wave numbers (inertia interval) in which the energy is not produced and is not dissipated, and is only transferred to higher wave numbers.[1] Analysis of the dimensions shows that for the inertia range[2] (Fig. 1.3):

$$E(k) \sim \Theta^{2/3} k^{-5/3}$$

Thus, it may be asserted that the spectral range of energy (I) and the dissipation range (II) at a high Reynolds number $\mathrm{Re}_\lambda = \lambda \left\langle u'^2 \right\rangle^{1/2} / \nu$ greatly differ in respect of frequency (Fig.1.3), and this also confirms the 'reality' of the previously made hypotheses on the dynamic independence of the large- and small-scale of structures. We shall discuss this in greater detail in §1, Chapter 3.

The spectral form of the Karman–Horvath equation has the form:

$$\frac{\partial E(k,t)}{\partial t} = T(k,t) - 2\nu k^2 E(k,t) \tag{1.1}$$

where $\left\langle u'^2 \right\rangle = \int_0^\infty E(k)dk$ is the total energy of the pulsating motion,

---

[1]According to Karman, a similar process occurs in the boundary layer: here we have an intermediate equilibrium layer between large-scale vortex motion (where energy is produced) and the viscous region of dissipation; analysis of the dimensions gives the well-known logarithmic law for flow in the boundary layer.

[2]The problem of the universality of this law remains unsolved

$T(k,t)$ is the function determining the distribution of energy in respect of the spectrum. The function $T(k,t)$ describes the variation of the energy of the 'spectral component' of turbulence with wave number k, produced by nonlinear 'inertia members' of the hydrodynamic equations. It is important to note that this change is reduced to the repartition of energy between the individual spectral components without any change of the total energy of turbulent motion as a whole.

The last term of the righthand part of equation (1.1) describes energy dissipation under the effect of viscosity forces. This shows that viscosity results in a decrease of the kinetic energy of perturbations with the wave number k, proportional to the intensity of these perturbations multiplied by $2\nu k^2$. Thus, the energy of long-wave perturbations (with low values of k) decreases under the effect of viscosity at a considerably lower rate in comparison with the energy of short-wave perturbations. This should be so because of the proportionality of the friction force in relation to the velocity gradient.

The behaviour of function $T(k)$ (negative value at low k and positive at high k) corresponds to the assumption according to which turbulent mixing should lead to both 'joining' of turbulent perturbations and to the transfer of the energy of large-scale components of motion to the energy of small-scale vortices which use their energy already immediately for overcoming 'viscous friction'. The fact that viscosity plays a significant role only for the relatively small-scale components of motion (characterised by high

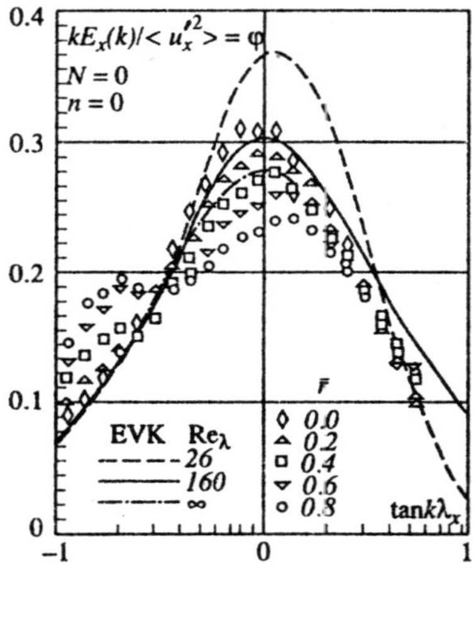

Fig.1.4. Spectral distribution in the coordinates $\varphi = kE_x(k)/\langle u_x'^2\rangle$ and $k\Lambda_x$ in the flow without rotation ($N = 0$; 0, Δ,... are experimental data obtained by the MFTI for different positions along the radius; lines of the EVKmodel for different values of Re$_\lambda$).

local velocity gradients), is expressed in the fact that in Fig.1.4 the maximum of the dissipation spectrum $2\nu k^2 E(k)$ is situated on the axis of the wave numbers far more to the right than the maximum of the energy spectrum $E(k)$ [96].

The energy range I (Fig.1.3) is characterised by the concentration of the main fraction (let us say, 80 or 90%) of total energy

$$\left\langle u'^2 \right\rangle = \int_0^\infty E(k) \quad dk \cong \int_0^{k_0} E(k)\,dk \tag{1.2}$$

and the main fraction of total dissipation is concentrated in the dissipation range II:

$$\Theta = 2\nu \int_0^\infty k^2 E(k,t) \quad dk \cong 2 \int_{k_0}^\infty k^2 E(k,t)\,dk \tag{1.3}$$

where k_0 is the intermediate value of the wave number, situated in the inertia range (behind the energy range and in front of the dissipation range). Equation (1.1) may be into integrated in respect of the entire frequency range:

$$\frac{\partial \left\langle u'^2 \right\rangle}{\partial t} = -2\nu \int_0^\infty k^2 E(k,t)\,dk \equiv -\Theta,$$

because for isotropic turbulence

$$\int_0^\infty T(k,t)\,dk = 0. \tag{1.4}$$

However, if we integrate equation (1.1) in the ranges $(0, k_0)$ and (k_0, ∞), then taking into account (1.2) and (1.6), we obtain

$$\frac{\partial \left\langle u'^2 \right\rangle}{\partial t} \cong \int_0^{k_0} T(k,t)\,dk,$$

$$0 \cong \int_{k_0}^\infty T(k,t)\,dk - \Theta,$$

but taking into account (1.4)

$$\frac{\partial \left\langle u'^2 \right\rangle}{\partial t} \cong \int_0^{k_0} T(k,t)\,dk = -\int_{k_0}^\infty T(k,t)\,dk = -\Theta. \tag{1.5}$$

Equation (1.5) indicates that the value of the rate of energy dissipation Θ is already determined within the energy range (this was utilised, in particular, in developing semiempirical Prandtl theory).

We shall now determine the order of the value of energy dissipation energy in turbulent motion [85]. This energy is taken from large-scale motion from where it is gradually transferred to smaller and smaller scales and until it dissipates in the viscous range. Although dissipation is, after all, due to viscosity, the order of Θ may be determined by means of the quantities characteristic of large-scale motion. They include density ρ, dimensions l and speed $\Delta\bar{u}$ (the variation of the mean velocity in the characteristic size of large-scale vortices l). From these quantities it is possible to produce only one combination characterised by the same dimensionality as the mean energy dissipation (in unit time in unit mass):

$$\Theta \sim \frac{(\Delta\bar{u})^3}{l}. \tag{1.6}$$

Characterising the properties of the turbulent flow by 'turbulent viscosity',

$$\nu_t \sim \Delta\bar{u}\, l,$$

we obtain

$$\Theta \sim \nu_t \left(\frac{\Delta\bar{u}}{l}\right)^2. \tag{1.7}$$

It should be mentioned that if molecular viscosity ν determines the dissipation of energy by the derivatives of the true velocity in respect of the coordinates, then ν_t links the dissociation with the gradient $\Delta\bar{u}\, l$ of the mean velocity of motion [85]. These relationships (6, 7) may be used when constructing numerical models.

Thus, the main features of the turbulent flow for high Reynolds numbers are determined in the energy-containing (non-viscous) range, i.e., in the range of long-wave perturbations (low values of k).[1] It also should be mentioned that this is the region of integral scales Λ.

3. It is also interesting to estimate the distribution of the energy range on the scale of the vortex scale (the data published by Onufriev).

The energy range $\langle u'^2 \rangle$ in respect of the spectrum is 'collected'

[1]Traditional investigations of turbulence are usually linked with the viscosity range causing large technical (and principal) difficulties in numerical examination of large-scale structures.

mainly in the range from $k\Lambda \approx 0.1$ to $k\Lambda \approx 10$. Using the generalised Karman model (EVK model) for homogeneous isotropic turbulenxe, Driscoll and Kennedy [157] obtained approximating relationships for the integral scales Λ and the Taylor (transverse) microscale λ:

$$\frac{\Lambda}{\eta} = \mathrm{Re}_\lambda^{1/2}\left[2.47 + 0.081\mathrm{Re}_\lambda^{1/2}\left(\mathrm{Re}_\lambda^{1/2} - 1\right)\right] \rightarrow 0.091\mathrm{Re}_\lambda^{1/2},$$

$$\frac{\lambda}{\eta} = (15)^{1/4}\,\mathrm{Re}_\lambda^{1/2}. \tag{1.8}$$

here $\eta = (\nu^3/\Theta)^{1/4}$ is the Kolmogorov microscale, $\mathrm{Re} = \Lambda U/\nu$, $\mathrm{Re}_\lambda = \mathrm{Re}_\lambda \langle u'^2\rangle^{1/2}/\nu$.

Equation (1.8) gives: $\Lambda/\lambda \approx 0.1\ \mathrm{Re}_\lambda$. It is now possible to make the following estimate:

$$\frac{\mathrm{Re}}{\mathrm{Re}_\lambda} \sim \frac{U\Lambda}{\lambda\langle u'^2\rangle^{1/2}} \sim \frac{U}{\langle u'^2\rangle^{1/2}\lambda} \approx 1000 \cdot 0.1\ \mathrm{Re}\lambda \approx 10\mathrm{Re}_\lambda. \tag{1.9}$$

Consequently, $\mathrm{Re} \approx 10\mathrm{Re}_\lambda^2$ and at $\mathrm{Re} = 10^7$, $\mathrm{Re}^\lambda = 10^3$:

$$\Lambda/\eta \approx \left(10^3 \div 10^4\right),\quad \Lambda/\lambda \approx 10^2,$$

i.e. the size of the longitudinal integral scale λ is two orders of magnitude (and more) greater than the size of the Taylor microscale λ and more than 3–4 times greater than the Kolmogorov microscale η. Thus, it is sufficient to 'work' in the low-frequency and initial zones of the inertia range of the spectrum. In order to determine ('collect') the required value of the energy $\langle u'^2\rangle$, in calculations it is important to have the scales 10–20 times smaller than Λ. The given estimates (1.9) indicate that in the calculations on 'real' grids we can obtain reliable results (Fig.1.2 shows the conventional value of the 'optimum' calculation range denoted by h^* – the universal part of the spectrum is 'cut off').

We shall now describe the approximate methods of determination of the longitudinal integral scale of turbulence. Onufriev et al [19, 57,244] carried out the experimental examination of the behaviour of the energy spectrum in a turbulent flow in a rotating (in relation to the longitudinal axis) pipe. Examination showed the relatively universal behaviour of the spectral distribution, represented in the dimensionless form, and the fact that the given universal distribution is similar to the generalised Karman model (EVK model) for uniform isotropic turbulence.

This situation (it will be referred to as the Onufriev rule) is formulated as follows: there is a similarity for the semiempirical isotropic EVK model of the spectrum: at high values of the Reynolds number, the maximum of the dimensionless quantity φ for the unidimensional spectrum (max $\varphi = \max[kE_1(k)/\langle u'^2\rangle]$) is situated at $k\Lambda \approx 1$. The value of this maximum differs only slightly (within the range of tens of percent), starting with $Re_\lambda > 100 \div 200$.

Figures 1.4 and 1.5 show the behaviour of the dependences of φ for the EVK model (line) and compare them with the experimental data obtained at the MFTI [57,244] for a developed turbulent flow in the pipe. On approaching the wall, the flow becomes highly nonuniform and the universal nature of the representation of the spectral distribution is disrupted (the systematic deviation in the spectrum may be taken into account, and is not very significant in the large part of the spectrum). The stipulation $\max[kE_1(k)/\langle u'^2\rangle])$ at $k\Lambda \simeq 1$" is also fulfilled for non-isotropic flows.

Consequently, if this assumption is regarded as the 'the rule of determination of the integral scale Λ', then it is sufficient to use a relatively narrow range of the scales in the low-frequency range where we can calculate $u'(t)$, $\langle u'^2\rangle$; in the range of these scales, there is a unidimensional spectrum $E_1(k)$ and this is followed by the determination of Λ.

After approximating the spectral Karman distribution which includes the low-frequency and inertia range of the spectrum, we determine (see, for example [119]) the rate of energy dissipation:

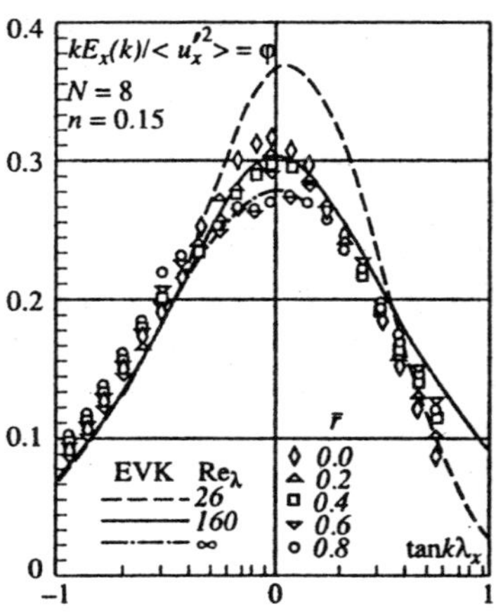

Fig.1.5. Spectral distribution in coordinates $\varphi = kE_x(k)/\langle u'^2\rangle$ and $k\lambda_x$ in a flow with rotation (number of revolutions of the outlet section of the channel N = 8 rev/s, other notations as in Fig.1.4.)

$$\Theta = \frac{(2/3)^{3/2}\left[\Gamma(5/6)/\Gamma(1/3)\right]^{5/2}}{(9/55)^{3/2}\pi^{1/4}(c_1)^{3/2}} \cdot \frac{E_1^{3/2}}{\Lambda} \sim \frac{0.71}{c_1^{3/2}} \cdot \frac{E_1^{3/2}}{\Lambda} \qquad (1.10)$$

or

$$\Theta \sim \frac{\left\langle u'^2 \right\rangle^{3/2}}{\Lambda}.$$

Here $c_1 = 1.4$ is the constant in the three-dimensional spectrum for the inertia range:[1] $E(k)=c_1\Theta^{2/3}k^{-5/3}$ (Fig.1.3). Subsequently, from the relationships for the uniform isotropic flow it is possible to determine the values of the Taylor λ and Kolmogorov η microscales, and also the value of Re_λ [96,119]:

$$\begin{cases} \lambda^2 = 10\nu\left\langle u_x'^2\right\rangle/\Theta, \\ \eta = \left(\nu^3/\Theta\right)^{1/4}, \\ \mathrm{Re}_\lambda = \left\langle u_x'^2\right\rangle\lambda/\nu. \end{cases} \qquad (1.11)$$

Finally, it is also necessary to take into account the effect of the non-isotropic nature and inhomogeneity of the flow on the values of the integral scale of the amplitude of spectral distribution, although in the first approximation this may be ignored.[2] Thus, the calculation of the low-frequency (long-wave) structures is important in examining turbulence for very high Reynolds numbers, and this 'large' structures should not be greatly distorted in the process of calculations.

Taking into account the previously mentioned assumptions and hypotheses (on the weak dependence of the large-scale formations on the nature of dissipation and on the distribution of the energy and dissipation ranges), it is possible to obtain fully satisfactory results in the calculation of the dynamics of large vortices using, in this case, very rough approximations ('averaging') subgrid turbulence (to a certain extent, this is a non-trivial problem in the given approach).

Taking this into account, we shall examine methods of 'rational' numerical determination in the calculation of large-scale turbulent

---

[1]In [216], the variance approach was used to calculation constant c equal to 1.2.

[2]The authors are grateful to Prof Onufriev for supplying material on this problem and very useful discussions.

structures, constructed on the basis of the integral laws of conservation for the ideal medium.

4. Recently, it has become more and more evident that the presence and dynamics of the structures of turbulent flows must be taken into account when examining these flows. The existence of the ordered structure in chaotic motion determines the formation of the regions of intensive concentration of vorticity – vortex tubes and vortex layers.

We shall regard the examination of the 'dynamics of large vortices' (DLV) as the main task of examination of large-scale formations in maximally developed turbulence. In this case, it is natural to use the concept of the 'division' of the flow into large ('oversized') and small ('undersized) vortices, but we shall attempt to develop such an approach without introducing semiempirical models of the closure of subgrid scales. We shall examine initially two important assumptions used as the basis for the methodology which we have developed [16,19,27].

As asserted in [85], "... at high Reynolds numbers Re, the Reynolds numbers Re_λ of large-scale motions are also high $Re_\lambda \sim v_\lambda \lambda / \nu$, where v_λ is the order of the value of the velocity of the scale λ). However, high Reynolds numbers are equivalent to low viscosities... Consequently, for large-scale motion, being the main type of motion in any turbulent flow, the viscosity of the fluid is not important and can be assumed to be equal to zero so this motion is described by the Euler equation.[1] In particular, this shows that there is no significant dissipation of energy in large-scale motion.

Previously, it was shown that for the low-frequency energy range (where energy is generated), the effect of viscosity is very small. Similarly, for the inertia range of motion (the scales of the vortices: $l > \lambda \gg \lambda_0$, where λ_0 is the internal scale of turbulence: $Re_{\lambda_0} \approx 1$), the main process will be the transfer of kinetic energy in the cascade of vortices with velocity Θ and, in this case, the breakup of vortices takes place only under the effect of inertia forces, and viscosity is insignificant (the second Kolmogorov similarity hypothesis of conservation of the energy flux). This type of motion is recorded at very high Reynolds numbers, and for the scales of the vortices $\lambda \gg \lambda_0$ it is convenient to use the relationship of local turbulence [72,85,96,119].

In fact, for free turbulence, modelling should be started from the scales of local turbulence, and the calculations of DLV (ordered

[1]The applicability of these equations to turbulent flows is determined by distancers of the order of the distances of the order of λ_0 [85].

structures and large-scale motions) can be carried out directly on the basis of complete dynamic equations for the ideal medium, i.e. without taking into account the effects of molecular viscosity (viscosity effects, naturally, play a significant role in the vicinity of the body).

Finally, we shall discuss the position relating to the problem of 'rational' averaging of the subgrid scales. In this case, two approaches are normally used. In the first of these approaches (as mentioned previously), we carry out averaging of the equations of motion in respect of the elementary volumes of the grid (see, for example [127,154]. etc,) and the effect of turbulence with the scales $\lambda < h$ is reflected in the form of Reynolds stresses, requiring the application of semiempirical closure models.

However, another approach based on the theory of filtration can also be used [160,187]. At first, we initially introduce a smoothing operator designed for separating small-scale motion in these cases. The application of this operator-filter to the equations of motion makes it possible to determine the effect of stresses, resulting from small-scale of turbulence, through the mean values in respect of volume. The approximation of the nonlinear members (of type $\overline{u}_i \overline{u}_j \neq \overline{u_i u_j}$, remaining after filtration, leads to the non-zero interaction of the velocities of the so-called Leonard stresses. Subsequently, any approximate method can be used for calculating the resultant equations.

As reported in [127], "the ideal case would be the one in which the filter excludes all contributions of small-scale motion above the selected wave number, without changing the mode in the range of wave numbers smaller than this limiting number. These filters have not as yet been used in finite-difference approximations...".

This property ('extinction of fine scale pulsations and having almost no effect on large-scale structures) are typical of the dissipative finite-difference schemes are based on 'oriented' approximations. In this case, it is very efficient to examine, instead of differential equations of motion, balance equations (in the form of integral laws of conservation) for the elements of the volume– the cells of the calculation grid. This method of preliminary 'integral smoothing' of the fine vortices appears to be most natural for the numerical calculation of large-scale turbulence. The 'splitting' scheme, used in organising the computing cycle, should approximate the mechanism of turbulent transfer.

5. Taking the above considerations into account, we shall note here the general assumptions regarding the approaches of 'rational'

numerical averaging used in the proposed approach (following the investigations by Tolstykh [28]).

In numerical modelling of large vortex structures of the turbulent flow using different schemes there are unavoidable errors which can be divided into two groups: amplitude and phase. Since the accurate and difference solutions can be represented in the form of superposition of harmonics with different frequencies and amplitude (in the form of a series or a Fourier integral), it is natural to conclude that relatively large structures correspond to 'long-wave' and 'medium-wave' harmonics, and 'short-wave' harmonics correspond to fine structures.

Instead of 'wave' exponent e^{ikx}, the difference solution (k is the wave number, the wavelength $\Lambda = 2\pi/k$) contains the representation $e^{ikx_n} = e^{ikhn}$ (h is the step of the grid, n is the number of the node). Because of periodicity, it may be concluded that $0 \leq kh < 2\pi$. In this case, the 'short difference' waves correspond to $\alpha = kh \approx \pi$ (at $kh = \pi\Lambda = 2\pi/k = 2h$, i.e. we have the shortest waves resolved by the grid). The values $kh \ll 1$ correspond to the 'long' waves, i.e. to the waves which we must describe with sufficient accuracy.

We examine the simplest linear analogue of the equation of transfer of vortices:

$$u_t + cu_x = 0, \quad c = \text{const.} \tag{1.12}$$

The amplitude errors ($\alpha_{dif}/\alpha_{exact}$, α is the amplitude) and phase errors (C_{dif}/C_{exact}, C in the interfacial velocity) can be written in the form

$$\frac{\alpha_{dif}}{\alpha_{exact}} = 1 - d(\alpha), \quad \frac{C_{dif}}{C_{exact}} = 1 - \tilde{l}(\alpha), \quad \alpha = kh,$$

where d is dissipation, $\tilde{l}$ is a function characterising the dispersion.

In most cases, all the schemes give small for 'long' waves ($kh \ll 1$) and large errors for short waves ($kh \approx \pi$). In this case, the dependence characterising the difference between C_{dif} and C_{exact} usually has the form shown in Fig.1.6a (solid curve). From the viewpoint of description of 'large' vortices (long and medium waves, where $\alpha = kh$ is not large) the errors in the description of 'fine, structures' (short waves, $\alpha \approx 1$) are far from harmless. The point is that these errors (small-scale perturbations) propagate with some 'difference' group velocity C^{gr}_{dif} which differs even more than C_{exact} from the 'accurate' value C of the velocity of propagation of perturbations (the dotted line in Fig.1.6a). At $kh \approx \pi C^{gr}_{dif}$, it may generally become negative and, consequently, the perturbations,

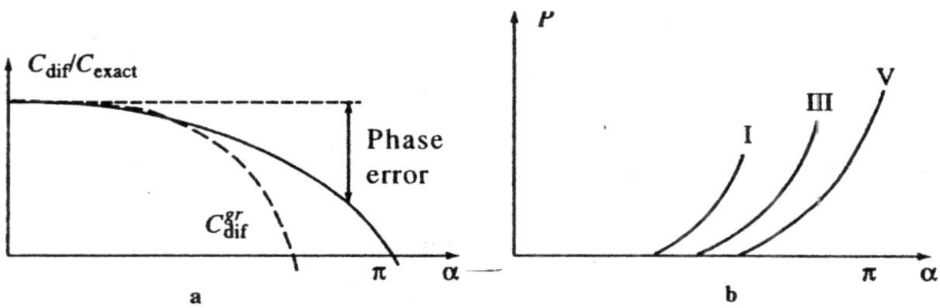

Fig.1.6. Examination of the modelling equation: a) C_{dif}/C_{exact} dependences characterising the difference between the 'difference' phase velocity and the exact values (solid curve) and the approximate law of variation of the 'difference' group velocity C^{gr}_{dif} of development of small-scale perturbations (broken curve); b) typiucal dependence of the function $P(\alpha)$, characterising the dissipation of difference schemes of different order of accuracy ($\alpha = kh$, k is the wave number, h is the step of the grid).

propagating upwards in the flow, may in certain cases greatly distort the structure of large vortices.

In a nonlinear case (for example in the case of an Euler equation), the situation is even worse: it can be shown that the 'pumping' of the energy of small-scale perturbations into larger scales is possible. This sometimes leads in the final analysis to the scheme nonlinear instability.

The situation in maybe improved (or even 'saved') by dissipation d. We shall examine 'non-dissipative' ($d = 0$) and dissipative ($d \neq 0$) schemes. The typical dependences for the dissipative schemes with different accuracy are presented in Fig.1.6b. It follows from here that the dissipative schemes are characterised by relatively low dissipation in the range of large scales ($kh \ll 1$) and by high dissipation in the range of small scales ($kh \approx \pi$). However, dissipation characterises the decay of the amplitude of perturbations. Thus, the dissipation-stable schemes describe efficiently the large structures (for example, large vortices) and weaken (or completely extinguish) the small-scale perturbations which we do not require.

When using the schemes with symmetric approximations, there is no dissipation and, in addition, there is no mechanism of weakening harmful (for describing large vortices) 'short-wave errors' (the only hope in this case is filtration, but this is equivalent to introducing dissipation). In addition, there may be a situation in which the dissipation for the given scale l of a large vortex is so high (h/l is too small) that this vortex itself will be extinguished ('blurred') by the scheme. Therefore, the schemes with large dissipation (for example, the schemes of the first order of accuracy) require the accurate selection of the grid (the ratio h/l).

We shall examine these problems in greater detail. The role of the dissipative mechanism, present in the difference schemes for the convective problems, can be examined most efficiently, as already mentioned, on an example of the transfer equation (1.12). In the equation, cu_x is replaced by the difference approximation $c\delta_x$, leaving u_t without changes (in order to eliminate the effect of the method of time discretisation). As any operator, $c\delta_x$ may be represented in the form of the sum of the self-conjugated $c\delta_x^{(0)}$ and skew-symmetric $c\delta_x^{(1)}$ parts, i.e.

$$c\delta_x = c\delta_x^{(0)} + c\delta_x^{(1)},$$

where

$$\left(\delta_x^{(0)}u, v\right) = \left(u, \delta_x^{(0)} v\right),\ \left(\delta_x^{(1)}u, v\right) = -\left(u, \delta_x^{(1)} v\right),$$

and (*,*) denotes the scalar products of grid functions. At first sight, it may appear that since $c\frac{\partial}{\partial x}$ is the skew-symmetric operator, the operator $c\delta_x$ should also be skew-symmetric (i.e. $\delta_x^{(0)} = 0$). For example, it may be assumed that δ_x is equal to the normal central difference $\delta_x u_j = (u_{j+1} - u_{j-1})/2h$. In particular, these approximations were also used in early stages (in the sixties) in the solution of relatively simple problems. However, in many cases, the presence of regions with steep gradients complicates the application of these 'natural' approximations, because the high-frequency 'parasitic' (scheme) oscillations, formed in this case, usually prevent obtaining a stable solution or greatly distort it.

One of the methods of overcoming this difficulty is, as is well known, the introduction of additional members of the type of artificial viscosity, carrying out in particular the role of smoothing filters. However, the shortcoming of this method is that it is necessary to select filter parameters for a specific problem (or classes of problems). In some cases, this cannot be carried out efficiently (in particular, this approach is difficult in spatial and nonstationary problems).

Another method is the application of schemes in which the dissipative mechanism (smoothing filters) is present in the operator itself δ_x, approximating convective transfer. In this case, the self-conjugated component of the operator $c\delta_x^{(0)}$ should be positive (i.e.

$(c\delta_x^{(0)} u,\ u) > 0$). In fact, after discretisation in respect of x, equation (1.1) may be presented in the following form:

$$\frac{\partial u}{\partial t} + c\delta_x u = 0. \tag{1.13}$$

Multiplying equation (1.30) in the scalar fashion by u, we obtain

$$\frac{\partial}{\partial t}(u,u) = \frac{\partial}{\partial t}\|u\|^2 = -(c\delta_x u,u).$$

Since $cd_x^{(1)}$ is the skew-symmetric operator, i.e. $((\delta_x^{(1)} u,\ u) = 0)$, then

$$\frac{\partial}{\partial t}\|u\|^2 = -\left(c\delta_x^{(0)} u,u\right) < 0.$$

This means that the norm of the solution decreases and dissipation takes place. However, this assertion does not yet indicate which components of the function (harmonics) attenuate at the highest rate. In order to explain this question, we shall assume that the initial data for (1.30) are presented in the form

$$u(x,0) = \exp(ikx).$$

and, in this case, the exact solution of (1.0) has the form $u(x,\ t) = \exp\ (ik(x-ct))$. The solution of (1.13) will be determined in the form $x(x,\ t) = v(t)\ \exp(ikx)$, $x = jh$, $j = 0, \pm 1, \pm 2, \ldots$, and, consequently, we obtain

$$\frac{\partial v}{\partial t} + (p + iq)v = 0.$$

where the actual values of $p(k)$ and $q(k)$ satisfy the relationships:

$$c\delta_x^{(0)} \exp(ikx) = p\ \exp(ikx), \quad c\delta_x^{(1)} \exp(ikx) = iq\ \exp(ikx);$$

and in this case, if $c\delta_x^{(0)} < 0$, then $p > 0$.

Consequently, we obtain

$$v(t) = \exp(-p(k)t) \exp(-iq(k)t).$$

The co-multiplier exp $(-iq(k)t)$ is the oscillatory component of the difference solution, determining the dispersion. The component $\exp(-p(k)t)$ is the coefficient of attenuation of the harmonics with the wave number k during time t (it may be regarded as a dissipation characteristic). In this case, as the value of $p(k)$ increases, the rate of attenuation of the harmonics (dependence $p(k)$ for the dissipative schemes of the different order of accuracy are shown qualitatively in Fig.1.6b). This shows that in the dissipative schemes the short waves show a higher rate of attenuation (higher kh), in particular the 'non-physical' scheme oscillations ($kh \approx \pi$). In this case, as the order of the scheme increases, the size of the region kh in which the harmonics do not change their amplitude becomes larger (as in the case of the exact solution). Thus, the dissipative schemes of a high order contain a natural filter which 'transmits' only the harmonics that describe the physical process and 'suppress' small-scale or 'non-physical' oscillations of the seesaw type (Fig.1.6 b).

The schemes with dissipation (with positive operators) use 'oriented', i.e. directional (non-centred) differences. By changing their direction in relation to the sign of c in (1.0) it is possible to retain the operator $c\delta_x$ positive also when the sign of c changes. In addition to the filtration effect, the 'positiveness' of the operator increases the margin of stability of the schemes. This is especially important in the solution of stationary problems and also when examining the dynamics of larger vortices in free developed turbulence.

However, it should be remembered that in a nonstationary case in the calculation for long times (for example, in weather forecasting), even at low values of k the quantity e^{-pt} may also extinguish the 'physical' harmonics. From this viewpoint, the advantages of the schemes with high order of accuracy are mainly the fact that this situation starts at a considerably later stage in comparison with the schemes of low order (Fig.1.6b).

At the present time, we use only dissipation-stable schemes: the method of large particles, the flow method, the approximation of viscosity of the equations, the compact schemes with a high order of accuracy, and others ([17,19,24] etc). In this case, the dissipative nature is achieved by 'orienting' the differences in accordance with the direction of the flow ('non-symmetric schemes) without introducing explicit members with artificial viscosity.[1] These methods, based on the splitting schemes, make possible to carry out calculations of complicated spatial-nonstationary flows in medium-power machines.

6. Thus, here we propose a completely new (in comparison with [94,104,154,160], etc) method of 'rational' averaging, based on constructing dissipation-stable and divergent-conservative difference schemes. We shall mention briefly the main assumptions of the approach which we have developed for calculating the free shear maximumally developed turbulence and 'transition' phenomena.

1. The initial equations in the examination of large-scale coherent structures are represented by balance relationships – the integral laws of conservation for the ideal medium.

2. We use the 'oriented' difference approximations in averaging convective members in respect of the volume of the elementary cell (this also leads to the formation, in the difference equations, of the dissipative mechanism which, ensuring the stability of the solution, should reflect in the approximate form, generally speaking, the effect of small-scale 'subgrid' vortices).[2]

3. By means of the direct calculation of macrostructures to the 'establishment of the results' on different approximation grids and for long periods of time we 'select' the energy integral and it is thus possible (according to the results of a large number of calculations) to determine the 'effective' value of such a dissipative mechanism, using the principle of 'the stability of the solution at the necessary resolution'. In this approach, we satisfy the requirements of the 'invariance of separation' into large and small scales and the solution is transferred to the steady ('equilibrium') regime in the determination of the non-stationary macrostructures of the flow.

4. In the calculation of the dynamics of large structures, we do not use the semi-empirical models of subgrid 'closure' (the introduction of the approximate models of 'effective' viscosity is

---

[1]See, for example, difference representations in the large particle method (Chapter 2), in the flow method (Chapter 3).

[2]According to calculations, the characteristics of the large-scale flow depend only slightly on the structure of the dissipative mechanism. This is the reason why in DLV calculations, the small-scale high-frequency turbulence can be extinguished or its effect can be 'roughly' taken into account, of mainly the universal part of the spectrum is positioned 'below the grid' (Fig.1.3.).

necessary only in the zones of local turbulence or when estimating the 'stochastic component' – the turbulence background).

5. The algorithms developed for this purpose make it possible to carry out 'natural' expansion also to special problems (the dissipative mechanism forms 'automatically' because of the previously described averaging process and it is not necessary, as for example in [228], to examine various types of coefficient of turbulence viscosity for two-dimensional and three-dimensional flows).

6. Using the proposed method of direct numerical modelling, it is possible using the 'smoothed' equations to determine the spatial-time structures of the dynamics of large vortices and by statistical processing of the pulsation-nonstationary regime determine their average 'fluctuation' characteristics: Reynolds stresses and turbulent energy (the correctness of this process is discussed below).

7. The numerical examination of the nominal-turbulent transition is carried out on the basis of complete Navier–Stokes equations.

8. 'The transition to chaos' (numeral experiment), carried out on the basis of complete dynamic models, can be carried out only in the presence of external perturbing forces, 'intensifying' the inherent nonlinear inertia mechanism of the system.

Finally, depending on the type of problem, it is convenient to use the schemes with a high order of accuracy. In this case, larger and larger part of the 'long-wave' harmonics can be 'resolved' and is not 'extinguished' by dissipation. The realisation of this approach greatly reduces the requirements on computer resources.

This is the principle of the proposed general methodology and its main difference from the approaches used by other authors for the direct numerical modelling of turbulence ([94,104,107,125,127,154, 187,228] and others). We shall now explain the proposed approach in greater detail ([15,16,19] and others, see also studies published by Babakov, Gushchin, Kon'shin and Yanitskii [128]).

1.3. SOME EXPERIMENTAL ASSUMPTIONS

If in the first years of examination of turbulence, these phenomena were interpreted as completely stochastic processes (determined by random distributions of pulsation quantities) then we believe that a principal change has taken place now in understanding these phenomena. It has been explained that turbulence includes, as an element, the organised motion of 'almost' coherent structures and intensive studies have started into the problems of the relationship of determined and chaotic elements ([46,71,113,231] and others).

Many experiments and theoretical investigations have shown (see, for example [71,113,236] and other) that the wide range of the turbulent flows with transverse shear (both in free flows and in boundary flow) is characterised by the existence of non-stationary organised motion of large-scale formations ('large vortices') with slight pulsation–ordered motion of 'stochastic structures'[1] which are characterised by a highly stable and typical (for the given problem) spatial–time form (Fig.1.7–1.9, experiments [236]). The internal zone of these flows, for example, for stream problems, is of the turbulent (stochastic) nature and consists of non-ordered small-scale pulsations with a relatively high intensity. but with an approximately homogeneous structure[2] (see [71,113, 119,236]). Figures 1.7–1.9 show photographs characterising the types of all the structures for various types of turbulent flows.[3]

For example, in the free shear flows (accompanying wakes, streams) examination showed (according to Townsend [113]) a dual ('alternating') structure of turbulence. The main zone of the turbulent fluid is characterised by a relatively small scale and is homogeneous. The external system of stable 'slow' (non-turbulent) large vortices is superimposed on the main zone, and the external system transfers completely the turbulent liquid from one part of the flow to another (Fig.1.10). Thus, the ordered structures are characteristic and form the natural basis of shear motion.[4]

In accordance with [46], we shall classify these turbulent structures:

– dynamic structures exist in the vicinity of the point of transition from the laminar to turbulent flow and are characterised by the bifurcation origin (structural stochasticity: 'chaos forms from order');

– quasi-equilibrium structures form in regions with developed turbulence, when random motion is so extensive that the system is

[1]When discussing the formation of ordered stochastic (organised, coherent, collective, etc) structures, it should be mentioned that this class of motion is stochastic both in respect of time and space [46,71].

[2]Cantwell [71] accepts that in a number of cases (for example, in flat mixing layers), "even very small-scale motions may be highly ordered".

[3] Fig.1.9 shows the small influence of molecular effects on the large-scale structure of the wake (in both cases, the structure of the wake is approximately the same, although Reynolds numbers differ greatly).

[4]New results of studies by Russian scientists in the theory of structural turbulence have been published in [46].

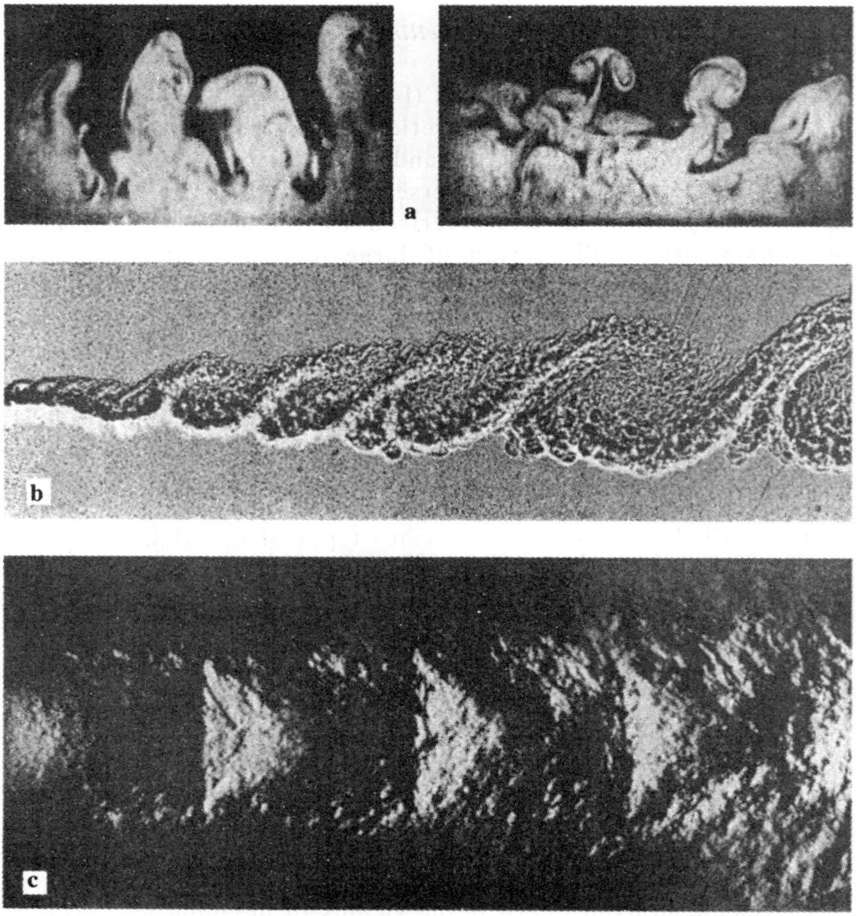

Fig.1.7. Different types of ordered turbulent structures (experiments in [236]): a) turbulent boundary layer; b) mixing layer (top – nitrogen, bottom – helium–argon mixture); c) supersonic jet (slow motion photography).

closely to thermodynamic equilibrium (the structural nature of this stochastic movement: formation of order from chaos');

– the flow structures occupy an intermediate area between the first two structures.

In free shear turbulent motions – streams, wakes, mixing layers – ordered large-scale formations are detected at any large Reynolds number. The structures of this type are also characterised by the two-dimensional (or quasi-two-dimensional) form [46]. The intensity, scale and form of this low-frequency ordered motion are quasi-determined (i.e. individual) for the given type of flow, and they can

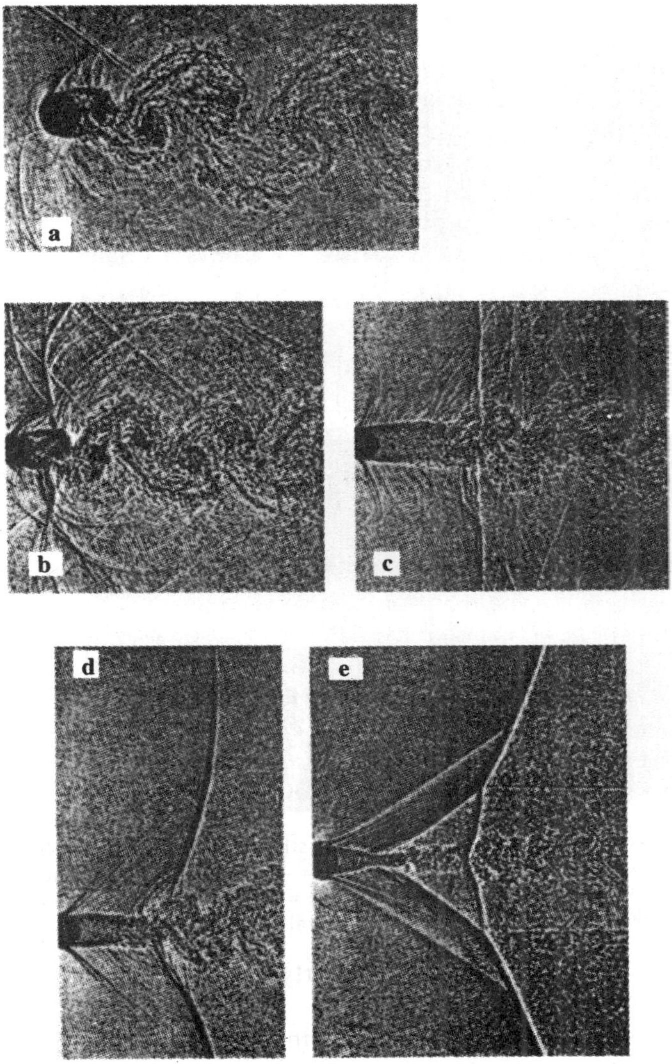

Fig.1.8. Different types of ordered turbulent structures: a wake behind a circular cylinder (experiments in [236]): a) $M_\infty = 0.64$, $Re = 1.35 \cdot 10^6$; b) $M_\infty = 0.80$, c) $M_\infty = 0.90$; d) $M_\infty = 0.95$; e) $M_\infty = 0.98$.

be described efficiently using hydrodynamics equations (and not statistical approaches). The dimensions of large vortices are comparable with the characteristic size of the flow and are considerably greater than the scales of vortices forming the natural turbulent motion ([71,113]). At the same time, the turbulence in the core of the wake is characterised by a completely sufficient degree of local isotropy (in relation to the intensity of pulsations of velocity

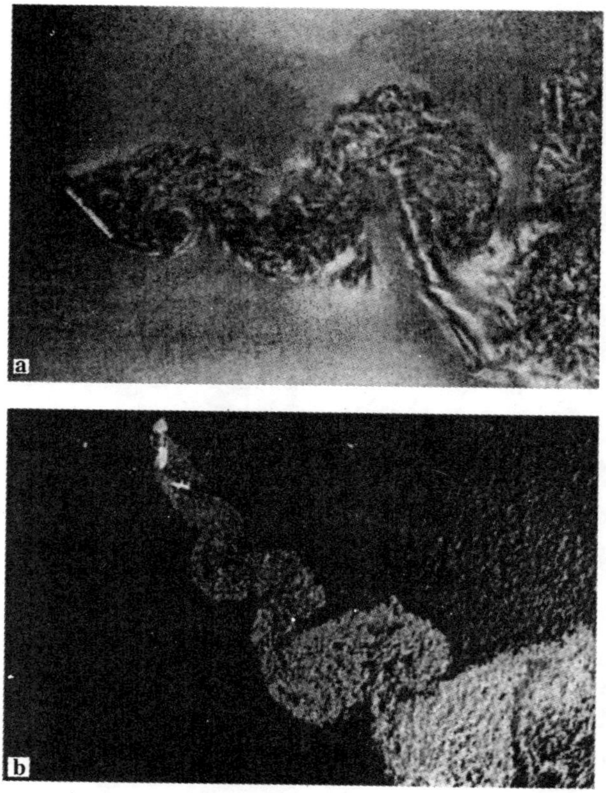

Fig.1.9. Different types of ordered turbulent structures (experiments in [236]): a) a wake behind an inclined sheet ($\alpha = 45°$, Re = 4300); b) a wake of oil discharged from a tanker which ran aground ($\alpha = 45°$, Re = 10^7). "In the last two cases the structures of the wake are surprisingly identical" [236].

in different directions and, consequently, in relation to the energy characteristics [119]).

In numerical investigations, it is important to model accurately the transfer processes. There are three types of energy phenomena, characteristic of real turbulent motions [46]: 1. The generation of large-scale vortices, which depends on the specific properties of the examined flow as a whole; 2. Break-up of these vortices as a result of nonlinear nature into smaller-scale vortices and the transfer of energy without any extensive loss downwards in the spectrum (the Kolmogorov 'cascade' process); 3. Viscous dissipation of energy on the smallest scales. Until recently, analysis by statistical methods was carried out only on types 2 and 3 and a theory was developed without taking into account the generation mechanism (the required pulsation energy was introduced into the flow from the outside by the introduction of a random external

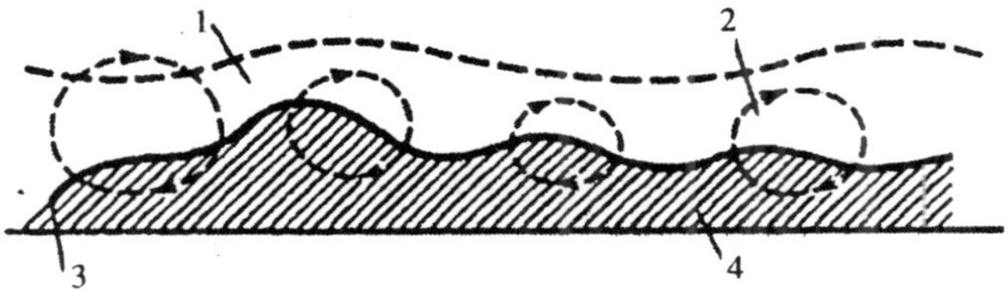

Fig.1.10. Diagram of flow in freely-developed turbulence (according to Townsend [113]): 1) non-turbulised fluid; 2) motion of large vortices; 3) boundary of turbulised fluid; 4) homogeneous turbulence (stochastic component – 'turbulent background').

force). At the same time, the generation aspect of the problem of large-scale turbulence, in particular (where the deterministic origin is dominant) is most important because it includes the reasons for the formation of turbulence and the mechanism of sustenance of turbulence.

The large-scale transfer of a turbulent liquid is evidently carried out mainly by organised motion of a group of large vortices causing the distortion of the boundary of the turbulence field and transfer of the turbulent fluid across the flow [71,113]. Consequently, the motion of ordered and large-scale turbulent structures determines mainly the dynamic, kinematic and energy characteristics of the flow as a whole which naturally also determines the properties of 'deep' turbulence where energy dissipation takes place.

The reversed effect of the small-scale turbulence (and molecular diffusion) on the main characteristics of a large-scale flow for freely developed shear turbulent flows in the case of high Reynolds numbers is evidently insignificant because these are effects of different orders, the small-scale structures are 'almost' universal (Fig.1.1), and the process of energy transfer is one-sided.

1.4. GENERAL FORMULATION OF THE PROBLEM

1. The above considerations also determine in principle the ideology of organising the computing process in direct numerical modelling in the wake of large-scale flows with freely developed shear turbulence for very high values of the Reynolds number. In this case, the general cycle of investigations breaks up into two related tasks.

Task 1: direct calculation of the nonstationary motion of ordered and large-scale turbulent structures.

The large-scale and organised nature of such a motion make it possible to describe the motion by means of numerical schemes based on nonstationary hydrodynamics equations (integral laws of conservation) for the model of the ideal medium ('difference Euler') and characterised by the approximate (effective) dissipation mechanism. The main task of such a dissipative mechanism is to 'smooth out' ('extinguish') small-scale pulsations ensuring the stability of the solution and the required resolution for the flow macrostructures. As shown by investigations and calculations, the application of the 'oriented' (non-symmetric) differences in the integral laws of conservation without the application of semi-empirical models of turbulence satisfies these conditions. The point is that the properties of large-scale motions are determined basically by volume convection (i.e. they have the 'wave', dynamic nature) and depend on the solution as a whole. Consequently, calculations should be carried out directly in the entire field of the flow on real ('large') difference grids with subsequent determination of the necessary average characteristics of the turbulent flow (for example, moments of different order) by means of appropriate statistical processing of the results.

Here, it should be mentioned that the possibility of determining the statistical characteristics of the non-stationary ('pulsation') large-scale turbulent flow on the basis of smoothed-out equations is far from evident because there is a problem of the correctness of the formulation of such a task (this will be discussed later).

In the given stage, as a result of solving the problem we simulate the movement of all ordered structures, large (energy-carrying) turbulent vortices and also the process of generation, and so on (based on large-scale transfer). Here, it is also possible, but already with the appropriate representation of the dissipative mechanism, to investigate the 'cascade' process of energy transfer, i.e. the 'descent' to small-scale locally isotropic turbulence. With appropriate initiation (and determination) on large structures, the energy characteristics of turbulence in the 'descent' will be modelled evidently quite accurately because the given process is sufficiently conservative.

This 'deterministic' approach makes it possible to separate the ordered and large-scale formations for structural turbulence, and in this case it is necessary to specify accurately the method of averaging pulsation fields.

Task 2: numerical modelling of the stochastic component of the turbulent shear flow (small-scale turbulence).

In this case, we model the local 'resolution' – the process of dissipation, the nature of propagation of energy in the turbulence core, and so on. Calculations of the flow of this type should be carried out, in our view, using statistical methods (or by a phenomenological procedure, introducing appropriate coefficients of turbulent viscosity). We can also use algorithms, based on the examination of pulsation equations (for example [112], where the the parameters of the mean flow are determined from the solution of task 1.

In this case, it is important to mention that the calculations of these flows in the given stage should be carried out only in limited sub-regions, 'cutting out' zones of high gradients from the general pattern of the flow. This approach is based on the assumption according to which on the turbulence scale small in comparison with the scale of mean motion, the local structure of turbulence is sufficiently universal for different flows and is determined only by local conditions (turbulent transfer is characterised in this case by gradient diffusion) [113]. Consequently, in the calculations we can use fully determined models and relatively fine calculation grids and the level of requirements on the computer resources is greatly reduced.

2. It is well-known that the averaging of the Navier–Stokes equations in respect of Re is carried out immediately on all scales of turbulence for long periods of time (separation of 'regular' mean motion takes place) and this requires modelling all structures at the same time. Thus, it is not realistic to construct a model. of turbulence that is universal for different classes of motion. In contrast to this approach, the concept described previously is based on 'splitting' general motion into large- and small-scale structures (the non-random' ordered motion of macrostructures is distinguished here). At the same time, motion of the ordered and large-scale turbulent vortices (with the size $\lambda \geq h^*$, where h^* is the step of resolution of the difference grid) is determined by direct integration of the complete equations of hydrodynamics, and modelling (smoothing) is applied only to small-scale pulsations which are not resolved in the explicit form in numerical integration and, as already mentioned, are characterised by relatively universal properties.

This concept is extremely important in the methods of direct numerical modelling of turbulence. It is interesting to note that the need for such division was mentioned as early as in 1948 by Drayden [158] who on the basis of analysis of the data obtained in the boundary layer using a thermoanemometer, concluded that the 'large masses of the liquid move more or less like coherent structures'.

This approach is fully adequate to the so-called models of subgrid closing (for example, [154,160], etc). However, the point is how to organise this process of splitting (averaging) and which models should be used to set up approximate (smoothed-out) equations describing the motion of ordered and large-scale turbulent structures. Of course, this results in problems associated with the correctness of formulation of the task, the selection of approximation for large-scale vortices and evaluation of their influence on large-scale motion. Finally, a central position should be occupied by the aspects, associated with the realisation of these approaches.

1.5. THE MODELLING OF COHERENT STRUCTURES IN TURBULENCE

1. In order to construct a numerical algorithm of the calculation of large-scale structures ('difference Euler'), the initial system should be represented by the approximation of the laws of conservation written, for example, for the simplest case in the integral form (the method of large particles, the flow method ([9] and others):

$$\frac{\partial F}{\partial t}=-\oint_{S_\Omega}\mathbf{Q}_F\,ds,\quad F=\{M,\mathbf{P},E\},\tag{1.14}$$

where S_Ω is the side surface of volume Ω; $\mathbf{Q}_F$ is the vector of the density of the flow with strength F (M is the mass, $\mathbf{P}=\rho\mathbf{v}$ is the pulse, E is total energy).

The equations (1.14) are presented (both for a viscous compressible gas and in the case of the ideal medium) taking into account the boundary conditions and are solved numerically for each calculation grid of the computing region.

Splitting the density vector $\mathbf{Q}_F$ into 'convective' and 'diffusion' component (and using appropriate approximations), it is possible to construct conservative schemes of the calculations of the second order of accuracy, taking into account correctly the domains of the influence and the nature of interaction of the investigated quantities.

In the numerical solution of the problem, the integrals in (1.14) are calculated from the quadrature formulas

$$\frac{F^{n+1}-F^{n}}{\tau}=-L_n\left(\oint_{S_\Omega}\mathbf{Q}_F\,ds\right),\tag{1.15}$$

where operator L_n carries out averaging (smoothing) of the characteristics of the medium for the elementary (but finite) size of the calculation grid.

The density of distribution can be determined using, for example, non-symmetric equations, 'retaining' the positive value of the transfer operator. In the schemes of the second order of accuracy [9]:

$$\rho_{m+1/2,n} = \begin{cases} 1.5\rho_{m,n} - 0.5\rho_{m-1,n} & u_{m+1/2,n} \geq 0, \\ 1.5\rho_{m+1,n} - 0.5\rho_{m+2,n} & u_{m+1/2,n} < 0, \end{cases}$$

The homogeneous finite-different schemes, constructed in this manner, are divergently conservative and dissipatively stable. This makes it possible to carry out, using a single algorithm, continuous calculations both in the region of smoothness of the solution and in the zones of breaks. In this case, the stability of calculations is ensured only by internal dissipation (approximation viscosity) so that these schemes can be used in the presence of curvilinear boundaries and also for spatial-nonstationary problems [19].

For example, the equations of the 'difference Euler' (obtained by means of differential approximation for the simplest schemes of splitting in the method of large particles) have the following form [24]:

$$\begin{aligned}
&\frac{\partial \rho}{\partial t} + \Delta(\rho\,\mathbf{v}) = 0, \\
&\frac{\partial \rho u}{\partial t} + \Delta(\rho u\,\mathbf{v}) + \frac{\partial p}{\partial x} = \frac{\partial}{\partial x}\left(\rho\varepsilon_x \frac{\partial u}{\partial x}\right) + \frac{\partial}{\partial y}\left(\rho\varepsilon_y \frac{\partial u}{\partial x}\right), \\
&\frac{\partial \rho v}{\partial t} + \Delta(\rho v\mathbf{v}) + \frac{\partial p}{\partial y} = \frac{\partial}{\partial x}\left(\rho\varepsilon_x \frac{\partial v}{\partial x}\right) + \frac{\partial}{\partial y}\left(\rho\varepsilon_y \frac{\partial v}{\partial y}\right), \qquad (1.16) \\
&\frac{\partial \rho E}{\partial t} + \nabla\left((\rho E + p)\mathbf{v}\right) + \frac{\partial p}{\partial y} = \frac{\partial}{\partial x}\left(\rho\varepsilon_x \frac{\partial E}{\partial x}\right) + \frac{\partial}{\partial y}\left(\rho\varepsilon_y \frac{\partial E}{\partial y}\right),
\end{aligned}$$

where p is pressure, $\varepsilon_x = |u|\ \Delta x/2$, ε_x y=$|v|\Delta y/2$ are the coefficients of approximation viscosity.

Here, we present expansions with the accuracy to O (Δt, h^2). It is interesting to note that in (1.16) the structure of the coefficients of approximation viscosity $\varepsilon \sim |v|h$ resembles the structure of the coefficient of turbulent viscosity for the λ-scale, $v_{T\lambda} \sim v_\lambda \lambda$ (evidently, this is explained by the single reason for their formation – the non-linearity of processes taking place and the adequacy of the applied schemes of splitting and the mechanism of turbulent mixing). If a turbulent vortex with scale λ, moving with velocity v_λ, is related with a large particle with size h, moving with velocity v_h, the expression for ε and $v_T\lambda$, will be identical. We assume that $\varepsilon_x \sim \varepsilon_y \sim v_{T\lambda}$ and, consequently, equations (1.16) at ρ = constant (incompressible fluid) acquire the form of the exact Navier–Stokes equations where molecular viscosity ν is replaced by the coefficient of effective turbulence ν_T (see [15,16,19]).

The lefthand part of (1.16) contains the exact expressions of the initial Euler differential equations, and the righthand part contains dissipative members, i.e. the 'perturbation background' formed as a result of smoothing (1.15) of subgrid fluctuations and breaks in the approximation of the initial system of the differential equations by finite-difference systems depending on the internal nature of the representations used. The dynamics of this 'background' is the source of fluctuations. As indicated by these equations, the smoothed equations (1.16), used in specific calculations, are dissipative (although the model of the ideal medium was used as the initial model). The specific type of the 'oriented' difference scheme was described by Babakov [9,128]. The type of dissipative mechanism depends on the nature of introduced approximations, and its structure can be controlled.

The dissipative mechanism in (1.16) is an analogue of 'Reynolds' stresses in joining with (1.15) in respect of the scales $\lambda < h$ and reflects in a generalised form the contribution of small ('subgrid') is vortices. For the schemes of the first order of accuracy, the coefficient of scheme ('effective') viscosity $\varepsilon \sim |v|h$ depends on the local velocity of the flow and the size of the difference grid h (and is independent of molecular viscosity). The resultant approximation viscosity in the schemes of the second order of accuracy [9] (in the main members of asymptomatic expansion) has the form $\varepsilon \sim \left|\frac{\partial v}{\partial y}\right| h^2$ which also corresponds to our approximate representation of turbulent viscosity. In fact, this shows that to obtain the values of 'effective' viscosity, satisfying the stability condition, calculation should be carried out using sufficiently 'large' calculation grids (or 'lumps' of molecules, clusters). Consequently, the common approach determining the nature and role of the dissipation mechanism may be the principle 'the stability of the solution for the necessary resolution'.

For comparison with the difference Euler (1.16), we shall present the equation of the 'difference Navier–Stokes'. The equation of the pulse for a compressed gas, obtained in [24], in the splitting schemes of the method of large particles for the Navier–Stokes model has the following form:

$$\frac{\partial \rho u}{\partial t}+\frac{\partial\left(p+\rho u^2\right)}{\partial x}=\left[\mu(A-2)+\rho|u|\frac{\Delta x}{2}\right]\frac{\partial^2 u}{\partial x^2}+O\left(\Delta t,\Delta x^2\right),$$

where $A = \lambda/\mu$, $\mu = \rho\nu$, λ is the second viscosity coefficient, ν is molecular viscosity. This shows that the approximation of 'initial' viscosity members is also affected by the scheme effects. In the calculations of flows with high Reynolds numbers, the stability of the calculation procedure can be ensured at a high effective viscosity and the real viscosity should be 'forgotten'.

It may be seen that the numerical solution of the boundary problems based on the Navier–Stokes equation at high Reynolds numbers encounters not only technical problems but also problems of principal nature (the approaches, associated with increase of the density of the grids cannot always be used to solve these problems). Thus, if the equation of the ideal gas includes the dissipation terms, then (as shown by investigations on calculations) in the presence of relatively wide assumptions regarding the nature of dissipation, the generalised equation in the examination of the macrostructures of the flows for the limiting conditions ($\mathrm{Re}\to\infty$) can be obtained with a specific accuracy from the equations with the approximate dissipation mechanism, and not from the Navier–Stokes equations.

It should again be stressed that: the semiempirical models of turbulence are not used in the described approach to the examination of the macrostructure of vortex motions. The role of the filter cutting off (or, more accurately, smoothing out) subgrid pulsations is played in this case and by the dissipative mechanism which is 'present' in the averaging operator (approximating convective transfer), which 'formed' as a result of the application of 'oriented' finite-difference representations. This dissipative mechanism in the average form also reflects the contribution of small-scale 'subgrid' vortices. In this case, direct calculations 'in the grid' of complete averaged-out dynamic macroequations should result in a stable solution, the 'collection' of the energy integral and the establishment of the parameters of the non-stationary flow. This makes it possible to obtain the effective value of the dissipative mechanism and satisfy the requirements of the invariance of the division into 'large' and 'small' scales. In the region of the large vortices we also determine important characteristics of the turbulent flow such as the rate of dissipation of energy and the integral scale (using, for example, Onufriev's rule). Thus, the described 'large scale' approach makes it possible to investigate the main properties of free maximally developed turbulence.

On the basis of the smoothed-out equations, it is possible to evaluate the characteristics also on smaller scales, investigate the 'cascade' process of energy transfer in respect of the hierarchy of the vortices, and so on. However, the phenomena of this type already require the introduction of semiempirical values of turbulent viscosity. We shall explain a possible scheme of such an approach.

2. The accurate modelling of nonstationary nature allows for the examination of the dynamics of development of the phenomenon. The approximate mechanism with 'fast' dissipation in the difference Euler equations can be used in principle to determine the average characteristics of the flow from the stability condition. The given dissipative mechanism can be controlled in such a manner as to ensure that it will model to a certain degree (for the appropriate resolution scales) the process of 'cascade' transfer for different values of λ. If the calculations of the DLV-motion of the non-stationary ordered macrostructures (which depends on the entire range of motion) is carried out using the difference Euler schemes with the viscosity of the type

$$\varepsilon \sim |v| h, \tag{1.17}$$

then in the examination of the average characteristics of the large scale 'energy-carrying' turbulent vortices in the region of high gradients it is possible to introduce the turbulent viscosity of the following type into the approximate dissipation mechanism:

$$\hat{\nu}_T \sim \Delta \bar{u} l, \tag{1.18}$$

where $\bar{u}$ is the variation of the mean velocity at distances of the order of l, and for the zones of local turbulence, determined by local gradients, it is efficient to use in the difference Euler effective viscosity $\hat{\nu}_{m,\lambda}$, satisfying the well-known Kolmogorov–Obukhov law (see [72,85]):

$$\nu_\lambda \sim (\Theta\lambda)^{1/3}\,\Theta \sim \frac{\nu_\lambda^3}{\lambda} \sim \hat{\nu}_{m,\lambda}\left(\frac{\nu_\lambda}{\lambda}\right)^2, \tag{1.19}$$

and, consequently,

$$\hat{\nu}_{m,\lambda} \sim \Theta\lambda^2/\nu_\lambda^2. \tag{1.20}$$

Here the Θ is the rate of dissipation of turbulent energy in the cascade of vortices (this quantity can be obtained directly from the calculations of large-scale formations, see, for example, approximation (1.8)–(1.10), ν_λ is the variation of velocity (mainly the pulsation component of velocity) at distances λ. In the zones of local turbulence ($\lambda \gg \lambda_0$), we examine the relative motion of fluid particles in the section, and not the absolute motion, when the entire section –'the cell'– moves as an integral unit, which is characteristic of large scales and follows from (1.17).

The introduction, into the calculation scheme (1.14)–(1.16), of relationships of the type (1.20), makes it possible to model accurately the law of attenuation of turbulence for different scales λ. In this case, in regions of the viscous motion range, characterised by the complete dissipation of energy (at distances of $\sim h$ at $Re_h \approx 1$), the effect of molecular viscosity becomes considerable and, in this case, it is already necessary to transfer to the calculation of complete Navier–Stokes equations with the molecular dissipation mechanism. (The author is grateful to Academician A.M. Obukhov for his attention to this work. The main concept of this approach was developed after our seminar on March 30, 1978).

Thus, the new coefficients of turbulent viscosity, corresponding to 'its' scale of motion, are gradually introduced.

As the resolution scales become finer, the scale of subgrid vortices in it becomes smaller and the estimates of subgrid pulsations become 'coarser' (larger and larger part of the spectrum is involved in direct resolution). The change of the type of coefficient of effective viscosity ('gluing' of the solution) takes place when fulfilling the condition of conservation of the energy flux in the cascade of the vortices in the regions in which the 'old' solution becomes unstable and it is necessary to transfer to finer steps (scales) of the calculation grid. This 'gluing' can also be used to determine the required constants in Eqs. (1.18) and (1.20).

Evidently, the proposed cascade model reflects with sufficient adequacy the process of turbulent transfer: at finer and finer resolution, the problems of stability dictate the transition to a more complicated solution in the conditions of continuously increasing dissipation.

In this case, it should be remembered that in the determination of the spatial–time field of ordered structures in organised motion, the role of the dissipative mechanism in (1.16) is reduced to the regularisation of the solution (the detailed structure of such a dissipative mechanism is, generally speaking, not significant and its optimum form is determined, as already mentioned, from the conditions of stability by calculations using different approximation grids). However, in the investigations of non-stationary (pulsation) properties of turbulence using equations (1.18)–(1.20) one can hope to obtain only to mean-statistical characteristics of the flow because also in this case the requirements on the level of the model for subgrid turbulence are fully acceptable. It may be seen that in this approach it is not necessary to solve the Navier–Stokes equations for high Reynolds numbers.

1.6. CORRECTNESS OF FORMULATION OF THE PROBLEM

If the construction of the flow pattern of 'slow' large vortices should not evidently lead to any serious problems (this is also confirmed by calculations), then the possibility of determining the averaged-out characteristics of non-stationary ('pulsation') large-scaloe turbulent motion using smoothed-out equations is far from obvious. Calculations using smoothed-out equations with the given dissipation mechanism are carried out in time periods up to the start of formation of a steady flow, i.e. appearance of stable characteristics (in a general cases non-stationary structures). Subsequently, to determine the averaged-out characteristics of the pulsation turbulent flow (moments) it is essential to carry out appropriate statistical processing of the results. This can be carried out directly using the calculation results or probability approaches.

We shall discuss this in greater detail, taking into account Ievlev's studies [62,63,100]. The general system of equations of mechanics of solid media for hydrodynamics problems can be written in the following form:

$$\frac{\partial v_i}{\partial t}+v_k\frac{\partial v_i}{\partial x_i}=F_i,\quad i=1,2,3,$$
$$\frac{\partial \rho}{\partial t}+v_k\frac{\partial \rho}{\partial x_i}=-\rho\frac{dv_i}{dx_i},\tag{1.21}$$
$$\frac{\partial \xi(m)}{\partial t}+v_k\frac{\partial \xi(m)}{\partial x_i}=\Pi(m),\quad m=1,2,\dots,N.$$

Here the symbol $\xi(m)$ denotes all independent parameters determining the state and motion of the medium, with the exception of velocity **v** and density ρ; v_k is the component of total velocity **v** along axis x_k (repeating indices indicate summation in respect of all coordinate axes); F_i are the righthand parts of equations of the amount of motion. For example, for an incompressible fluid:

$$F_i=-\frac{1}{\rho}\frac{\partial \rho}{\partial x}+\nu\frac{\partial^2 v_i}{\partial x_k\,\partial x_k}.$$

In a turbulent flow, instantaneous values of v_i, ρ, $\xi(m)$ pulsate and are random quantities. We shall select a group of n points in the flow and use f_n to indicate the probability of different values of random quantities v_i, ρ, $\xi(m)$ at these points at the same moment of time t. Let it be that A_n denotes the set of values of v_i, ρ, $\xi(m)$ at all selected points; quantities, relating to some point γ, will be given the index (γ). Knowing the probability density f_n, we can determine the average values (mathematical expectations) of any functions of A_n, in particular, we can determine the averaged-out values of the quantities v_i, ρ, $\xi(m)$ at every examined point, one- and multipoint moments of a higher order, etc.

The equation describing the time dependence of f_n has the following form [62,63]:

$$\frac{\partial f_n}{\partial t}=\sum_{\gamma=1}^{n}\left\{-v_k^{(\gamma)}\frac{\partial f_n}{dx^{(\gamma)}}+f_n\left\langle\left(\frac{\partial v_l}{\partial x_l}\right)^{(\gamma)}\right\rangle_{A_n}-\frac{\partial}{\partial v_i^{(\gamma)}}\left[f_n\left\langle F_i^{(\gamma)}\right\rangle\right]+\right.$$
$$\left.+\frac{\partial}{\partial \rho^{(\gamma)}}\left[f_n\rho^{(\gamma)}\left\langle\left(\frac{\partial v_l}{\partial x_l}\right)^{(\gamma)}\right\rangle_{A_n}\right]-\sum_{m=1}^{N}\frac{\partial}{\partial \xi^{(\gamma)}(m)}\left[f_n\left\langle\overset{(\gamma)}{\Pi}(m)\right\rangle_{A_n}\right]\right\}.\tag{1.22}$$

Here the symbol $\langle ... \rangle$ denotes conventional mathematical expectations in the braces of the quantities for the given values of all arguments A_n of the coordinates of points and time.

The equation for f_n is exact but not closed because the conventional mathematical expectations of different quantities cannot be determined only on the basis of values of f_n (to determine them, it is also necessary to know f_{n+1} and sometimes also more multipoint distributions of probabilities).

In addition to the true values of the quantities v_i, ρ, $\xi(m)$, we can also examine their approximate values $\tilde{v}_i, \tilde{\rho}$ and $\tilde{\xi}(m)$, determined from equations with smoothed-out righthand parts:

$$
\begin{aligned}
&\frac{\partial \tilde{v}_i}{\partial t} + \tilde{v}_k^{(\gamma)} \frac{\partial \tilde{v}_i^{(\gamma)}}{\partial x_k^{(\gamma)}} = \left\langle F_i^{(\gamma)} \right\rangle \tilde{A}_n, \quad i = 1,2,3, \\
&\frac{\partial \tilde{\rho}^{(\gamma)}}{\partial t} + \tilde{v}_k^{(\gamma)} \frac{\partial \tilde{\rho}^{(\gamma)}}{\partial x_k^{(\gamma)}} = -\tilde{\rho}^{(\gamma)} \left\langle \left(\frac{\partial v_l^{(\gamma)}}{\partial x_l^{(\gamma)}} \right)^{(\gamma)} \right\rangle, \\
&\frac{\partial \tilde{\xi}^{(\gamma)}(m)}{\partial t} + \tilde{v}_k^{(\gamma)} \frac{\partial \tilde{\xi}^{(\gamma)}(m)}{\partial k_k^{(\gamma)}} = \left\langle \overset{(\gamma)}{\Pi}(m) \right\rangle_{\tilde{A}_n}, \quad m = 1,2,\ldots N,
\end{aligned}
\tag{1.23}
$$

where A_n denotes the range of values of $\tilde{v}_i, \tilde{\rho}$ and $\tilde{\xi}(m)$ at all points n.

Let us assume that $\tilde{v}_i, \tilde{\rho}$ and $\tilde{\xi}(m)$ in equation (1.23) have random values (because of the instability of solutions of smoothed-out equations or because of random initial and boundary conditions). Then, as shown by Ievlev, the equations for the probability density $\tilde{v}_i, \tilde{\rho}$ and $\tilde{\xi}(m)$ of the range $\tilde{A}_n\{\tilde{v}_i, \tilde{\rho}, \tilde{\xi}(m)\}$ will have the form identical with (1.22). If we specify identical initial and boundary values for f_n and $\tilde{f}_n$, then the functions f_n and $\tilde{f}_n$ will be completely identical, i.e. all the statistical characteristics of the flow determined from A_n and $\tilde{A}_n$ will also be identical (although the time dependences of $\tilde{v}_i^{(\gamma)}, \tilde{\rho}^{(\gamma)}$ and $\tilde{\xi}^{(\gamma)}(m)$ will not be in agreement at the same initial conditions with true distributions $v_i^{(\gamma)}$, $\rho^{(\gamma)}$ and $\xi^{(\gamma)}(m)$).

Thus, calculations carried out using smoothed-out equations may yield accurate statistical characteristics of the flow which depends on large-scale turbulence, although the detailed spatial–time pattern of this pulsation motion will not reproduce any real process.

This fundamental result will be referred to as Ievle's principle. It provides a positive answer to the question of the possibility of direct numerical modelling of non-stationary (pulsation) large-scale turbulent motion using smoothed-out motion equations. In this case it is of course necessary to define correctly the conventional mathematical expectations in the righthand parts of the smoothed-out equations (1.23). It should also be mentioned that righthand parts of the smoothed-out equations (1.23) are determined unambiguously by the quantities $v_i^{(\gamma)}$, $\rho^{(\gamma)}$ and $\xi^{(\gamma)}(m)$ in the nodes of the calculation grid (in contrast to exact equations (1.21) where the righthand parts also include additional random oscillations). Thus, 'closure' takes place here by appropriate averaging (smoothing) of the effects of subgrid small-scale turbulence. This principle also confirms the hypothesis on dynamic independence of large- and small-scale motions of turbulence at high Reynolds numbers.

The application of the method of numerical modelling is especially important in problems of mathematical physics, plasma physics, mechanics of solids (gas dynamics, elasticity theory, etc.) [19,24,45,55,91,109,114,129]. This is explained by a number of circumstances.

Difficulties in carrying out experiments

In examining phenomena taking place, for example, at hypersonic flight speeds there are high temperatures which lead to the effects of dissociation, ionisation in the flow and, in a number of cases, even in the 'glow' of the gas. In these cases, it is very difficult to model the phenomenon in laboratory and full-scale conditions because to ensure similarity between the actual situation and the modelling experiments it is no longer sufficient to satisfy only classic criteria of similarity – equality of the Mach and Reynolds numbers for the model and the nature. It is also important to ensure the equality of absolute pressures and absolute temperatures; this is possible only under the conditions of equality of the dimensions of the model and the natural object. All this leads to considerable technical difficulties and expensive experiments, not mentioning the fact that the results of experimental measurements are often limited. An identical situation also forms in the design of large tonnage vessels – the selection of the optimum shape, the prediction of the velocity in the wake behind the stern, calculation of the dynamic and strength characteristics of propellers, etc.

The active application of the method of numerical modelling [19, 24,26,45,91,109,114,129,142] and image recognition [56] in the development, on their basis, of expert systems and automatic control systems makes it possible to reduce greatly the duration of scientific and design studies. In cases in which actual experiments are difficult to carry out, and the information on the process is unclear, and indirect, mathematical modelling is practically the only investigation tool. However, the principal role of physical experiments must not be underestimated. Experiments will always be the basis of any investigation, confirming (or rejecting) a scheme and a solution in the theoretical approach.

Complexity of equations

The deep penetration of numerical methods into the mechanics of solid media and plasma physics is also explained by the fact that the equations, describing the phenomena taking place here, represent the most complex (in comparison with other areas of mathematical physics) system of integral differential equations in partial derivatives.

In a general case, this is a nonlinear system of the mixed type with the unknown form of the transition surface (where the equations change their type) and 'moving boundaries': the boundary conditions of the problem are defined on surfaces or on lines which themselves are determined during calculations. The range of variation of the initial functions is so wide that the conventional methods of analytical investigation (linearisation of equations, expansion into series, separation of a small parameter, etc.) are not suitable in a general case for a complete solution of the problem.

It should be mentioned that for the large majority of gas dynamics problems no mathematical theorems of existence and uniqueness have been proved and, in addition, it is even not believed that such theorems can be obtained. In most cases, the mathematical formulation of the problem is not defined and there is only a physical formulation which is far from the actual situation. As already mentioned, the mathematical difficulties of examining problems of this type are connected with 'strong' nonlinearity and nonstationarity of equations and also with a large number of independent variables (spatial-nonstationary problems). Of special interest are numerical methods which make it possible to investigate quantitatively some phenomena and processes which did not have previously qualitative description, for example, in economics, medicine, etc. [24,26,105].

As regards the problem of constructing systems of automation of design, there are usually three conceptual levels in this case, the problem of synthesis of complex technical systems (CTS) of the given target application (comparative analysis of alternatives of the solution and their rejection, 'compression' of the set of alternatives, etc.); the problem of analysis of CTS of the given design response (determination of the characteristics of designed CTS on the basis of verification calculations, determination of the adequacy of proposed functional characteristics of CTS, etc.); finally, the problem of formalisation and the preparation of general system programming. If the first two levels form a problem-orientated subsystem, i.e. 'producing' a branch of ADS, then the last task is the 'infrastructure' of the ADS.

We believe that the main task in the problem of 'optimum ADS' is the task of algorithmic facilities for ADS (for example, the Boeing company carried out a thorough analysis of the results of approximately 700 calculation studies carried out on computers in the 70s and 80s; it was found that 70% of the final calculation results were inaccurate because of the application of inadequate mathematical models).

Thus, the development of mathematical models and, on the basis of these models, of intellectual expert systems (IES) and the mechanics of solids, medicine, etc. adequate to the examined physical process or objects, is the main factor of the reliability of developed systems of automation of design [24,26,51,56,105,116]. We have knowingly separated the mechanics of solids as the subject area because, firstly, it covers an extremely wide sphere of applications of ADS, secondly, the mathematical models available in this area are most complicated from the viewpoint of application in a computer, and, thirdly, regardless of the existing experience in solving applied problems of the mechanics of solid media, this subject area contains a number of fundamental problems whose solution will greatly affect the success of development of ADS in related subject areas [51]. We shall mention another important special feature of the investigated problems.

It is well known that modern supercomputers carry out millions and billions of operations per second. This has opened completely new possibilities for quantitative processing of information. Computers have penetrated into almost all regions of human activity and this has greatly expanded the range of problems that can be solved by mathematical methods. It has been shown that the formulation of many problems of scientific and technical advances does not fit the formulation of the problems of theoretical mathematics associated with the 'principle of absolute accuracy' which is the basis of Aristotle's formal logics. Typical examples include the problems of processing scientific investigations, optimum planning and many others [114].

In order to use computers, it is important to have: a mathematical model of the formulated task, a stable method for solving the appropriate mathematical task, a stable algorithm, constructed on the basis of this method, and a computer programme corresponding to this algorithm.

The solutions of unstable problems with inaccurate data greatly change even in the presence of very small changes in the initial data. The algorithm reproduces a solution corresponding to the initial data. In this case, its application in a computer will give unstable results. Thus, the realisation in a computer of solutions of unstable problems within the framework of 'the concept of accuracy' of the formulation of problems of theoretical mathematics does not guarantee obtaining stable results when examining unstable problems. The methods of computer solution of unstable problems with inaccurate data relate to the class of mathematical problems which are outside the range of theoretical mathematics and are often found in formulation of the problems of technical and scientific advances [114]. In this case, the problem of finding a stable solution, adequate to the examined process, is central.

Analysis of the evolution of mathematical models, used in the mechanics of solids, shows that the evolution takes place mainly along the path of quantitative accumulation of factors determining the examined process of phenomenon. The prediction constructed on the basis of extrapolation of traditional approaches to mathematical modelling, for example, the flow of a viscous compressible glass around aircraft, shows that the numerical realisation of the problem requires a computing system with a productivity of 10^{12}–10^{14} (or more) operations per second and with a memory volume of 10^{12} layers. Not discussing the question of technical feasibility of these requirements, it is important to mention another extremely important circumstance. Analysis of the effect of the error of calculations in these volumes of arithmetic operations shows that the traditional approach is to the construction of numerical methods cannot ensure the required reliability of calculation results in these cases. A rational increase in the length of the computer word is not a solution in this case. Now it is necessary to process very large volumes of information, and the problem of 'arithmetic redundancy' associated with the build up of computation errors (especially in calculations over a long period of time) is practically insoluble.

The catastrophic increase of the volume of arithmetic losses is associated not only with the arithmetic complexity of calculations but, in a number of cases, the physical nature of the solved problems and the imperfection of applied 'traditional' algorithms. It is quite clear that the controlling quality of the designed components (complex technical systems) is the reliability of the functioning during the maximally long period of time. From the viewpoint of mathematical (numerical) modelling this indicates that it is necessary to ensure the possibility of examining the behaviour of the model for some suitable long period of time. To this it is necessary to add the greatly 'dynamic' nature of functioning of the design models, determined by the time-dependence of external influences.

From the mathematical viewpoint, the above requirement indicates that the applied 'imitation' model (the system of determining equations with boundary conditions) for an element should 'in the mean' have the property of asymptotic stability (or it should be maximally close to such a model). For asymptotically stable algorithms, the main condition of obtaining a solution is the existence of dissipative nature (i.e. no increase in the error of calculations over long periods of time) [19,24]. In this case, it is important to note that many of the currently available numerical models in the mechanics of solids do not generally have this property.

The solution to this situation is seen in the development of the principle of rational numerical modelling. Obviously, in near future, successes in the solution of the development of ADS in different subject areas will be achieved not as much by increasing the operating speed and memory of computing systems (this requires huge material expenditure) as by the development of rational models and IES.

Non-stationarity, multidimensionality, dissipative nature and nonlinearity are the characteristic features of the large number of practical problems being solved at the present time ([19,24,45,91,109,114, 129], etc.). In this connection, it is important to mention problems of the examination of dynamics of solids moving in gas or liquid and also the dynamics of oceans and atmospheres, examination of stalled flows, turbulence, the development of hydrodynamic instabilities, speed interaction of solids, filtration processes, etc. The existence of different scales, nonlinearity and nonstationary nature of the processes comprising these complex phenomena, requires a new approach to constructing numerical models. In this case, the traditional 'classical' approaches are characterised by low efficiency and require extremely long computing times. In many cases, these models cannot be applied.

One of the possible solutions of this situation is to abandon the generally accepted metric (point) evaluation of the results and transfer to a structural evaluation [19,24,26,142]. It is important to note that the transition to a structural correspondence opens new prospects in constructing a stable solution and determining its adequacy to the examined phenomenon (this situation is in complete agreement with Tikhonov's theory [114] in computer solution of 'unstable problems with inaccurate data'. In the presence of only information on the initial operator and the right-hand part of the equation there is, generally speaking, no numerical method for obtaining a stable solution over long periods of time for the examined problem. However, in transition to a structural description of the medium (in introducing 'parametric' expansion) such a solution can be obtained and the stability parameter depends on the expansion parameter – the step of the calculation grid, etc. [32, 114]). This circumstance is especially important when we discuss the modelling of physical processes on the basis of conservation laws in the most general discrete representation for large classes of functions or using statistical approaches.

Thus, modelling on the basis of the laws of conservation using a structural description of the system (large particles, finite elements, discrete vortices, etc.) with the application of the principle of asymptotic (structural) stability forms a basis for constructing methods of rational numerical modelling [19,24]. In fact, here it is important to build simulation models (macro-equations for a structural element) adequate to the examined nonlinear processes.

2.2. SOME NUMERICAL METHODS FOR CALCULATION OF MULTIDIMENSIONAL NONLINEAR PROCESSES

In this case, various numerical algorithms are constructed using, in fact, the same approach which is highly physical and adequate to the examined transition phenomena, the so-called splitting method [19]. As is always the case with computers, the integration range is covered with a fixed (Euler) calculation grid, and the modelled media is replaced with a set of the so-called liquid (Lagrange) particles present inside every elementary cell or, if this is a discrete medium (for example, rarified gas) by a set (ensemble) of some particles and correlated molecules for which the Monte Carlo method is used to construct a simulation model reflecting the investigated stochastic process.

The general principle of splitting will now be formulated [19,24]. The ideology of the given method makes it possible to carry out successively calculations in Euler–Lagrange variables so that it is possible to use 'stronger' aspects of these approaches. A specific iterative process is constructed in each time step: initially, we examine 'adaptation', i.e. we examine the motion of a subsystem only inside every cell without its displacement (without exchange with adjacent cells for a solid medium or collisional relaxation in a spatially-homogeneous case for a discrete model of a medium) and this is followed by modelling 'shear' which reflects exchange processes and displacement of the subsystems (collisionless relaxation) but already without changes of the initial condition.

The equilibrium distribution of all parameters of the medium is calculated after stabilising the process in time.

For numerical models of the method of 'large particles', developed together with Yu.M. Davydov [24] on the basis of Euler equations, instead of the set of particles in the cells we examine the mass of the entire fluid (Euler) cell as a whole – 'large particle' (hence the name of the method). Subsequently, on the basis of finite-difference or integral presentations of the laws of conservation, we examine non-stationary (and continuous) flows of these 'large particles' through an Euler grid. In fact, we use the laws of conservation written in the form of balance equations for a cell of finite dimensions (as is usually carried out in the process of deriving gas-dynamic equations, but without further limiting transition from a cell to a point).

In the FLUX method, developed by A.V. Babakov (see [19]), integral laws of conservation of the following type are used:

$$\frac{\partial}{\partial t}\oiint\limits_{\Omega}\int F\,d\Omega=-\oint\limits_{S_\Omega}\int(\mathbf{Q}_F\mathbf{n})dS,$$

where S_Ω is the surface restricting finite volume Ω; $F = \{M, X, Y, Z, E\}$; M; X, Y, Z; E is the mass, the components of the pulse and energy Ω of the volume, respectively; $\mathbf{Q}_F$ is the vector of the density of the flow of each of these quantities.

The general assumptions of the flux method and the difference approximation are shown schematically in Table 2.1. Non-symmetric 'oriented' approximations for the density of distributions of components F generate a dissipative mechanism and ensure the second order of accuracy. As a result, we obtain divergent-conservative and dissipative-stable numerical schemes which enable us to examine a wide range of complex problems of gas dynamics (supercritical conditions, turbulent flows in the wake behind a solid, diffraction problems, transitions through the speed of sound, etc.). These schemes are characterised by internal dissipation (scheme viscosity). Thus, they enable us to carry out stable computations without preliminary definition of breaks (homogeneous schemes of continuous computing).

The results of calculations of a block of problems in examination of multi-dimensional flows in aerodynamics are presented in Figs.2.1–2.6.

Statistical methods of 'particles in cells' are used widely for

Table 2.1

Main assumptions of the 'FLUX' method

$$\frac{\partial}{\partial t}\oiiint_{\Omega} F\, d\Omega = -\oiint_{S_\Omega}(\mathbf{Q}_F\,\mathbf{n})\,dS,\quad \oiiint_{\Omega} F\, d\Omega \Rightarrow M, X, Y, Z, E$$

$$\Omega\frac{F^{n+1}-F^{n}}{\tau} = -L_h\left[\oiint_{S_\Omega}(\mathbf{Q}_F\mathbf{n})dS\right].$$

$$(\mathbf{Q}_M)_i = \rho V_i,\ (\mathbf{Q}_{X_J})_i = P\delta_{ij} + (\rho V_j)V_i - \sigma_{ij},$$

$$(\mathbf{Q}_E)_i = \varepsilon V_i + PV_i - \sum_j \sigma_{ij}V_j - \lambda\frac{\partial T}{\partial X_i},\ \sigma_{ij} = (-2/3)\mu\,\mathrm{div}\,\mathbf{V}\,\delta_{ij} + \mu\left(\frac{\partial V_i}{\partial X_j} + \frac{\partial V_j}{\partial X_i}\right).$$

Difference approximation

$$\frac{F^{n+1}-F^{n}}{\tau} = -\frac{1}{\Omega}\sum_{S_i}(\mathbf{Q}_F\mathbf{n})_{S_i}S_i,$$

$$(V_x)_{l+1/2,k} = 0.5\left[(V_x)_{l+1,k} + (V_x)_{l,k}\right],$$

$$(\rho V_i)_{l+1/2,k} = \begin{cases} 1.5(\rho V_i)_{l,k} - 0.5(\rho V_i)_{l-1,k}, & (\rho V_x)_{l+1/2,k} \ge 0, \\ 1.5(\rho V_i)_{l+1,k} - 0.5(\rho V_i)_{l+2,k}, & (\rho V_x)_{l+1/2,k} < 0, \end{cases}$$

$$P_{l+1/2,k} = 0.5(P_{l+1,k} + P_{l,k}),$$

$$\left(\frac{\partial f}{\partial x}\right)_{l+1/2,k} = (f_{l+1,k} - f_{l,k})/\Delta x,$$

$$\left(\frac{\partial f}{\partial y}\right)_{l+1/2,k} = (f_{l+1,k+1} - f_{l+1,k-1} + f_{l,k+1} - f_{l,k-1})/(4\Delta y).$$

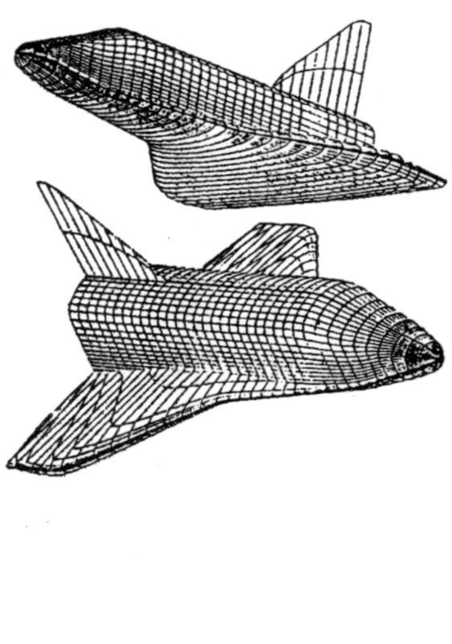

Fig.2.1. Generation of difference grids on bodies of complicated configuration (Yu.D. Shevelev).

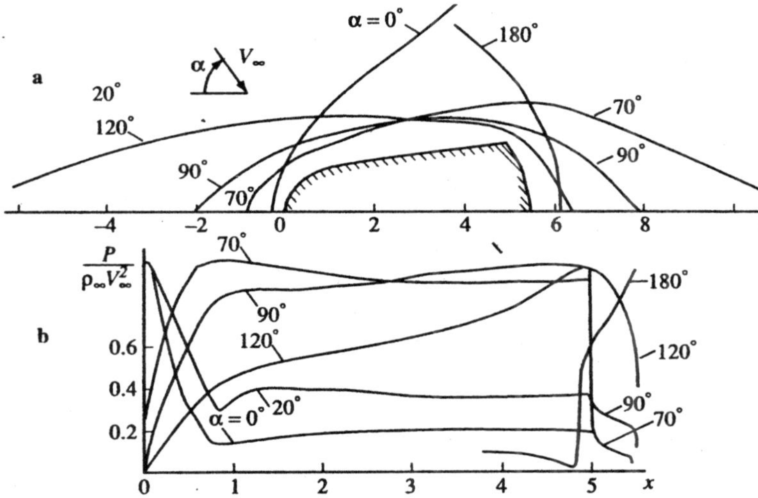

Fig. 2.2. Supersonic flow-around ($M_\infty = 2$) of descending space systems at different incidence angles α: a) position of shock waves, b) distribution of pressure in the solid (A.S. Kholodov).

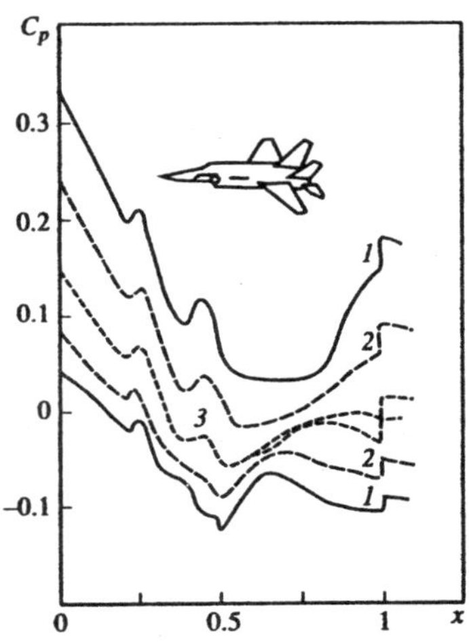

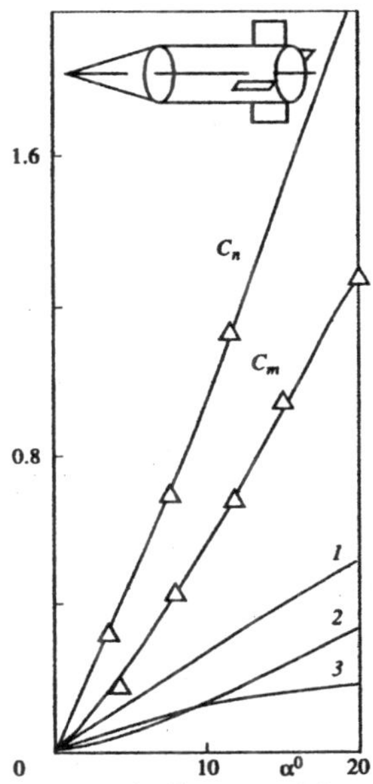

Fig. 2.3. Distribution of pressure coefficient in supersonic flow-around an aircraft ($M_\infty = 3$) and different incidence angles α: 1) for α = 20°, 2) for α = 5°, 3) for α = 0° (G.P. Voskresenskii).

Fig. 2.4. Calculation of aerodynamic coefficients of normal force and momentum in supersonic ($M_\infty = 3$) flow-around of a body with tail panels, installed symmetrically under angles: 1) for 90°; 2) for 135° and 3) for 45° (G.P. Voskresenskii).

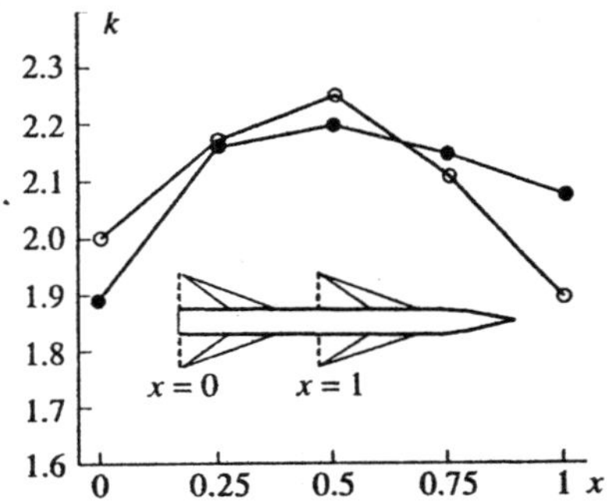

Fig. 2.5. Calculation of interference coefficients in supersonic flow-around of a configuration with complicated distribution of wings: • – calculations, ○ – experiments (Yu.D. Shevelev).

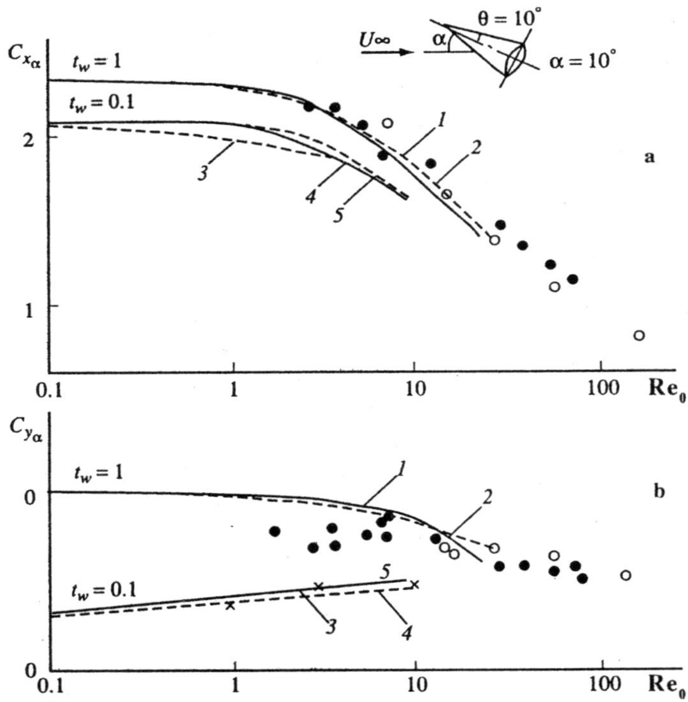

Fig. 2.6. Flow of a rarefied gas around a cone (C_x on (a) and C_y on (b)) under an incidence angle at height of $\simeq 100$ km: solid and broken lines – calculations ($\gamma = 5/3$; $\mu = T^{\omega}$; 1) for $\{M_\infty = 10, \omega = 1, t_w = 1\}$; 2) for $\{10, 0.5, 1\}$; 3) for $\{10, 1, 0.1\}$; 4) for $\{1, 0.5, 0.1\}$; 5) for $\{20, 0.5, 0.1\}$), points – experiments ($\gamma = 1.4$, $M_\infty = 5 \div 10$; ○ – $t_w = 1$, • – for $t_w = 0.5$) (A.I. Erofeev).

solving the problems of dynamics of rarefied gas (and now also turbulence phenomena). They are based on the principles of the splitting method [143] and the method of direct numerical modelling developed by V.A. Yanitskii [33] (on the basis of Monte Carlo methods), one of the most complicated equations in mechanics, i.e. the Boltzmann equation. In fact, in this case we are concerned with the computer imaging of physical processes forming the basis of investigated phenomena. We shall discuss this problem in greater detail [29,33].

Let us assume that in the region Ω with the boundary Γ we examine the flow of a Boltzmann gas from spherically symmetrical molecules. Each molecule is characterised by position **x** and by velocity **c** and they collide with each other and with boundary Γ. The laws of their elastic collisions are defined by means of differential and total sections $d\sigma_{ij}$ and $\sigma_{ij} = \int d\sigma_{ij}$, and the Boltzman equation for the molecules $f(t,\mathbf{x},\mathbf{c})$ at the moment of time t at point $(\mathbf{x},\mathbf{c})$ of the six-dimensional space which has the following form:

$$\frac{\partial f}{\partial t}+c\frac{\partial f}{\partial x}+\frac{F}{M}\frac{\partial f}{\partial c}=\int\left(f'f_1'-f\,f_1\right)g\,d\sigma\,dc_1.$$

Here **F** is the external force acting on the molecule, M is the mass molecule and $f_1 = f(t,\mathbf{x},\mathbf{c}_1)$ is the distribution of molecules for which integration is carried out, $g = |\mathbf{c}-\mathbf{c}_1|$. The apostrophe at f indicates f and f_1 calculated from velocities $\mathbf{c}'$ and $\mathbf{c}_1'$ after their collision. The task is to find a solution of the Boltzmann equation corresponding to the given initial $f(t = 0,\mathbf{x},\mathbf{c}) = f_0(\mathbf{x},\mathbf{c})$ and boundary solutions $f(t,\mathbf{x}_\Gamma,\mathbf{c}) = \int K(\mathbf{c},\mathbf{c}_1)\,f(t,\mathbf{x}_\Gamma,\mathbf{c}_1)d\mathbf{c}_1$. The approach is based on the simulation of the Boltzmann gas by a system consisting of a finite number of N particles (of the order of 1000 or 10000). For this purpose, in space Ω we introduce a relatively fine grid which divides the space into cells of the given volume. Continuous time t is replaced by discreet time $t_\alpha = \alpha\Delta t$. At the initial moment, space Ω is filled by particles. Their co-ordinates and velocities are defined by the Monte Carlo methods in accordance with the initial distribution of co-ordinates and velocities $n_0(\mathbf{x}) = \int f_0(\mathbf{x},\mathbf{c})d\mathbf{c}$ and $\varphi_0(\mathbf{x},\mathbf{c}) = f_0(\mathbf{x},\mathbf{c})/n_0(\mathbf{x})$.

The calculation of the evolution of this system of particles over a short period of time Δt is split into the following two physical stages: in the first stage we model paired collisions of particles in cells (only the velocities change in this case) and in the second stage the free-flight of particles from cell to cell and their interaction with the boundary Γ. The alternation of these stages of collisions and displacements leads to the trajectories of movement of the examined system in space whose points are represented by the numbers of particles in the cells, their position and speed. Having the set of realised trajectories, we can calculate any of the macroparameters of the initial system through appropriate independent estimates [33] of the corresponding integrals of the type

$$n(t,\mathbf{x}) = \int f \, d\mathbf{c}.$$

The main difficulty in constructing a numerical algorithm for the Boltzmann equation is the calculation of the collision integral present in the right-hand part of the equation. The splitting scheme in this approach is shown in Table 2.2.

The principal moment of our approach [33] is the fact that in the next collisional stage of the splitting scheme modelling of the Boltzmann equation without a convective derivative is replaced by introducing Katz's model which is asymptotically equivalent (at $N \gg 1$) to the Boltzmann equation in the spatially-heterogeneous

Table 2.2

Stage I	Stage II
Complete Boltzmann equation	
$\frac{\partial f}{\partial t} + \mathbf{c}\frac{\partial f}{\partial x} = I(f \cdot f_1)$	
Splitting	
Stage I	Stage II
$\frac{\partial f_\beta^*}{\partial t} = I\left(f_\beta^* \cdot f_{1\beta}^*\right),\ f_\beta^*\left(t = t_\alpha\right) = f_\beta^\alpha$	$\frac{\partial f_\beta}{\partial t} = \mathbf{c}L\left(f_\beta\right) = 0,\ f_\beta\left(t = t_\alpha\right) = f_\beta^{*(\alpha+1)}$
Markovian approximation in a cell	
$\frac{\partial f_\beta^*}{\partial t} = I\left(f_\beta^* \cdot f_{1\beta}^*\right) -$ Boltzmann equation without $\mathbf{c}\frac{\partial f}{\partial \mathbf{x}}$	
equivalent (at $N \gg 1$)	
$\frac{\partial}{\partial t}\varphi\left(t, c_1, \ldots, c_N\right) = K[\varphi] -$ Katz model	

case. The Katz model (and contrast to the Boltzmann equation) is linear and enables constructing a strictly Markovian process of collision of particles [33] for a multidimensional case [19]. Figure 2.6 shows an example of calculation of the flow of a rarefied gas around a cone under a given incidence angle carried out by A.I. Erofeev using the scheme described previously [19].

This approach was also generalised by V.E. Yanitskii for the case of examination of problems of stochastic turbulence [29].

One of the promising directions for numerical solutions of many current problems of the mechanics of solids, described by multidimensional systems of equations of the hyperbolic type, are the grid–characteristic methods developed in [88].

Of special interest is the development of effective numerical methods of solving multi-dimensional evolution equations of the purely hyperbolic type or of parabolic equations containing a hyperbolic part. These mathematical model describe many spatially-nonstationary problems of the mechanics of solid and plasma physics. The construction of the computing algorithm for the problems of this type is a very complicated problem which is usually solved in stages. Here, we shall formulate several general assumptions of the mathematical technology of constructing these schemes.

Hyperbolic nature of equations

The hyperbolic part of parabolic equations is most 'unfavourable' in the computing plan because it is a source of the formation of steep gradients in narrow zones (discontinuities in solutions are found in purely hyperbolic problems). The effective methods for solving hyperbolic equations are therefore used in a large number of applications. In fact, the method of splitting with respect to the physical processes enable a relatively formal and efficient introduction of almost any previously developed (for solving hyperbolic equations) method into the general algorithm of solving a parabolic problem, containing the hyperbolic part.

As examples, one can mention splitting in the method of the type of particles in cells (into convective and non-convective parts); McCormack schemes (non-viscous flow and a boundary layer); modification of the grid-characteristic method for problems of plasma physics (the hyperbolic part and members associated with electronic thermal conductivity), etc.

Multidimensionality

The most universal and effective approaches to solving multi-dimensional parabolic and hyperbolic equations (for explicit and implicit schemes) are various methods of splitting with respect to spatial variants. This includes the well-known scheme proposed by Pisman and Rackford, Yanenko, Samarskii, Godumov, Magomedov, Kholodov, etc. Using this approach, it is possible to generalise naturally almost any unidimensional numerical algorithm into a multidimensional case. Thus, the initial problem is greatly simplified and reduced to finding 'good' unidimensional schemes.

Nonlinearity. Conservative nature

The difference schemes, constructed for linear equations, can be generalised, generally speaking, also for the case of quasi-linear and even nonlinear equations. For explicit schemes this includes, for example, the integral–interpolation methods, for implicit schemes the Runge–Kutta method. In this case, the non-conservative scheme can 'made' conservative.

Characteristic properties of hyperbolic equations

These concept are fundamental for the given type of problem because perturbations in the medium propagate along the characteristic manifolds. By taking them into account and introducing into the numerical scheme it is possible to determine accurately the range of the dependence of the solution. This is important when preparing efficient calculation algorithms.

In addition to this, the application of the characteristic properties of the hyperbolic equation makes it possible to 'split' the initial hyperbolic system into considerably simpler conditions of compatibility along some characteristic directions (in particular, in a linear unidimensional case into mutually independent simpler transfer equations of the type $u_t \pm \lambda u_x = 0$, where $\lambda > 0$). However, for a separate condition of compatibility, it is possible to construct and analyse a very wide range of difference schemes, optimise them and generalise for the case of the initial system of equations [88]. As shown by a large number of examples, the properties included in 'elementary' scheme are also retained for the initial system of the hyperbolic equations. This type of splitting is an important element when constructing numerical methods for complicated (including nonlinear and multidimensional) systems of hyperbolic equations [88].

The method of indeterminate coefficients

It is very efficient to use the method of indeterminate coefficients (with introduction of linear spaces of these coefficients) in the stage of analysis of difference schemes for the simplest transfer equations. This approach makes it possible to construct schemes of different classes for arbitrary grid templates with positive approximation (monotonic or majorant schemes according to another terminology), playing an important role in computational mathematics. In a more general case, it is possible to confirm the absence of different schemes with positive approximation with a higher (higher than the first) order of accuracy on solutions of initial equations. The same approach was used for the most widely used grid template (both explicit and implicit) to construct grid-characteristic schemes of the second and third order of accuracy which are closest, in the examined space of the coefficients, to the schemes with positive approximation. In particular, new, more effective modifications of the Lax-Wendroff, McCormack, Rusanov, etc., different schemes, employed widely in computing practice, were obtained. This approach is highly promising also in constructing the so-called hybrid schemes for the effective regularisation of discontinuous numerical solution.

Thus, the efficiently developed apparatus of generalisation (unidimensional different schemes to multidimensional equations, linear to nonlinear cases, the method of solving hyperbolic systems – to parabolic equations, etc.) enabled us to develop a specific set of numerical schemes differing both in the accuracy and realisation (explicit or implicit, conventional or absolutely stable, etc.) and suitable for solving a wide range of nonlinear multidimensional problems of the mechanics of solid and plasma physics. In the final analysis, the basis of this 'pyramid', constructed using the given mathematical technology, consists of a specific elementary scheme. The selection of the scheme controls the properties of the computing algorithm as a whole [17,88].

These methods have proved to be highly effective in examining the characteristics of flying systems of complex shapes, investigating the problems of mechanics of deformed media, laser compression of shells [17,88], etc.

2.3 AN EFFECTIVE QUASI-MONOTONIC HYBRID SCHEME FOR GAS-DYNAMIC CALCULATIONS

The proposed method is a generalisation (for a compressible gas) of an explicit hybrid scheme described in [23] on the basis of the grid-characteristic formalism [87]. The concept of 'hybridity' was proposed for the first time by Fedorenko in 1962 [118]. In fact, this idea is used in many current schemes (TVD, ENO, UNO, etc.). This generalisation was carried out by Oparin [102]. The scheme has proved to be simple to apply, reliable and highly economical.

This scheme has the second order of accuracy on smooth solutions and, being monotonic, does not use artificial viscosity nor smoothing, nor the flux limiter procedure, used often in modern schemes of computing dynamics of the liquid. At the same time, the method makes it possible to carry out failure-free calculations in a wide range on the basis of the Mach number (practically from zero to several tens) over any period of time and enables large discontinuities (up to several orders) to be used in the initial data (density, pressure, etc.).

2.3.1 Scheme for the modelling equation of transfer

For better understanding, it is important to mention briefly the finite-difference scheme for the linear modelling equations [23]

$$f_t + af_x = 0, \quad a = \text{const}. \tag{2.1}$$

We shall consider a space-uniform grid $\Omega = \{x_i = ih, h > 0, i = 0, 1, \dots\}$ and time step τ. On Ω we determine the grid function f_i^n coinciding with the nodes of the grid with the required function f. We introduce the conservative finite-difference approximation of the transfer equation

$$\frac{f_i^{n+1} - f_i^n}{\tau} + a\frac{f_{i+1/2} - f_{i-1/2}}{h} = 0. \tag{2.2}$$

We shall examine the following two-parameter family of different schemes which depend on parameters α and β:

$$f_{i+1/2} = \begin{cases} \alpha f_{i-1}^n + (1-\alpha-\beta) f_i^n + \beta_{i+1}^n, & a \ge 0, \\ \alpha f_{i-2}^n + (1-\alpha-\beta) f_{i+1}^n + \beta_i^n, & a < 0. \end{cases} \tag{2.3}$$

The first differential approximation for equation (2.2):

$$f_t + af_x = \frac{h}{2}|a|(1+2\alpha-2\beta)f_{xx} - \frac{\tau}{2}f_{tt} =$$

$$\left[\frac{h}{2}|a|(1+2\alpha-2\beta) - \frac{\tau a^2}{2}\right]f_{xx}. \tag{2.4}$$

For a scheme with orientated differences, i.e. at $\beta = 0$, the requirement of the minimum of the approximated viscosity imposes, as indicated by (4), the following condition on α:

$$\alpha = -0.5(1-C). \tag{2.5}$$

where $C = |a|\ \tau/h$. However, for schemes with central differences ($\alpha = 0$), the identical condition is given by the equation:

$$\beta = 0.5(1-C). \tag{2.6}$$

It is also assumed that there is a monotonic grid function, for example $f_i^n, \Delta f_{i+1/2}^n \equiv f_{i+1}^n - f_i^n \geq 0$. Function f_{i+1}^n is also monotonic if the following conditions are fulfilled:

a. for schemes with $\beta = 0$ and α from the relationship (5)

$$\Delta f_{i+1/2}^n \geq \zeta(C)\Delta f_{i-1/2}^n, \quad \text{where} \quad \zeta(C) = 2\cdot\frac{(1-C)}{2-C};$$

b. for schemes with $\alpha = 0$ and β from equation (2.6)

$$\Delta f_{i+1/2}^n \geq \sigma(C)\Delta f_{i-1/2}^n, \quad \text{where} \quad \sigma(C) = 2\cdot\frac{(1+C)}{C}.$$

It may be seen that the monotonic regions of the investigated homogeneous schemes have a non-hollow intersection. Thus, there is a whole class of monotonic hybrid schemes differing from each other by the condition of switching from one homogeneous scheme to another. The general form of this condition is as follows:

$$\Delta f_{i+1/2}^n = \delta\Delta f_{i-1/2}^n, \quad \text{where} \quad \zeta(C) \leq \delta \leq \sigma(C).$$

The selection $\delta = 1$ corresponds to the points of variation of the sign of the second difference f_i^n and enables us to obtain the estimate $f_{xx} = O(h)$ for the required function f at the contact points, consequently, it is possible to retain the second order of approximation with respect to spatial variables in smooth solutions.

It is convenient to use the conditions of switching in the following form: if $a\left(\left|f_{i+2}^n - f_{i+1}^n\right| - \left|f_i^n - f_{i-1}^n\right|\right) > 0$, then we use the scheme with $\beta = 0$, otherwise the scheme with $\alpha = 0$ is used.

For smooth solutions, the given scheme has the second order of approximation with respect to time and spatial variables. It is stable if the Courant condition is fulfilled and is monotonic [23].

2.3.2. Generalisation for the system of equations of unidimensional gas dynamics

We shall write the Euler equation for the unidimensional planar case in the following form:

$$\frac{\partial \mathbf{W}}{\partial t} + \frac{\partial \mathbf{F}}{\partial x} = 0, \tag{2.7}$$

or

$$\frac{\partial \mathbf{W}}{\partial t} + \mathbf{A}\frac{\partial \mathbf{F}}{\partial x} = 0 \tag{2.8}$$

where

$$\mathbf{W} = \begin{pmatrix} \rho \\ \rho u \\ \rho E \end{pmatrix}, \quad \mathbf{F} = \begin{pmatrix} \rho u \\ \rho u^2 + p \\ (\rho E + p)u \end{pmatrix}.$$

$$\mathbf{A} = \frac{\partial \mathbf{F}}{\partial \mathbf{W}} = \begin{pmatrix} 0 & 1 & 0 \\ p_\rho - u^2 & 2u + p_{(\rho u)} & p_{(\rho E)} \\ -uE + p_\rho u - \frac{p}{\rho}u & E + \frac{p}{\rho} + up_{(\rho u)} & u\left(1 + p_{(\rho E)}\right) \end{pmatrix}.$$

The system of equations (7) is hyperbolic. Characteristic relationships [87] are used. Let it be that $\mathbf{l}_k$ and $\mathbf{r}_k$ are the left and right eigenvectors of the matrix **A**, corresponding to the eigennumbers λ_k ($\lambda_1 = u + c$, $\lambda_2 = u$, $\lambda_3 = u - c$): $\mathbf{l}_k^T\mathbf{A} = \lambda_k \mathbf{l}_k^T$, $\mathbf{A}\mathbf{r}_k = \lambda_k \mathbf{r}_k$; or in the matrix form: $\Omega\mathbf{A} = \Lambda\Omega$, $\mathbf{A}\Omega^{-1} = \Omega^{-1}\Lambda$, where $\Omega^T = \{\mathbf{l}_1\ \mathbf{l}_2\ \mathbf{l}_3\}$, $\Omega^{-1} = \{\mathbf{r}_1\ \mathbf{r}_2\ \mathbf{r}_3\}$, $\Lambda = \{\lambda_k\}$ is the matrix with the eigennumbers distributed along the main diagonal of the matrix, and the remaining elements are equal to 0. Here $p = p(\rho, \rho u, \rho E)$ is the equation of state in the general form; p_ρ, $p_{(\rho u)}$ and $p_{(\rho E)}$ are the appropriate derivatives; $c^2 = p_\rho + u p_{(\rho u)} + (E + p/\rho)\, p_{(\rho E)}$. In a partial case of the ideal gas $p_\rho = 0.5(\gamma-1)\, u^2$, $p_{(\rho u)} = -(\gamma-1)\, u$, $p_{(\rho E)} = \gamma-1$ and $c^2 = \gamma p/\rho$.

The matrices of the right and left eigenvectors have the form:

$$\Omega^{-1} = \begin{pmatrix} 1 & 1 & 1 \\ u - c & u & u + c \\ E + \dfrac{p}{\rho} - cu & E + \dfrac{p}{\rho} - \dfrac{c^2}{p_{(\rho E)}} & E + \dfrac{p}{\rho} + cu \end{pmatrix}$$

and

$$\Omega^T = \frac{1}{2c^2}\begin{pmatrix} p_\rho + uc & 2(c^2 - p_\rho) & p_\rho - uc \\ p_{(\rho u)} - c & -2p_{(\rho u)} & p_{(\rho u)} + c \\ p_{(\rho E)} & -2p_{(\rho E)} & p_{(\rho E)} \end{pmatrix}.$$

If **A** = const, the system of equations (2.8) after multiplication from the left by Ω can be written in the following form:

$$\frac{\partial g_k}{\partial t} + \lambda_k = \frac{\partial g_k}{\partial x} = 0, \quad \text{where} \quad g_k = \mathbf{l}_k^T\mathbf{W}, \quad k = 1, 2, 3.$$

This means that system (2.8) breaks up into three linear equations. The system can be solved by the same method as the previously examined linear modelling equation. This justifies the application of the solution of the nonlinear system (2.8) of the following conservative different scheme:

$$\frac{\mathbf{W}_i^{n+1} - \mathbf{W}_i^n}{\tau} + \frac{\mathbf{F}_{i+1/2}^n - \mathbf{F}_{i-1/2}^n}{h} = 0,$$

where

$$\mathbf{F}^n_{i+1/2} = \Omega^{-1}\{\alpha_k\lambda_k\}\Omega\mathbf{W}^n_{i-1} + \Omega^{-1}\{\beta_k\lambda_k\}\Omega\mathbf{W}^n_i +$$
$$+\Omega^{-1}\{\gamma_k\lambda_k\}\Omega\mathbf{W}^n_{i+1} + \Omega^{-1}\{\delta_k\lambda_k\}\Omega\mathbf{W}^n_{i+2}.$$

The matrices of the eigenvectors and eigenvalues are calculated to the point $i + ½$. The coefficients α_k, β_k, γ_k, δ_k are selected from the conditions of switching separately for each characteristic: $G_k = \lambda_k\left(\left|(g_k)^n_{i+2} - (g_k)^n_{i+1}\right| - \left|(g_k)^n_i - (g_k)^n_{i-1}\right|\right)$:

Table 2.3

$\lambda_k \geq 0$		$\lambda_k < 0$	
$G_k \geq 0$	$G_k < 0$	$G_k \geq 0$	$G_k < 0$
$\alpha_k = -0.5(1-c_k)$ $\beta_k = 1-\alpha_k$ $\gamma_k = 0$ $\delta_k = 0$	$\alpha_k = 0$ $\beta_k = 1-\gamma_k$ $\gamma_k = 0.5(1-c_k)$ $\delta_k = 0$	$\alpha_k = 0$ $\beta_k = 0$ $\gamma_k = 1-\delta_k$ $\delta_k = -0.5(1-c_k)$	$\alpha_k = 0$ $\beta_k = 0.5(1-c_k)$ $\gamma_k = 1-\beta_k$ $\delta_k = 0$

where $c_k = \Delta t\,|\lambda_k|\,/h$ is the Courant number corresponding to the k characteristic.

As a test of the given numerical procedure, we shall examine the well-known problem of breakdown of a discontinuity. The integration region is a unit section which is divided into halves and each half has its own values of density, pressure and speed at the initial moment of time. In the given case, the initial values for the gas with $\gamma = 1.4$ are as follows: $\rho_L = 1$, $\rho_R = 0.125$, $P_L = 100$, $P_R = 0.1$ and $u_L = u_R = 0$. In this case, the exact solution consists of a shock wave moving to the right, and a rarefaction wave moving to the left. A contact discontinuity, displaced to the left, is situated between them. This example was specially made more complicated by the presence of a pressure gradient of three orders of magnitude. In addition to this, the rarefaction wave contains a 'dangerous' region in which the sound characteristic changes the sign $|u| = c$ (u and c are the local values of the speed of the flow and the speed of sound). In the vicinity of this region, certain numerical methods are characterised by an error leading to a non-physical rarefaction jump. Figures 2.7–2.9 show the results of calculation of this problem for a sequence of doubling grids (the grid with the lowest density consists of 100 cells) for the moment of time t =

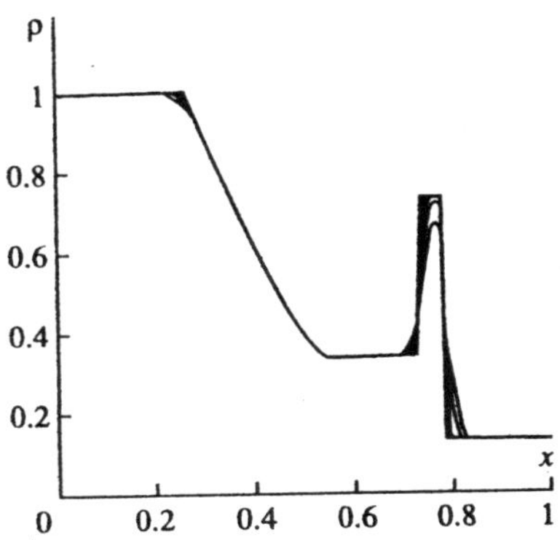

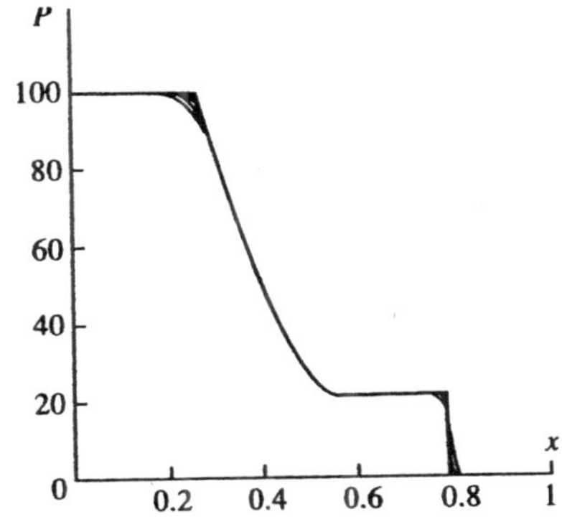

Fig.2.7. Breakdown of a discontinuity. Calculated profiles of density for a sequence of finer and finer grids (100, 200, 400,...).
Fig.2.8. Breakdown of a discontinuity. Calculated profiles of pressure for a sequence of finer and finer griods (100, 200, 400,...).

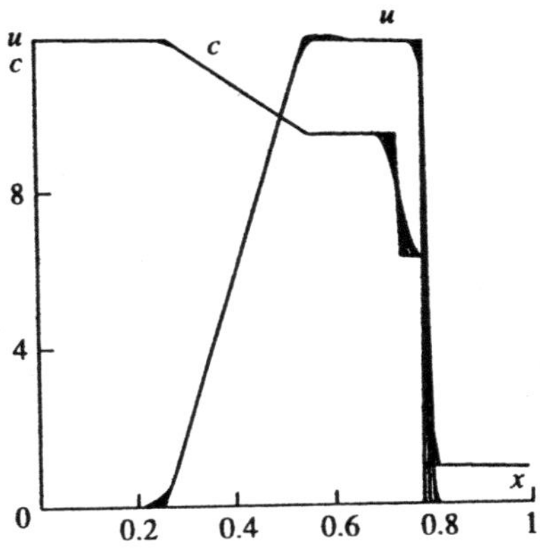

Fig.2.9. Breakdown of a discontinuity. Calculated profiles of the velocity of sound (c) anf flow velocity for a sequence of finer and finer grids (100, 200, 400...).

0.02 (the actual step with respect to time was determined using the condition of stability with a Courant number of 0.4). A rapid convergence to the exact solution is observed. The quality of the numerical solution is good. The fronts of the shock wave and the contact discontinuity (to a slightly lower degree) are relatively steep and contain no oscillations.

The numerical method was generalised for planar and spatial cases using a relatively standard technology of splitting with respect to the spatial variables.

2.4 SUBSTITUTION OF GRID FUNCTIONS OF DEPENDENT VARIABLES IN FINITE-DIFFERENCE EQUATIONS

A method has been proposed for constructing conservative difference schemes for weak solutions of the Euler equations in non-conservative variables. The method is based on the application of identical transformations of the finite-difference equations for the grid functions of conservative variables in equations for the grid functions of arbitrary variables, including: (p, **v**, h), (p, **v**, e), (ρ, **v**, e), (p, **v**, ρ). Rules have been formulated for the substitution of the finite differences of the grid functions of dependent variables in finite-difference equations, leading to equivalent finite-difference equations. It has been shown that for a numerical method based on the local-characteristic approach, the proposed method of replacement of variables in the discrete form greatly simplifies the finite-difference equations whilst maintaining conservative nature. The results are presented of testing the schemes in the variables (p, **v**, h), (p, **v**, e), (ρ, **v**, e), (p, **v**, ρ) equivalent to Harten (TVD2) and Young (UNO3) schemes showing the correctness and computing efficiency of the new methods.

It is well known that the laws of conservation in the mechanics of solids lead to differential equations in the divergent form:[1]

$$\partial_t \mathbf{U} = -\nabla_k \mathbf{F}^k(\mathbf{U}), \tag{2.9}$$

where **U** is the vector of conservative variables, representing the densities of the conserved quantities (mass, pulse, energy), ∇_k is the derivative with respect to direction k, and $\mathbf{F}^k$ is the vector of the fluxes of the appropriate quantities. The equations may be written in the non-divergent form

$$\partial_t \mathbf{U} = -A^k \nabla_k \mathbf{U}, \tag{2.10}$$

where $A^k = \partial \mathbf{F}^k / \partial \mathbf{U}$ is the Jacobi matrix of the vector of fluxes $\mathbf{F}^k$ in relation to the vector of conservative variables. It is assumed that **V** is the vector of variables, linked with the components of vector **U** by some mutually-unambiguous, continuously differentiated functional dependence[2] $\mathbf{U} = U(\mathbf{V})$ which is such that det[∂U/

[1]Equations are written in the Cartesian coordinate system. The rule of summation with respect to the repeating index $k = 1, 2, 3$ is used:

[2]Usually it is assumed that $\mathbf{U} = (t, \mathbf{r})$ and $\mathbf{V}$ $(t, \mathbf{r})$ are continuously differentiated functions of their arguments in some spatial–time region.

$\partial \mathbf{V}] \neq 0$. The requirement for the mutual unambiguity of the transformation corresponds to the equivalence of describing the state of the medium by means of the vectors **U** and **V**.

The transition to non-conservative (sometimes referred to as 'simplest') variables leads to a system of equations for **V** in the non-divergent form:

$$\partial_t \mathbf{V} = -B^k \nabla_k \mathbf{V}, \tag{2.11}$$

where $B^k = M^{-1}A^k M$, $M = \partial \mathbf{U}/\partial \mathbf{V}$.

Whilst the integral form of the laws of conservation permits discontinuities in the solutions of the functions, the differential description based on equations (2.9)–(2.11) usually assumes the continuous differentiability of the latter. The generalisation of the differential description for the case of solutions permitting discontinuities leads to the concept of weak solutions. The weak solutions correspond to the equations (2.9)–(2.11) in the regions of smoothness and to relationships linking the jumps of different variables at discontinuities, where the latter can usually be obtained using the appropriate differential equations or it can be obtained on the basis of more complete physical models. It is natural to assume that a correct discrete model should approximate (2.9) or (2.11) in the range of smoothness of the solution and to a certain degree should 'approximate' the appropriate solutions at discontinuities. The models using the separation discontinuities within the framework of a numerical algorithm, and the so-called methods of continuous computing are used in most cases. In the latter case, to 'approximate' relationships at discontinuities, it is necessary to use special methods of calculating the fluxes at the edges of the cells based on the exact (by Godunov's method [44]) and approximate (for example, the Roe method [22]) solution of the problem of breakdown of an arbitrary discontinuity. At present, conservative numerical methods for solving the equations of the mechanics of solids, based on the approximation of equations in the divergent form (2.9) have been developed and are used widely. In this case, using the concept of a reference volume, the approximation of equations (2.9), expressing the finite-difference analogues of the laws of conservation for the vector **U** in the integral form is constructed in such a manner that the increment of the vector **U** in the computing cell is represented in the form of a sum of fluxes through its edges [44]. In studies by Harten et al. (see, for example, [168–171,227]) methods were proposed of calculating the fluxes at the edges of the cells leading to the schemes of a high order of approximation in a one-dimensional case, schemes TVD [168], UNO [170], etc. In the solution of the spatial problems, the latter also ensure high accuracy in the region of smoothness and good resolution of discontinuities. The difference methods for solving the equations in the simplest variables are not used widely because of the fact that the direct approximation of equations (2.10) results in non-conservative schemes. We shall show how to obtain conservative numerical methods in non-conservative variables. The concept is based on the fact that, using as the finite-difference equations for grid vector functions $\mathbf{U}_h^\tau$ as the basis, the equation for $\mathbf{V}_h^\tau$ is obtained as a result of identical transformations of discrete equations for $\mathbf{U}_h^\tau$ and not by approximation of equations (2.10). The existence of an equivalent conservative difference scheme guarantees the conservative properties of the method, and the set of the variables will be determined for a specific problem by the suitable form of the equation of state, numerical efficiency, etc.

The replacement of the dependent variable and differential equations (2.10) is carried out using an equation describing the relationship of infinitely small increments of the vectors $\mathbf{U}$ and $\mathbf{V}$: $d\mathbf{U} = M\ d\mathbf{V}$. We shall examine transformations of the finite-difference equations, based on the application of an exact relationship for finite differences of the grid vector function $\mathbf{U}_h^\tau$ and $\mathbf{V}_h^\tau$:$\Delta\mathbf{U}_h^\tau = M\Delta\mathbf{V}_h^\tau$.

Let us assume that the vectors $\mathbf{U}\in\Omega_U\subset\Re^m$ and $\mathbf{V}\in\Omega_V\subset\Re^m$ are linked by the transformation $\mathbf{U} = U(\mathbf{V})$. Here m is the number of independent parameters of the state (thermodynamic variables, components of pulse or velocity, etc.), determining unambiguously the thermomechanical state of the medium. The regions of permissible values of Ω_U and Ω_V are determined by the conditions of physical feasibility of the appropriate states; for example, temperature, density, internal energy cannot have negative values. We introduce the matrix operator acting on the set Ω_V of permissible values of vector $\mathbf{V}$ which is such that the differences of any two vectors υ_1, $\upsilon_2\in\Omega_V$ are related to the difference of their images in Ω_U, i.e. the following equality is valid

$$U(\upsilon_2)-U(\upsilon_1)=M(\upsilon_2,\upsilon_1)(\upsilon_2-\upsilon_1),\quad \det(M(\upsilon_2,\upsilon_1))\neq 0. \tag{2.12}$$

Example 1. Let it be that $u=U(\upsilon)=\upsilon^n\ \forall\upsilon_2,\upsilon_1\in\Re^1$,

$$u(\upsilon_2)-u(\upsilon_1)=(\upsilon_2^{n-1}+\upsilon_2^{n-2}\upsilon_1+...+\upsilon_2\upsilon_1^{n-2}+\upsilon_1^{n-1}\ (\upsilon_2-\upsilon_1),$$
$$M(\upsilon_2,\upsilon_1)=\upsilon_2^{n-1}+\upsilon_2^{n-2}\upsilon_1+...\upsilon_2\upsilon_1^{n-2}+\upsilon_1^{n-1}.$$

Since we have exactly n terms of the order $n-1$, it can be written that: $M(\upsilon_2,\upsilon_1)=n\bar{\upsilon}^{n-1}$ is the derivative $\partial u/\partial\upsilon$ at point $\bar{\upsilon}$; $\bar{\upsilon}=\left(\upsilon_2^{n-1}+\upsilon_2^{n-2}\upsilon_1+...\upsilon_2\upsilon_1^{n-2}+\upsilon_1^{n-1}\right)^{1/(n-1)}/n$ is the mean of the values of υ_1 and υ_2, i.e., the point at which the value of the derivative coincides with the tangent of the angle of the secant:

$$\frac{\partial u}{\partial\upsilon}\left(\bar{\upsilon}\right)=\frac{u(\upsilon_2)-u(\upsilon_1)}{\upsilon_2-\upsilon_1}.$$

Example 2. Roe proposed a method for intepreting grid differences of the conservative vector of the flux $\Delta\mathbf{F}$ and the vector of conservative variables $\Delta\mathbf{U}$ as a sum of jumps $\Delta\mathbf{F}=\Delta\Sigma\mathbf{F}_i$, $\Delta\mathbf{U}=\Delta\Sigma\mathbf{U}_i$, on which the relationships are fulfilled, $\Delta\mathbf{F}_i=\lambda_i\,\Delta\mathbf{U}_i$, where λ_i is the eigenvalue of the Jacobi matrix of the vector of the fluxes $\mathbf{A}=\partial\mathbf{F}/\partial\mathbf{U}$. In the method proposed by Roe, this is achieved by introducing a special procedure for averaging the vector of the solution at the edges of the cells (we shall return to it later) which is such that

$$\Delta\mathbf{F}=\tilde{A}\Delta\mathbf{U}.$$

$\tilde{A}=A\left(\tilde{\mathbf{U}}\right),\Delta\mathbf{U}=\mathbf{U}_2-\mathbf{U}_1,\tilde{\mathbf{U}}$ is the mean (according to Roe) of the derivative states $\mathbf{U}_1$ and $\mathbf{U}_2$. For more details, please see studies by Roe [219,220]. For our purposes, it is clear that the matrix $\tilde{A}$ gives an example of the previously determined operator M and in the given specific case $\mathbf{V}$ is a vector of conservative variables, and the transformation of U is determined by the functional dependence of the vector of the fluxes from the vector of the conservative variables $\mathbf{F}=U(\mathbf{U})$.

Any difference scheme, approximating equation (2.9) for the grid function $\mathbf{U}_h^\tau$, determined on the spatial–time grid $\omega_\tau\times\omega_h$ (ω_τ and ω_h are respectively the time and spatial division of the calculation region), can be written in the vector form [1]

---

[1] It is assumed that the boundary conditions are included in the given system

$$\Phi(\mathbf{U}_h^\tau)=0. \tag{2.13}$$

This equation expresses only that we have N nonlinear algebraic equations with respect to the number of components of the vector Φ for N unknown components of the grid function $\mathbf{U}_h^\tau$, representing the required solution in the nodes of the calculation grid. As regards the vector form of the finite-difference equations, the reader should refer to [110]. For identical transformations of nonlinear vector finite-difference equations from some grid functions to others, it is necessary to replace finite differences. Therefore, in addition to the vector $\mathbf{U}_h^\tau$ with dimension N, we introduce vector $\mathbf{k}_h^\tau$ with the dimension ½(N – 1) N in respect to the number of possible combinations, with the elements of the vector being the finite differences

$$\Delta_{qr}^{sp}U=U_q^s-U_r^p\,,\quad \text{where } \left(s,p\in\omega_t,q,r\in\omega_k\right) \tag{2.14}$$

of the grid function U_h on the grid $\omega_\tau\times\omega_h$. Instead of $\Phi(\mathbf{U}_h^\tau,\mathbf{k})=0$, we shall write

$$\Phi(U_j^k,\Delta_{qr}^{sp}U)=0, \tag{2.15}$$

indicating for which elements of the vector there is a dependence for the specific methods. Equation (2.15) will be regarded as the representation of the finite-difference equation (2.13) having in mind the relationship (2.14). In all cases, the expression of the type $U(\mathbf{V}_h^\tau)$ or $M\mathbf{k}$ will denote the element by element application of given operations. It should be stressed that the elements $\mathbf{V}_h^\tau$ and $\mathbf{k}$-vectors from $\Re^m$, where m is the number of equations (2.1), and the region of the permissible value is determined by the conditions of realisability, resulting from the physical sense of the individual components, for example, the density and internal energy should be positive, etc. We shall write a nonlinear vector finite-difference equation $\Phi^U(\mathbf{U}_h^\tau)$ = 0 for the system (2.9) in the form of the dependence of the values of the grid function and the vector of grid differences, determined on the space–time grid $\omega_\tau\times\omega_h$

$$\Phi(U_j^k,\Delta_{qr}^{sr}U)=0,\quad \text{where } (k,s,p\in\omega_\tau,q,r,j\in\omega_h). \tag{2.16}$$

It should be mentioned that taking into account the relationships $\Delta^{sp}_{qr}U = U^s_q - U^p_r$, the system is closed. Using the definition of the operator M (2.12), we carried out identical transformations of the equations for the grid functions $\mathbf{U}^\tau_h$ in finite-difference equation for the grid functions $\mathbf{V}^\tau_h$:

$$\Phi^V\left(\mathbf{V}^\tau_h\right)=\Phi^{'}\left(V^k_j,\Delta^{sp}_{qr}V\right)=\Phi(U(V^k_j);\quad M\Delta^{sp}_{qr}V)=0, \tag{2.17}$$

where $M\Delta^{sp}_{qr}V=M(V^s_q,V^p_r)(V^s_q-V^p_r)$ and $(k,s,p\in\ \omega_\tau,\ j,q,r\in\ \omega_h)$.

Because of the large numbers of methods of writting equations (2.13) in the form (2.14) which results from the definition of the vector of grid differences, the proposed method can be used to obtain different finite-difference equations $\Phi^V(\mathbf{V}^\tau_h) = 0$. For example, the explicit difference scheme for the determination of grid functions U_h^{n+1} may be written in the form

$$U^{n+1}_h=\Phi\left(U^k_j,\Delta^{sp}_{qr}U\right),$$

$$\text{where}\quad \left(k,s,p\leq n\in\omega_\tau,k,s,p\in\omega_\tau,q,r,j\in\omega_h\right), \tag{2.18}$$

or in the δ-form:

$$U^{n+1}_h-U^n_h=\Phi\left(U^k_j,\Delta^{sp}_{qr}U\right). \tag{2.19}$$

The transition to grid functions $\mathbf{V}^\tau_h$ in equation (2.10) results in the explicit scheme:

$$V^{n+1}_h=u^{-1}\left(\Phi\left(U\left(V^k_j\right),M\Delta^{sp}_{qr}V\right)\right). \tag{2.20}$$

The transition to grid functions $\mathbf{V}^\tau_h$ in equation (2.18) leads to equations in finite differences

$$M\left(V^{n+1}_h,V^n_h\right)\left(V^{n+1}_h-V^n_h\right)=\Phi\left(U\left(V^k_j\right),M\Delta^{sp}_{qr}\right). \tag{2.21}$$

The given finite-difference equation is very similar to the corresponding differential equation $M\partial_t\mathbf{V} = (A_kM)\nabla_k\mathbf{V}$, obtained in transition from (2.10) to (2.11). The previously defined matrix operator $M(\upsilon_1,\upsilon_2)$ is identical to the Jacobi matrix M in the following respect. Whereas the Jacobi matrix M links infinitely small increments of the functions **U** and **V**, the operator $M(\upsilon_1,\upsilon_2)$ links their finite differences. For a large number of sets of functions, characterising the thermomechanical state of the medium, we can construct, and not by one method only, simply analytical equations for the operator $M(\upsilon_1, \upsilon_2)$. In addition, it appears in many cases that $M(\upsilon_1,\upsilon_2) = M(\tilde{\upsilon})$, where M is an analytical expression for the appropriate Jacobi matrix, and the vector of state $\tilde{\upsilon}$ is obtained as a result of averaging states υ_1,υ_2. The Roe method provides a suitable example. Like in the case of equation (2.11) which when presented in the simplest variables **V** are simpler than the quasi-linear form of the equations in conservative variables (2.10), for an entire family of currently available numerical methods, based on the application of the local-characteristic approach, the finite-difference equations for grid functions $\mathbf{V}_h^\tau$ are simpler than equations in the finite differences for grid functions $\mathbf{U}_h^\tau$. Equations (2.17) are obtained from equations (2.16) using identical transformations, and we consequently obtain a simple relationship between the solutions on the basis of equations (2.17) and (2.16), and for the same initial data these solutions coincide to a certain degree. We introduce the relationship of the equivalence of the finite-difference schemes for different grid functions $\mathbf{U}_h^\tau$ and $\mathbf{V}_h^\tau$, linked by the function of dependence $\mathbf{U} = U(\mathbf{V})$, and we shall assume that this scheme is not degenerated.

Determination. The finite-difference equation for the grid vector functions $\mathbf{U}_h^\tau$ and $\mathbf{V}_h^\tau$ will be regarded as equivalent if from the condition $U_h^k = U(V_h^k)$ for all $k \leq n$ the equations give $U_h^{n+1} = U(V_h^{n+1})$, where U is some generally nonlinear vector function.

It is obvious that the solutions of the equivalent finite-difference equations coincide in the sense $U_n^k = U(V_h^k)$, for all $k \in \omega_\tau$ for the corresponding selection of the initial data and boundary conditions and on the condition that calculations are carried out on an 'ideal' computer carrying out arithmetic operations with an infinite number of significant numbers.

It may easily be shown that the equations (2.20) and (2.21) are equivalent to the initial finite-difference equation for **U** in terms of the previously described definition. For equation (2.20), this follows immediately because if $U_h^k = U(V_h^k)$ $\forall k \leq n$, then $U(V_i^{n+1})$ satisfies the equation for U_h^{n+1}. In equations (2.21) in the same conditions $U_h^k = U(V_h^k)$ $\forall k \leq n$ and $\forall i \in \omega_h$, the value $M(V_i^{n+1}, V_i^{n+1} - V_i^n)$ satisfies the equations for $U_i^{n+1} - U_i^n$. From this we obtain:

$$U_i^{n+1} = U_i^n + M\left(V_i^{n+1}, V_i^n\right)\left(V_i^{n+1}, V_i^n\right) =$$
$$= U_i^n + U\left(V_i^{n+1}\right) - U\left(V_i^n\right) = U\left(V_i^{n+1}\right),$$

which uses the definition of the operator M (2.12) and the initial assumptions $U_i^k = U(V_i^k)$ for all $k \leq$ n.

Since the equations (2.20) are already resolved in relation to unknown grid functions on the $(n + 1)$-th time layer, two methods of integration with respect to time can be investigated for equation (2.21).

Method 1 (introduction of internal iterations)

Step 1: $\mathbf{k}_h^{n+1n} = \Phi\left(U\left(V_j^k\right),\ M\Delta_{qr}^{sp}V\right),$

Step 2: $V_h^{k+1} = V_h^n + \left(M\left(V_h^k, V_h^n\right)\right)^{-1}\mathbf{k}_h^{n+1n}$ (2.22)

Method 2 (application of the property of equivalents)

Step 1: $\mathbf{k}_h^{n+1\,n} = \Phi\left(U\left(V_j^k\right), M\Delta_{qr}^{sp}V\right),$

Step 2: $M\left(U^{-1}(U\left(V_h^n\right)\left(+\mathbf{k}_h^{n+1\,n}\right), V_h^n\right)\left(V_h^{n+1} - V_h^n\right) = \mathbf{k}_h^{n+1\,n},$

$$V_h^{n+1} = V_h^n + \left(M\left(U^{-1}\left(U\left(V_h^n\right) + \mathbf{k}_h^{n+1\,n}\right), V_h^n\right)\right)^{-1}\mathbf{k}_h^{n+1\,n}. \quad (2.23)$$

The meaning of the vector $\mathbf{k}_h^{n+1n}$ introduced here is very simple: for $\forall\, i \in \omega_h$ $k_i^{n+1n} = U_i^{n+1} - U_i^n$ and equivalent equations for grid functions $\mathbf{U}_h^\tau$.

It is important to note that the introduced transformations of the equations in the finite differences according to the rules

$$\Delta\mathbf{U}_h^\tau \to M\Delta\mathbf{V}_h^\tau,\quad \mathbf{U}_h^\tau \to U\left(\mathbf{V}_h^\tau\right), \quad (2.24)$$

where M is determined by the condition (2.12), do not change the graph of the initial equations on the spatial–time grid $\omega_\tau \times \omega_h$ (spatial division may be structurised, partly structurised or non-structurised;

interpretation of division – points in space, finite volumes, particles – also has no meaning for the application of the given approach, because we transform relationships written in discrete form).

The proposed mathematical method can be regarded as a method of solving the initial system of nonlinear algebraic equations for $\mathbf{U}_h^\tau$ by the method of replacement of variables according to the rules (2.24). From the computational viewpoint, this procedure with the appropriate selection of the variables results in simplification of the equations and greater computing efficiency of the method.

Example 3. We examine a one-dimensional equation with the characteristic form

$$U_t + \lambda U_x = 0$$

and the difference scheme of the type 'explicit corner':

$$U_i^{k+1} - U_i^k = -\sigma\left(U_i^k - U_{i-1}^k\right), \tag{2.25}$$

where $\sigma = \lambda \Delta t / \Delta x$. There are also variables $\upsilon : \upsilon^2 = U$. Evidently, the differential equation for υ has the form:

$$\upsilon_t + \lambda \upsilon_x = 0. \tag{2.26}$$

The equation for υ in the finite differences can be obtained directly approximating (2.26):

$$\upsilon_i^{k+1} - \upsilon_i^k = -\sigma\left(\upsilon_i^k - \upsilon_{i-1}^k\right).$$

From condition $U(x,0) = \upsilon^2(x,0)$ in the differential formulation we obtain:

$$U(x,t) = U\left(x - \lambda t, 0\right) = \upsilon^2\left(x - \lambda t, 0\right) = \upsilon^2\left(x,t\right) \text{ at } t > 0.$$

In this case, we are not really concerned with the equations in finite differences. If for all j from the regional definition $U_j^k = (\upsilon_j^k)^2$ at $k = n$, then $U_j^{n+1} \neq (\upsilon_j^{n+1})^2$, which can be easily confirmed by direct verification. If the equation for U expresses the law of conservation, the difference equations for υ are non-conservative, and vice versa. Let us assume that U is a conservative variable and we obtain equations in the differences for υ directly from (2.17) by replacement of the finite differences, as proposed previously. Initially, it is necessary to construct the operator M such that

$$(\upsilon_2)^2-(\upsilon_1)^2=M(\upsilon_1,\upsilon_2)(\upsilon_2-\upsilon_1);$$

and evidently

$$M(\upsilon_1,\upsilon_2)=\upsilon_2+\upsilon_1.$$

Carrying out the replacement of the finite differences using the operator of the equivalent substitution M, we obtain difference equations in the variables υ:

$$M\left(\upsilon_i^{n+1},\upsilon_i^n\right)\left(\upsilon_i^{n+1}-\upsilon_i^n\right)=-\sigma M\left(\upsilon_i^n,\upsilon_{i-1}^n\right)\left(\upsilon_i^n-\upsilon_{i-1}^n\right)$$

or

$$\begin{aligned}&\left(\upsilon_i^{n+1}+\upsilon_i^n\right)\left(\upsilon_i^{n+1}-\upsilon_i^n\right)=-\sigma\left(\upsilon_i^n+\upsilon_{i-1}^n\right)\left(\upsilon_i^n-\upsilon_{i-1}^n\right),\\&\left(\upsilon^{n+1}\right)^2-\left(\upsilon^n\right)^2=-\sigma\left[\left(\upsilon_{i+1}\right)^2-\left(\upsilon_{i-1}\right)^2\right].\end{aligned}\tag{2.27}$$

The integration method, determined previously, are written in the following form.

$$\left(\upsilon_i^k+\upsilon_i^n\right)\left(\upsilon_i^{k+1}-\upsilon_i^n\right)=-\sigma\left(\upsilon_i^n+\upsilon_{i-1}^n\right)\left(\upsilon_i^n-\upsilon_{i-1}^n\right):$$

Method 1 (introduction of internal iterations)

Step 1: $\kappa=-\sigma\left(\upsilon_i^n+\upsilon_{i-1}^n\right)\left(\upsilon_i^n-\upsilon_{i-1}^n\right)$,

Step 2: $\upsilon_i^{k+1}=\upsilon_i^n+\kappa/\left(\upsilon_i^k+\upsilon_i^n\right)$, after reaching the required accuracy $\upsilon_i^{n+1}=\upsilon_i^{k+1}$.

Method 2 (direct)

Step 1: $\kappa = -\sigma\left(\upsilon_{i,}^{n} + \upsilon_{i-1}^{n}\right)\left(\upsilon_{i}^{n} - \upsilon_{i-1}^{n}\right)$,

Step 2: $\upsilon_{i}^{n+1} = \upsilon_{i}^{n} + \kappa / \left\{\upsilon_{i}^{n} + \left[\left(\upsilon_{i}^{n}\right)^{2} + \kappa\right]^{1/2}\right\}$.

Method 3 (see (2.20)): $\upsilon_{i}^{n+1} = \left[\left(\upsilon_{i}^{n}\right)^{2} - \sigma\left(\upsilon_{i,}^{n} + \upsilon_{i-1}^{n}\right)\left(\upsilon_{i}^{n} - \upsilon_{i-1}^{n}\right)\right]^{1/2}$.

We shall construct the operators of the equivalent substitution of the finite differences for different sets of gas-dynamic variables. It will be shown that for the entire family of conservative numerical methods for the Euler equations, the transition from the grid conservative variables to non-conservative ones results in a simplification of the difference equations. Finally, we obtain the conservative difference schemes in the variables $(p,\mathbf{v},h)^{\mathrm{T}}$, $(p,\mathbf{v},e)^{\mathrm{T}}$, $(\rho,\mathbf{v},e)^{\mathrm{T}}$, $(p,\mathbf{v},\rho)^{\mathrm{T}}$ for Euler equations, equivalent to the conservative local-characteristic methods, using the Roe solution and designed for calculations of flows with discontinuities. Let it be that $V = (p,u,h)^{\mathrm{T}}$, and it is necessary to construct the matrix operator $M(V_1, V_2)$ such that for arbitrary states $V_1 = (p_1,u_1,h_1)^{\mathrm{T}}$ and $V_2 = (p_2,u_2,h_2)^{\mathrm{T}}$ we would have $U_2 - U_1 = M(V_1,V_2)(V_2 - V_1)$. We shall use the following procedure: according to Roe, we introduce the parametric vector $\mathbf{z}$ in such a manner that the conservative vector of the flows $\mathbf{F} = (\rho u,\ p+\rho u^2,\ u(h_+ u^2/2)^{\mathrm{T}}$ and the vector of conservative variables $U = (\rho, \rho\mathrm{v}, \rho(h_{+1/2}u^2) - p)^{\mathrm{T}}$ depend in a quadratic manner on the components of vector $\mathbf{z}$:

$$\mathbf{z} = \left(\sqrt{\rho}, u\sqrt{\rho}, H\sqrt{\rho}\right)^{\mathrm{T}}, \quad \mathbf{U} = \begin{bmatrix} z_1^2 \\ z_1 z_2 \\ z_1 z_3 - p \end{bmatrix}, \quad \mathbf{F} = \begin{bmatrix} z_1 z_2 \\ z_2^2 + p \\ z_2 z_3 \end{bmatrix}.$$

It was shown in [203] that if the relationship:

$$\Delta p = P_{\rho}\big|_{m,E} \Delta\rho + P_{m}\big|_{\rho,E} \Delta m + P_{E}\big|_{\rho,m} \Delta E, \tag{2.28}$$

is fulfilled (here and the rest $\Delta(\cdot) = (\cdot)_2 - (:)_1$), and ρ, m, E are the components of vector $\mathbf{U}$), then taking into account expansions

$$\Delta\rho = 2\bar{z}_1 \Delta z_1, \quad \Delta(\rho u) = \bar{z}_2 \Delta z_1 + \bar{z}_1 \Delta z_2, \quad \Delta(\rho H) = \bar{z}_3 \Delta z_1 + \bar{z}_1 \Delta z_3, \tag{2.29}$$

$$\Delta p = \frac{1}{1+P_E}\left(P_\rho \Delta\rho + P_m \Delta(\rho v) + P_E \Delta(\rho H)\right) \tag{2.30}$$

we can write

$$\Delta \mathbf{U} = C(\bar{\mathbf{z}})\Delta \mathbf{z}, \quad \Delta \mathbf{F} = B(\bar{\mathbf{z}})\Delta \mathbf{z},$$

where $\bar{\mathbf{z}} = (\mathbf{z}_1 + \mathbf{z}_2)/2$,

$$B(\bar{\mathbf{z}}) = \begin{bmatrix} \bar{z}_2 & \bar{z}_1 & 0 \\ \dfrac{(2\bar{z}_1 P_\rho + \bar{z}_3 P_E - \bar{z}_2 P_m)}{1+P_E} & 2\bar{z}_2 + \dfrac{\bar{z}_1}{1+P_E} & \dfrac{\bar{z}_1 P_E}{1+P_E} \\ 0 & \bar{z}_3 & \bar{z}_2 \end{bmatrix}$$

$$C(\bar{\mathbf{z}}) = \begin{bmatrix} 2\bar{z}_1 & 0 & 0 \\ \bar{z}_2 & \bar{z}_1 P_m & 0 \\ \dfrac{(\bar{z}_3 - 2\bar{z}_1 P_\rho)}{1+P_E} & \dfrac{-\bar{z}_1 P_m}{1+P_E} & \dfrac{\bar{z}_1}{1+P_E} \end{bmatrix}$$

This leads to an equation for the matrix operator A which is such that the following condition is fulfilled for arbitrary states: $\Delta\mathbf{F} = A\Delta\mathbf{U}$:

$$A = B(\bar{\mathbf{z}})C(\bar{\mathbf{z}})^{-1}.$$

Taking into account the fact that the form of this matrix coincides with the analytical expression for the appropriate Jacobi matrix $A(\tilde{\mathbf{U}}) = \partial\mathbf{F}/\partial\mathbf{U}$, this leads immediately to Roe's averaging:

$$\tilde{\mathbf{U}} : \rho = \sqrt{\rho_1\rho_2},$$

$$\tilde{u} = \left(u_1\sqrt{\rho_1} + u_2\sqrt{\rho_2}\right)/\left(\sqrt{\rho_1} + \sqrt{\rho_2}\right),$$

$$\tilde{H} = \left(H_1\sqrt{\rho_1} + H_2\sqrt{\rho_2}\right)/\left(\sqrt{\rho_1} + \sqrt{\rho_2}\right),$$

We shall use the following notations of the mean states: the tilder above the letter – the mean according to the Roe method, the two dots above the letter – the mean in accordance with the equation:

$$\overline{\overline{\varphi}}=\left(\varphi_1\sqrt{\rho_1}+\varphi_2\sqrt{\rho_2}\right)/\left(\sqrt{\rho_1}+\sqrt{\rho_2}\right),$$

where φ is some function of the thermomechanical state. For example $\tilde{u}=\overline{\overline{u}}$ and $\tilde{H}=\overline{\overline{H}}$. It is required that for two not necessarily close states the condition $\Delta U = M(\mathbf{V}_1,\mathbf{V}_2)\Delta\mathbf{U}$ is fulfilled accurately.

$\Delta\mathbf{V}$ will be presented in the form $\Delta V=D(\mathbf{U},\bar{\mathbf{z}})\Delta\mathbf{z}$. The appropriate expansion for pressure is obtained as a result of substitution of (2.29) into (2.30). Identity $\Delta(\rho\varphi)=\tilde{\rho}\Delta\varphi+\overline{\overline{\varphi}}\Delta\rho$, because of which we can carry out immediately direct verification, taking into account $\tilde{u}=\overline{\overline{u}}$, gives the following equation for the speed:

$$\Delta u=\left(\Delta(\rho u)-\tilde{u}\Delta\rho\right)/\tilde{\rho}. \tag{2.31}$$

The identical relationship for total enthalpy

$$\Delta H=\left(\Delta(\rho H)-\tilde{H}\Delta\rho\right)/\tilde{\rho}$$

leads to the following expansion for enthalpy:

$$\Delta h=\left(\Delta(\rho H)-\tilde{H}\Delta\rho\right)/\tilde{\rho}-\tilde{u}\Delta u+\left(\tilde{u}-\bar{u}\right)\Delta u. \tag{2.32}$$

Gradually, we obtain an expansion for internal energy. For an arbitrary equation of state we obtain:

$$\Delta e=\frac{1}{\tilde{\rho}}\left(\Delta(\rho e)-e\Delta\rho\right)=\frac{1}{\tilde{\rho}}\left(\Delta(\rho H)-\Delta P-\tilde{e}\Delta\rho\right)+\left(\tilde{e}-\overline{\overline{e}}\right)\frac{\Delta\rho}{\tilde{\rho}}.$$

In particular, for the ideal gas

$$\tilde{e}=\frac{1}{\gamma}\left(\tilde{H}-\frac{\tilde{u}^2}{2}\right)=\overline{\overline{e}}+\frac{\overline{\overline{u^2}}-\tilde{u}^2}{2\gamma}.$$

Substituting (2.29) into (2.31), (2.29) and (2.31) into (2.32), and taking into account the equation for the difference of the means:

$$\bar{\bar{\varphi}}-\dot{\bar{\varphi}}=\frac{\varphi_1\sqrt{\rho_1}+\varphi_2\sqrt{\rho_2}}{\sqrt{\rho_1}+\sqrt{\rho_2}}-\frac{\varphi_1+\varphi_2}{2}=\frac{\Delta\varphi}{4\bar{z}_1}\Delta z_1,$$

in the final analysis we obtain

$$D(\mathbf{U},\mathbf{z})=\begin{bmatrix}\dfrac{\left(2\bar{z}_1P_\rho+\bar{z}_2P_m+\bar{z}_3P_E\right)}{1+P_E} & \dfrac{\bar{z}_1P_m}{1+P_E} & \dfrac{\bar{z}_1P_E}{1+P_E}\\ \dfrac{\left(\bar{z}_2-2\tilde{u}\bar{z}_1\right)}{\tilde{\rho}} & \bar{z}_1/\tilde{\rho} & 0\\ \dfrac{(\Delta u)^2}{4\bar{z}_1}+\bar{z}_3-\tilde{u}\bar{z}_2-2\bar{z}_1\left(\tilde{H}-\tilde{u}^2\right)/\tilde{\rho} & -\bar{z}_1\tilde{u}/\tilde{\rho} & \bar{z}_1/\tilde{\rho}\end{bmatrix}.$$

Using the matrices B, C, D, we obtain the matrices setting the relationship of the finite differences:

$$\Delta\mathbf{F}=B(\bar{\mathbf{z}})D\left(\tilde{\mathbf{U}},\bar{\mathbf{z}}\right)^{-1}\Delta\mathbf{V}\quad\text{and}\quad\Delta\mathbf{U}=C(\bar{\mathbf{z}})D\left(\tilde{\mathbf{U}},\bar{\mathbf{z}}\right)^{-1}\Delta\mathbf{V}.$$

In order to realise the explicit TVD and ENO schemes, we require only matrix M^{-1} which should be such that $\Delta\mathbf{V} = M^{-1}\,\Delta\mathbf{U}$:

$$M^{-1}=D(\tilde{\mathbf{U}},\bar{\mathbf{z}})\,C(\bar{\mathbf{z}})^{-1}=\begin{bmatrix}P_\rho & P_m & P_E\\ -\dfrac{\tilde{u}}{\rho} & \dfrac{1}{\tilde{\rho}} & 0\\ \dfrac{\delta+\tilde{u}^2-\tilde{H}+P_\rho}{\tilde{\rho}} & \dfrac{-\tilde{u}+P_m}{\tilde{\rho}} & \dfrac{1+P_E}{\tilde{\rho}}\end{bmatrix},$$

where $\delta=\dfrac{(\Delta u)^2\tilde{\rho}}{2\bar{z}_1}$.

Comparing the structure of the produced matrix with the

appropriate Jacobi matrix, linking the infinitely small increments of the vectors $d\mathbf{V} = M^{-1}d\mathbf{U}$:

$$M^{-1} = \frac{\partial \mathbf{U}}{\partial \mathbf{V}} - \begin{bmatrix} P_\rho & P_m & P_E \\ -\frac{\bar{u}}{\rho} & \frac{1}{\rho} & 0 \\ \frac{u^2 - H + P_\rho}{\rho} & \frac{-u + P_m}{\rho} & \frac{1 + P_E}{\rho} \end{bmatrix},$$

it may seen that the matrix, describing the relationship of the finite increments, is a sum of two matrices: the appropriate Jacobi matrix, calculated using averaging according to Roe, and the matrix having the second order in relation to the differences. In the present case, it has one element differing from zero.

Transforming the differences equations for the grid functions of the conservative variables using relationships of type (2.24), we obtain equivalent difference equations for the grid functions of the non-conservative variables. In a general case, this does not reduce the extent of calculations, but for the local-characteristic approaches, using the Roe method, the resultant equations are simplified.

We shall present a different scheme in the simplest variables, equivalent to schemes: Harten scheme TVD2 and Young scheme UNO3 using the Roe method. Both conservative schemes are written in the form

$$\mathbf{U}_j^{n+1} - \mathbf{U}_j^n = -\lambda\left[\mathbf{F}_{j+1/2}^n - \mathbf{F}_{j-1/2}^n\right], \quad \mathbf{F}_{j+1/2}^n = \left[\mathbf{F}_j^n + \mathbf{F}_{j+1}^n + R_{j+1/2}\Phi_{j+1/2}\right]/2.$$

The elements φ^l of vector $\Phi_{j+1/2}$ has the following form: for TVD2 (Harten, 1983)

$$\varphi_{j+1/2}^l = \sigma\left(a_{j+1/2}^l\right)\left(g_j^l + g_{j+1}^l\right) - \psi\left(a_{j+1/2}^l + \gamma_{j+1/2}^l\right)\alpha_{j+1/2}^l, \tag{2.25}$$

for UNO3 (Young, 1992)

$$\varphi^l_{j+1/2} = \alpha\left(a^l_{j+1/2}\right)\left(\beta^l_j + \beta^l_{j+1}\right)$$
$$\begin{cases} \tilde{\sigma}\left(a^l_{j+1/2}\right)\left(\tilde{\beta}^l_j + \tilde{\beta}^l_{j+1}\right) - \psi\left(a^l_{j+1/2} + \gamma^l_{j+1/2} + \tilde{\gamma}^l_{j+1/2}\right)\alpha^l_{j+1/2}, \\ \text{if } \left|\alpha^l_{j-1/2}\right| \le \left|\alpha^l_{j+1/2}\right| \\ \hat{\sigma}\left(a^l_{j+1/2}\right)\left(\hat{\beta}^l_j + \hat{\beta}^l_{j+1}\right) - \psi\left(a^l_{j+1/2} + \gamma^l_{j+1/2} + \hat{\gamma}^l_{j+1/2}\right)\alpha^l_{j+1/2}, \\ \text{if } \left|\alpha^l_{j-1/2}\right| > \left|\alpha^l_{j+1/2}\right| \end{cases}$$
$$\beta^l_j = m\left[\alpha^l_{j+1/2}, \alpha^l_{j-1/2}\right], \qquad (2.34)$$
$$\tilde{\beta}^l_j = \bar{m}\left[\Delta_-\alpha^l_{j-1/2}, \Delta_+\alpha^l_{j-1/2}\right], \text{ if } \left|\alpha^l_{j-1/2}\right| \le \left|\alpha^l_{j+1/2}\right|$$
$$\hat{\beta}^l_j = \bar{m}\left[\Delta_-\alpha^l_{j+1/2}, \Delta_+\alpha^l_{j+1/2}\right], \text{ if } \left|\alpha^l_{j-1/2}\right| > \left|\alpha^l_{j+1/2}\right|.$$

If $\alpha^l_{j+1/2} = 0$, then $\gamma^l_{j+1/2} = \tilde{\gamma}^l_{j+1/2} = \hat{\gamma}^l_{j+1/2} = 0$. Otherwise:

$$\gamma^l_{j+1/2} = \sigma\left(a^l_{j+1/2}\right)\left(\beta^l_{j+1} - \beta^l_j\right)/\alpha^l_{j+1/2},$$
$$\tilde{\gamma}^l_{j+1/2} = \tilde{\sigma}\left(a^l_{j+1/2}\right)\left(\tilde{\beta}^l_{j+1} - \tilde{\beta}^l_j\right)/\alpha^l_{j+1/2},$$
$$\hat{\gamma}^l_{j+1/2} = \hat{\sigma}\left(a^l_{j+1/2}\right)\left(\hat{\beta}^l_{j+1} - \hat{\beta}^l_j\right)/\alpha^l_{j+1/2}.$$

The following functions were used when writing the schemes:

$$\bar{m}(x,y) = \begin{cases} x, \text{ if } |x| \le |y|, \\ y, \text{ if } |x| > |y|, \end{cases}$$

$$\sigma(x) = \frac{1}{2}\left(\psi(x) - \lambda x^2\right), \; \tilde{\sigma}(x) = \frac{1}{2}\left(2|x| - 3\lambda|x|^2 + \lambda^2|x|^3\right), \qquad (2.35)$$
$$\hat{\sigma}(x) = \frac{1}{6}\left(\lambda^2|x|^3 - |x|\right),$$

$$m(x,y) = \begin{cases} s\min\left(|x|,|y|\right), \text{ if } \operatorname{sgn} x = \operatorname{sgn} y = s, \\ 0, \text{ otherwise} \end{cases}$$

$$\psi(x)=\begin{cases}|x|, & |x|\geq\varepsilon,\\ \dfrac{\left(x^2+\varepsilon^2\right)}{2\varepsilon}, & |x|<\varepsilon\end{cases}$$

The methods are described in detail in [45,91,114,241].

Vector $\boldsymbol{\alpha}_{j+1/2}$, approximating the variation of the local characteristics of the variables, is computed through the variation of the vector of conservative variables:

$$\alpha_{j+1/2}=R_{j+1/2}^{-1}\left(\mathbf{U}_{j+1}^{n}-\mathbf{U}_{j}^{n}\right). \tag{2.36}$$

The columns of the matrix R^{-1} are the left eigenvectors of the Jacobi matrix of the flow vector.

It should be mentioned that the approximation of the differences of the characteristic variables in half-integral points of the template is in these schemes the most cumbersome and laborious operation. The equivalent replacement of the finite differences in transition to equations for grid functions $\mathbf{V}_h^{\tau}$ is reflected mainly on the method of calculating $\boldsymbol{\alpha}_{j+1/2}$. Direct substitution gives

$$\alpha_{j+1/2}=S_{j+1/2}^{-1}\left(\mathbf{V}_{j+1}^{n}-\mathbf{V}_{j}^{n}\right),$$

where $S^{-1}_{j+1/2} = R^{-1}_{j+1/2}\, M^{n}_{j+1/2}$. If matrix R is constructed using the Roe method, then for an entire series of the simplest variables the resulting matrices S^{-1} are simpler than matrices R^{-1}.

In accordance with the rule of transformation of the finite differences, the equivalent schemes in the variables $\mathbf{V} = (p,\mathbf{v},h)^{\mathrm{T}}$ are written in the form

$$U\left(\mathbf{V}_{j}^{n+1}\right)-U\left(\mathbf{V}_{j}^{n}\right)=-\lambda\left[\mathbf{F}_{j+1/2}^{n}-\mathbf{F}_{j-1/2}^{n}\right],$$

or

$$M_{j}^{n+1/2}\left(\mathbf{V}_{j}^{n+1}-\mathbf{V}_{j}^{n}\right)=-\lambda\left[\mathbf{F}_{j+1/2}^{n}-\mathbf{F}_{j-1/2}^{n}\right],$$

where

$$\mathbf{F}_j^{n+1/2} = \left[\mathbf{F}_j^n + \mathbf{F}_{j+1/2}^n + R_{j+1/2}\Phi_{j+1/2}\right]/2,$$
$$\alpha_{j+1/2} = S_{j+1/2}^{-1}\left(\mathbf{V}_{j+1}^n - \mathbf{V}_j^n\right), \quad S_{j+1/2}^{-1} = R_{j+1/2}^{-1}M_{j+1/2}^n.$$

The following notations were used here: $M_j^{n+1/2}=M(\mathbf{V}_j^{n+1},\mathbf{V}_j^{n+1/2})$ $M_{j+1/2}^n=M(\mathbf{V}_{j+1}^n,\mathbf{V}_j^n)$. All functions and matrix operators are written through the variable $\mathbf{V} = (p,\mathbf{v},h)$, and we use the equation of state of the type $\rho = \rho(p,h)$, $\rho = \rho(p,e)$, $p = p(\rho,e)$, $h = h(p,\rho)$. We use one of the methods of integration in respect of time (2.20).

Method 1

$$\mathbf{V}_j^{n+1} = U^{-1}\left(U(\mathbf{V}_j^n) - \lambda\left[\mathbf{F}_{j+1/2}^n - \mathbf{F}_{j-1/2}^n\right]\right).$$

Method 2

Step 1: $\xi_j = -\lambda\left[\mathbf{F}_{j+1/2}^n - \mathbf{F}_{j-1/2}^n\right].$

Step 2: $\left(M^{-1}\right)_j^{n+1/2} = M^{-1}\left(\mathbf{V}_j^n, \upsilon\left(\mathbf{V}_j^n, \xi_j\right)\right).$

Step 3: $\mathbf{V}_j^{n+1} = \mathbf{V}_j^n + \left(M^{-1}\right)_j^{n+1/2}\xi_j.$

Here υ has the meaning of the preliminary value $\mathbf{V}_j^{n+1}$ and differs from the final results by the rounding error. For the variable $\mathbf{V} = (p,u,h)^{\mathrm{T}}$ and the equation of state in the form $\rho = \rho(p,h)$

$$\upsilon\left(\mathbf{V}_j^n,\xi_j\right) = \left[p,(\rho u+\xi_2)/(\rho+\xi_1),\left(p+\varepsilon^*\right)/(\rho+\xi_1)\right]^{\mathrm{T}},$$

where

$$\varepsilon^* = E + \xi_3 - 1/2(\rho u + \xi_3)^2/(\rho+\xi_1)$$

and ξ_1, ξ_2, ξ_3 are the components of vector ξ_j.

In a general case, we obtain an equation for p:

$$\rho\left(p,\left(p+\varepsilon^{*}\right)/\left(\rho+\xi_{1}\right)\right)=\rho+\xi_{1}.$$

In a partial case of the ideal gas, $p = (\gamma-1)\varepsilon^*$, γ is the exponent of the adiabate.

It should be mentioned that it is not compulsory to find the vector $\mathbf{\upsilon}$ in the computation stage of $(M^{-1})_j^{n+1/2}$ and it is sufficient to determine the quantities included in the equations used for determining the mean according to Roe and δ.

Method 3

Step 1: $\xi_j=-\lambda\left[\mathbf{F}_{j+1/2}^{n}-\mathbf{F}_{j-1/2}^{n}\right].$

Step 2: $\mathbf{V}_j^{(0)}=\mathbf{V}_j^{n},$

and subsequently we carry out iteration with respect to k of the type $\mathbf{V}_j^{(k+1)} = \mathbf{V}_j^{(k)} + M^{-1}(\mathbf{V}_j^{n}, \mathbf{V}_j^{(k)})\xi_j$ on reaching the required accuracy $\mathbf{V}_j^{n+1} = \mathbf{V}_j^{(k)}$.

At first sight, it may appear that method 2 is far more complicated than method 1. Firstly, the equations of method 2 are written in δ-form and when solving the stationary problems by the stabilization method, the resultant solution contains the conservative approximation of stationary equations. Secondly, method 2 is generalised in a natural manner for a case in which the equations contain rigid sources and must be linearised with respect to time. Iteration method 3 has an advantage when using complex equations of state.

When solving the problems in several spatial measurements in the right-hand part of step 1, we carry out summation with respect to the co-ordinate directions. In the finite volume method $\lambda = \Delta t/\Omega$, where Ω is the measure of the cell with index j. Matrix M^{-1} is computed as follows:

$$\mathrm{M}^{-1}=\left(\mathbf{V}_{1},\mathbf{V}_{2}\right)=\tilde{M}^{-1}\left(\tilde{\mathbf{V}},\delta\right),$$

where $\tilde{\mathbf{V}}$ is determined unambiguously by the averaging of states $\mathbf{V}_1$ and $\mathbf{V}_2$ by the Roe method, δ-correction having the second order of smallness away from the discontinuities of the solution. In the appendix, we present the specific type of matrices S^{-1} and M^{-1} for different vectors $\mathbf{V}$. All variables, including S^{-1} and M^{-1}

represent means of the corresponding state according to Roe.

The results of numerical testing

As in the replacement of dependent variables in differential equations (2.9), (2.10), the substitution of grid functions in explicit numerical local-characteristic methods is relatively simple. In the case of TVD and UNO schemes it is sufficient for this purpose to:

1 To change, by identical transformations, the computation procedure of all grid functions in accordance with the variation of the mean of the grid functions of the variables of the flow fields;
2 Change, if necessary, the method of averaging quantities at the edges of the cells for averaging by the Roe method;
3 Change of the procedure of calculating the approximations of the differences of the characteristic variables (2.36), using instead of the matrix of left eigenvectors R^{-1}, the matrix $S^{-1} = R^{-1}M$, which has a considerably simpler structure (see Appendix);
4 Use one of the integration schemes in respect of time (*method* 1, *method* 2 or *method* 3).

We shall verify the equivalency of the methods based on the substitution of the grid functions of the dependent variables in the Harten and Young schemes.

Comparison was carried out for the previously mentioned schemes in different variables. The parameter of enthalpy correction in the schemes UNO and TVD (2.35), and also the criteria of smallness of the variations of the characteristic variables in (2.33) and (2.34) were assumed to be identical in all calculations. The initial data for the methods in different variables have the form:

$$\mathbf{V}_i^0 = \begin{cases} \mathbf{V}_L, & i \le N/2, \\ \mathbf{V}_R, & i > N/2, \end{cases}$$

where i is the grid index, and N is the number of nodes of the calculation grid with a constant step.

The initial data represent the same solution expressed by means of different variables (Table 2.4).

In every case, we used the appropriate form of the equation of state of the ideal gas $p(\rho, e) = (\gamma-1)\rho e$, $p(\rho,h) = (\gamma-1)\rho h/\gamma$, $\rho(p,h) = \gamma p/[(\gamma-1)h]$ etc., where it was assumed that $\gamma = 1.4$.

The calculations were carried out on the identical spatial grid

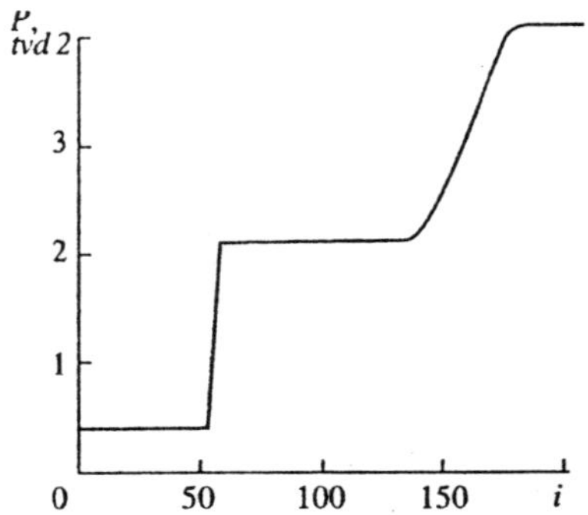

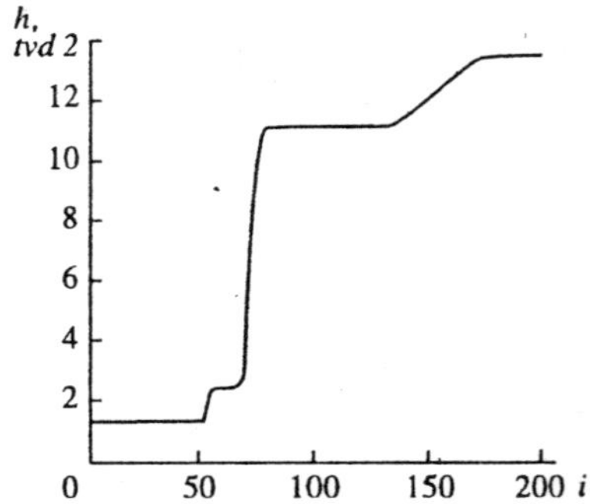

Fig.2.10. Breakdown of a discontinuity. Profiles of pressure determined using two 'equivalent' TVD2 schemes in conservative and non-conservative variables coincide.
Fig.2.11. (right) Breakdown of a discontinuity. Profiles of enthalpy determined using two 'equivalent' TVD2 schemes in conservative and non-conservative variables coincide.

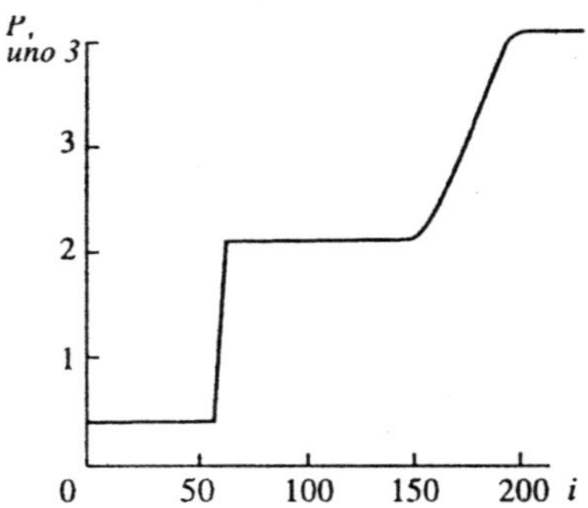

Fig.2.12. Breakdown of a discontinuity. Profiles of pressure determined using two 'equivalent' UNO3 difference schemes in conservative and non-conservative variables coincide.
Fig.2.13. Breakdown of a discontinuity. Profiles of enthalpy determined using two 'equivalent' UNO3 difference schemes in conservative and non-conservative variables coincide.

Table 2.4

Variables	$\mathbf{V}_L$	$\mathbf{V}_R$
$(\rho, \rho u, \rho(e+u^2))$	(1. 0. 1)	(1. 0. 10)
(ρ, u, h)	1. 0..14/10)	(1. 0. 14)
(p, u, h)	(4/10. 0. 14/10)	(4. 0. 14)
(p, u, e)	(4/10. 0. 1)	(4. 0. 10)
(ρ, u, e)	(1. 0. 1)	(1. 0. 10)
(p, u, ρ)	(4/10. 0. 1)	(4. 0. 1)

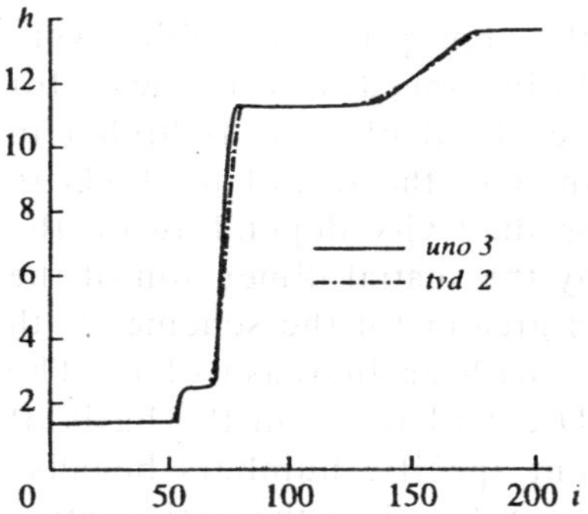

Fig.2.13. Breakdown of a discontinuity. Profiles of enthalpy determined using two 'equivalent' TVD2 and UNO3 difference schemes.

N = 200, with the same sequence of steps in time, at CFL = 0.1. The grid functions, representing a solution for each of the tested methods after 1000 integration steps with respect to time, were reduced to 'pressure–enthalpy' variables $p_i = p(V_i^{n=1000})$, $u_i = u(V_i^{n=1000})$, $h_i = h_i(V_i^{n=1000})$ and are shown in Figs.2.10 and 2.11 for the method equivalent to the TVD2 scheme, and in Figs.2.12, 2.13 for the UNO3 method. The numerical solutions, obtained by the TVD2 and UNO3 methods are shown in Fig.2.14. Differences in the form of the curves in Fig.2.14 are explained by the properties of TVD2 and UNO3 schemes. At the same time, the difference in the curves in Figs.2.10–2.13 cannot be resolved on the given scale, it is a difference in the norm max$|\Delta|$, where Δ is the difference of the appropriate quantities, obtained by calculations using the method in conservative **U** and simplest variables **V**, presented in Table 2.5 justified in the form in which they were derived as a result of calculations in the DOUBLE format.

It is important that the programs for both methods used statements of all variables of the real type such as REAL (KIND = 8), and also the function of double accuracy (DABS, DSQRT). The programmes were compiled using Digital Fortran and calculations were carried out under the control of a 32-bit operating system.

Table 2.5.

Method	$\max_i \left\lvert p_i^{(V)} - p_i^{(U)} \right\rvert$	$\max_i \left\lvert u_i^{(V)} - u_i^{(U)} \right\rvert$	$\max_i \left\lvert h_i^{(V)} - h_i^{(U)} \right\rvert$
V=(p, u, h) TVD2	5.960464832810430E--008	7.003605777301127E--015	2.384186501558361E--007
V=(p, u, h) UNO3	5.960464477539062E--008	0.0E+000	2.384185791015625E--007
V=(ρ, u, h) UNO3	5.960464388721221E--0.08	0.0E+000	2.384185791015625E--007

The programs of the methods in the appropriate variables were not optimised for the maximum operating speed, nevertheless, for the algorithms (*methods* 1, 2) the rate of calculations is higher in the simplest variables in comparison with the initial methods in conservative variables (to 10% of productivity depending on the number of equations m, determined by the spatial dimension of the problem), and the advantage becomes greater for the schemes with a high approximation order and also with an increase of m. The maximum effect was achieved for 3D calculations on the basis of the UNO3 scheme. We do not present specific numbers because it is necessary to employ a strict approach based on the optimisation of the programme code for methods in different variables. It should only be mentioned out of all the methods tested, the maximum computing efficiency was obtained in integration with respect to time using scheme (*method* 1) in variables (ρ, u, h)

APPENDIX

Here we present the matrices S^{-1} and M^{-1} for the Euler equations in two spatial measurements for different sets of the variables. The variables $V = (p, u, v, h)^T$:

$$M^{-1} = \begin{bmatrix} c^2 + \left(u^2 + v^2 - H\right)\Gamma & -u\Gamma & -v\Gamma & \Gamma \\ -\dfrac{u}{\rho} & \dfrac{1}{\rho} & 0 & 0 \\ -\dfrac{v}{\rho} & 0 & \dfrac{1}{\rho} & 0 \\ \dfrac{\delta + c^2 + \left(u^2 + v^2 - H\right)(1+\Gamma)}{\rho} & -\dfrac{u(1+\Gamma)}{\rho} & -\dfrac{v(1+\Gamma)}{\rho} & \dfrac{1+\Gamma}{\rho} \end{bmatrix}$$

$$S^{-1} = \begin{bmatrix} \frac{\Gamma}{c^2 - \Gamma\delta}\frac{c^2+\delta}{c^2} & 0 & 0 & -\rho\frac{\Gamma}{c^2-\Gamma\delta} \\ 0 & n_y & -n_x & 0 \\ \frac{1}{\rho c} & n_x & n_y & 0 \\ \frac{1}{\rho c} & -n_x & -n_y & 0 \end{bmatrix},$$

$$\delta = h - \bar{\bar{h}} = \rho^2 \frac{(\Delta u)^2 + (\Delta v)^2}{8(\bar{z}_1)^2}.$$

Variables $\mathbf{V} = (p, u, v, e)^{\mathrm{T}}$:

$$M^{-1} = \begin{bmatrix} c^2 + (u^2 + v^2 - H)\Gamma & -u\Gamma & -v\Gamma & \Gamma \\ -\frac{u}{\rho} & \frac{1}{\rho} & 0 & 0 \\ -\frac{v}{\rho} & 0 & \frac{1}{\rho} & 0 \\ \frac{\frac{1}{2}(u^2+v^2) - \bar{\bar{e}}}{\rho} & -\frac{u}{\rho} & -\frac{v}{\rho} & \frac{1}{\rho} \end{bmatrix},$$

$$S^{-1} = \begin{bmatrix} \frac{(h-\bar{\bar{e}})\beta}{c^2} & 0 & 0 & -\rho\beta \\ 0 & n_y & -n_x & 0 \\ \frac{1}{\rho c} & n_x & n_y & 0 \\ \frac{1}{\rho c} & -n_x & -n_y & 0 \end{bmatrix}, \quad \beta = \frac{\Gamma}{c^2 - \Gamma(h - \bar{\bar{e}})}.$$

Variables $V = (p,\ u,\ v,\ \rho)^T$:

$$M^{-1} = \begin{bmatrix} c^2 + \left(u^2 + v^2 - H\right)\Gamma & -u\Gamma & -v\Gamma & \Gamma \\ -\dfrac{u}{\rho} & \dfrac{1}{\rho} & 0 & 0 \\ -\dfrac{v}{\rho} & 0 & \dfrac{1}{\rho} & 0 \\ 1 & 0 & 0 & 0 \end{bmatrix},$$

$$S^{-1} = \begin{bmatrix} \dfrac{1}{c^2} & 0 & 0 & 1 \\ 0 & n_y & -n_x & 0 \\ \dfrac{1}{\rho c} & n_x & n_y & 0 \\ \dfrac{1}{\rho c} & -n_x & -n_y & 0 \end{bmatrix}$$

Variables $V = (p,\ u,\ v,\ e)^T$:

$$M^{-1} = \begin{bmatrix} 1 & 0 & 0 & 0 \\ -\dfrac{u}{\rho} & \dfrac{1}{\rho} & 0 & 0 \\ -\dfrac{v}{\rho} & 0 & \dfrac{1}{\rho} & 0 \\ \dfrac{\frac{1}{2}\left(u^2 + v^2\right) - \bar{\bar{e}}}{\rho} & -\dfrac{u}{\rho} & -\dfrac{v}{\rho} & \dfrac{1}{\rho} \end{bmatrix},$$

$$S^{-1}=\begin{bmatrix} \beta & 0 & 0 & -\rho\dfrac{\Gamma}{c^2} \\ 0 & n_y & -n_x & 0 \\ \dfrac{(1-\beta)c}{\rho} & n_x & n_y & \dfrac{\Gamma}{c} \\ \dfrac{(1-\beta)c}{\rho} & -n_x & -n_y & \dfrac{\Gamma}{c} \end{bmatrix}, \ \beta=\frac{\Gamma\left(h-\overline{\overline{e}}\right)}{c^2}.$$

The above methods correspond to the diagonalised form:

$$\Lambda=\mathrm{diag}\left[\sigma v_n,\sigma v_n,\sigma(v_n+c),\sigma(v_n-c)\right]$$

and the matrix of right eigenvectors:

$$R=\begin{bmatrix} 1 & 0 & \alpha & \alpha \\ u & \rho n_y & \alpha(u+cn_x) & \alpha(u-cn_x) \\ v & -\rho n_x & \alpha(v+cn_y) & \alpha(v-cn_y) \\ H-\dfrac{c^2}{\Gamma} & \rho(un_y-vn_x) & \alpha\left(H+c(un_x+vn_y)\right) & \alpha\left(H-c(un_x+vn_y)\right) \end{bmatrix}.$$

In the equivalent method in the conservative variables we use the following matrix of left eigenvectors and it is given here for comparison with matrix operators S^{-1}.

$$R^{-1}=\begin{bmatrix} 1-\dfrac{c^2-(H-u^2-v^2)\Gamma}{c^2} & \dfrac{\Gamma u}{c^2} & \dfrac{\Gamma v}{c^2} & -\dfrac{\Gamma}{c^2} \\ \dfrac{n_x-n_y u}{\rho} & \dfrac{n_y}{\rho} & -\dfrac{n_x}{\rho} & 0 \\ \dfrac{c^2(H-u^2-v^2)\Gamma-c(un_x-vn_y)}{\rho c} & \dfrac{-\Gamma u+n_x c}{\rho c} & \dfrac{-\Gamma v+n_y c}{\rho c} & \dfrac{\Gamma}{\rho c} \\ \dfrac{c^2(H-u^2-v^2)\Gamma+c(un_x+vn_y)}{\rho c} & \dfrac{-\Gamma u-n_x c}{\rho c} & \dfrac{-\Gamma v-n_y c}{\rho c} & \dfrac{\Gamma}{\rho c} \end{bmatrix},$$

$$\alpha = \frac{\rho}{2c}.$$

The following notations were used here:

$$\Gamma = P_E\big|_{\rho,m}, \quad c^2 = P_\rho\big|_{E,m} + \left(H - u^2 - v^2\right)P_E\big|_{\rho,m} \tag{A.1}$$

with additional relationships $P_{m_k}\big|_{\rho,E} = -m_k\, P_m\big|_{\rho,m_{i\neq k}}$.

The partial derivatives of pressure with respect to the components of the vector of conservative variables in (A.1) are determined using a specific equation of state, for example:

$$P_\rho\big|_{m,E} = \left.\frac{\partial p}{\partial \rho}\right|_e + \frac{1}{\rho}\left.\frac{\partial p}{\partial \rho}\right|_\rho \left(u^2 + v^2 - \frac{E}{\rho}\right), \quad P_E\big|_{\rho,m} = \frac{1}{\rho}\left.\frac{\partial p}{\partial \rho}\right|_\rho, \tag{A.2}$$

where arbitrary pressures in respect of the thermodynamic variables at the boundaries of the cells are determined using the equation of state in such a manner as to fulfil the relationship (2.20) [2.10], [2.12]. In the partial case the equation of state of the ideal gas equation (2.20) is fulfilled in the appropriate derivatives are expressed by a means of mean values in respect of Roe. Using the relationship between the arbitrary thermodynamic quantities [2.13], the equations (A.2) may be written in the form suitable for using the equation of state of any type.

TABLE OF NOTATIONS

p	–pressure
h	–enthalpy
$v=(u, v)^T$	–velocity
ρ	–density
e	–internal energy per unit mass
$E = \rho(e+1/2u^2)$	–total energy per unit volume
h	–specific enthalpy
$H = h+1/2u^2$	–total specific enthalpy
$\mathbf{U} = (\rho, \rho v, E)^T$	–vector of conservative variables
$\mathbf{V}$	–vector of arbitrary variables
∇_k	–derivative in respect of spatial direction k
$\mathbf{F}^k$	–conservative vector of fluxes corresponding to spatial direction

$\mathbf{F} = (\rho u, p+\rho u^2, uH)^T$	–the same, for a single spatial measurement
$A^k = \partial \mathbf{F}^k / \partial \mathbf{U}$	–Jacobi matrix of the vector of fluxes (matrix of coefficients of quasilinear form of the differential equations in conservative variables $\mathbf{U}$)
A	–the same for a single spatial measurement
M	–Jacobi matrix $\partial \mathbf{U}/\partial \mathbf{V}$
$B^k = M^{-1}A^kM$	– matrix of coefficients of quasilinear form of the differential equations in variables $\mathbf{V}$
$U : \mathbf{U} = U(\mathbf{V})$	–mutually unambiguous transformation $U: \Omega_U \to \Omega_V$, where $\Omega_U, \Omega_V \in \Re^m$
$\mathbf{U}^\tau_h, \mathbf{V}^\tau_h$	–net functions, determined on a generalised spatial–time net $\omega_\tau \times \omega_h$
$\mathbf{U}^i_j$	–component (i,j) of the net function $\mathbf{U}^\tau_h$, where $i \in \omega_\tau$, $j \in \omega_h$
$\mathbf{U}^n_h$	–set of values of the net function on time layer $i \in \omega_\tau$
Δ	–difference operator
$\boldsymbol{\kappa}$	–vector of net differences, determined on the generalised spatial–time net $\omega_\tau \times \omega_h$
$\mathbf{k}^{sp}_{qr} = \Delta^{sp}_{qr} U^\tau_h = U^s_q - U^p_r$	–component of vector $\boldsymbol{\kappa}$, where s, $p \in \omega_\tau$, $q, r \in \omega_h$
M	–operator (4), and also the appropriate net operator, acting on the vector of net differences
$M^{n+1/2}_j = M(\mathbf{V}^{n+1}_i, \mathbf{V}^n_j)$ $M^n_{j+1/2} = M(\mathbf{V}^n_{j+l}, \mathbf{V}^n_j)$	
$\Phi, \Phi^U, \Phi^V, \Phi'$	–nonlinear vector net functions
$\Phi^U(\mathbf{U}^\tau_h) = 0,\ \Phi^V(\mathbf{V}^\tau_h) = 0\ldots$	–nonlinear vector net equations,
$\Phi^U(\mathbf{U}^\tau_h) = 0$	–set of all equations of the discrete model, including discrete formulation of boundary conditions, whereas $\Phi(\mathbf{U}^n_h, \mathbf{U}^k_h) = 0$, where $k \le n$, k, $n \in \omega_\tau$,
$\Phi_{j+1/2}$	–part of the approximation of fluxes of local characteristic variables in TVD, UNO schemes

$\mathbf{m} = (m_1, m_2, m_3)^{\mathrm{T}} = (\rho u, \rho v, \rho w)^{\mathrm{T}}$ –momentum per unit volume

$\mathbf{z} = (z_1, z_2, z_3)^{\mathrm{T}} = \left(\sqrt{\rho}, u\sqrt{\rho}, H\sqrt{\rho}\right)^{\mathrm{T}} \left(\sqrt{\rho}, u\sqrt{\rho}, H\sqrt{\rho}\right)^{\mathrm{T}}$ –Roe parametric vector

B, D, C –matrices in quasilinear representation of the differences in respect of vectors **F**, **B**, **V**, through differences of parametric vector **z**

Λ –diagonal matrix of eigenvalues of the Jacobi matrix of the flux vector

$\mathbf{F}_{j+1/2}$ –numerical approximation of the vector at the face of the cell

$\mathbf{n} = (n_x, n_y, n_z)^{\mathrm{T}}$ –unit vector of the normal in the designation of the finite volume method

$v_n = (\mathbf{v}, \mathbf{n})$ –velocity in the direction of vector **n**

σ –area of the face in the designation of the finite volume method

c –adiabatic velocity of sound

γ –parameter of the adiabate in the equation of the state of the ideal gas

$P_\rho = P_\rho|_{m,E}$, $P_m = P_m|_{\rho,E}$, $P_E = P_E|_{\rho,m}$ –partial derivatives of pressure in respect of the components of the vector of conservative variables

2.5. DECOMPOSITION OF PROBLEMS OF NUMERICAL MODELLING, DESCRIBED BY HYPERBOLIC EQUATIONS, ON PARALLEL COMPUTERS

Here, we shall describe the technique of one-, two- and three-dimensional decomposition used for parallelization of problems of numerical modelling of hydro- and gas-dynamic flows. We present the dependence of the resultant speedup and the efficieny on the number of processes and the size of the grid for different types of decomposition. We present measurements of productivity on an example of the problems of numerical modelling of the uplift of a near-surface thermal in a stratified atmosphere.

Introduction

The majority of problems of hydro- and gas-dynamics can be solved explicitly in the quadratures. For an approximate solution of these problems, it is necessary to select a suitable formalisation of the problem within which the computing experiments are carried out [17]. Methods of numerical modelling have been used for a relatively long period of time, but only recently the rapid development of computing methods has made it possible to carry out demanding calculations and obtain physically substantiated results for three-dimensional problems. In addition to this, the application of supercomputers and parallel programs increases the productivity of calculations (and, consequently, accuracy) several tens of times.

Supercomputers increase the productivity by parallel execution of several processes of data processing simultaneously in different processors. Specialised parallel algorithms and programmes, realising these algorithms, have been developed for this purpose. In addition to this, not all algorithms permit parallelization, for example, if an algorithm consists of a consecutive modification of the same data, and the selection of the branch of the algorithm of the iteration step depends on the initial data, then such an algorithm is very difficult to parallelize. In cases in which we have a parallel algorithm, there are often several alternative variants and, consequently, it is difficult to select the most efficient algorithm.

It is possible to select two sources of parallelism in the problem: parallelism with respect to control and parallelism with respect to data. In the first case, there are several independent processes of data processing, for example, calculations and communications. In the second case, the data can be divided into independent parts and they can be processed independently (all of them or part of them). Parallelism with respect to the data is encountered more often in practice. In the numerical methods of solution of differential equations in partial derivatives, the parallelism forms in a natural manner because of the presence of a computing grid, and each node of the grid or a group of nodes are candidates for the data which can be processed partially and independently.

The procedure

The majority of systems of equations of hydro- and gas-dynamics can be represented in the form of a system of equations in partial derivatives of the hyperbolic type [107]

$$\frac{\partial \mathbf{W}}{\partial t}+\mathbf{A}\frac{\partial \mathbf{W}}{\partial x}+\mathbf{B}\frac{\partial \mathbf{W}}{\partial y}+\mathbf{C}\frac{\partial \mathbf{W}}{\partial z}=\mathbf{G}$$

where x, y, z are any independent co-ordinates in the region; **A**, **B**,

C are the Jacobi matrices (of the order n, where n is the number of components of vector **W**) of the functions determining the flows of the main calculated variables, in the general case, the matrices **A**, **B**, **C** may depend not only on the co-ordinates x, y, z but also on the solution vector **W**. We are not interested in the splitting scheme, a specific calculation method, the type of coordinate or even the grid template for realisation of this method, because the general procedure does not depend on these methods. The region of application of the decomposition procedure described below includes all grid methods of solving hyperbolic systems of equations.

The characteristic feature of the hyperbolic system of equations is that their solutions have a locality property, i.e. during finite time τ the variation manages to spread over the distance $\tau \cdot \max(|\lambda^A_k|)$, $\tau \cdot \max(|\lambda^B_k|)$, $\tau \cdot \max(|\lambda^C_k|)$ along the axes x, y and z, where λ^A_k, λ^B_k, λ^C_k are the eigen numbers of the matrices **A**, **B**, **C**. Thus, restricting the step with respect to time τ, corresponding to one iteration of the numerical method, using the Courant condition

$\tau \le \dfrac{h_x}{\max\left(\left|\lambda^{A}_k\right|\right)}, \tau \le \dfrac{h_y}{\max\left(\left|\lambda^{B}_k\right|\right)}, \tau \le \dfrac{h_z}{\max\left(\left|\lambda^{C}_k\right|\right)}$, where h_x, h_y and h_z are the dimension of the cell, we can guarantee that the effect of the change of the solution in one cell will propagate during this time τ only to its adjacent cells (Table 2.6). Thus, almost any division of the calculation region into groups of adjacent nodes will correspond to the property of parallelism with respect to the data.

Since the calculation region is presented in most cases in some co-ordinates of a rectangle, a parallelepiped or a region restricted by sections, parallel to the coordinate axes, the most suitable division for realisation is the division into regions representing identical rectangles/parallelepipeds. Figure 2.15 shows one-, two- and three-dimensional decomposition of a three-dimensional region, i.e. division into rectangular regions with respect to one, two or three co-ordinates, respectively.

Each such parallelepiped can also be processed by the same procedure, simultaneously with others, in a separate process. In reality, cutting out such a parallelepiped represents the separation of several terms of a large cycle of treatment of the points of the region on an individual processor.

The points directly in the vicinity of boundary S_1 of the division region U_1 (Fig.2.16) are calculated using a general scheme by

Table 2.6

Three-dimensional decomposition					
N	1	8	27	64	125
T (c)	32.75	4.305	1.578	0.716	0.437
S	1	7.61	20.76	45.75	74.95
E	1	0.951	0.768	0.715	0.600

Two-dimensional decomposition						
N	1	9	15	64	100	144
T (c)	32.75	3.951	1.596	0.890	0.682	0.563
S	1	8.29	20.52	36.80	48.03	58.18
E	1	0.921	0.821	0.575	0.480	0.404

One-dimensional decomposition					
N	1	8	24	60	120
T (c)	32.75	4.375	1.750	0.984	0.675
S	1	7.49	18.72	33.29	48.52
E	1	0.936	0.780	0.555	0.404

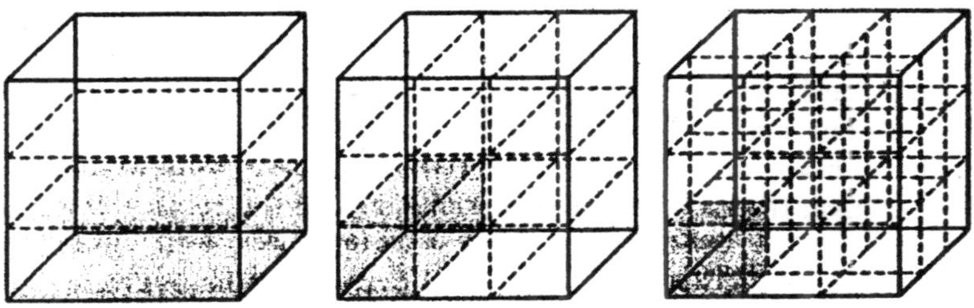

Fig.2.15. One-, two- and three-dimensional decomposition of the three-dimensional integration region.

adding additional points B_1^r, situated outside the region U_1, as it is carried out for calculation of the boundary points of the main divided region U by introducing additional points B_0^l. A difference

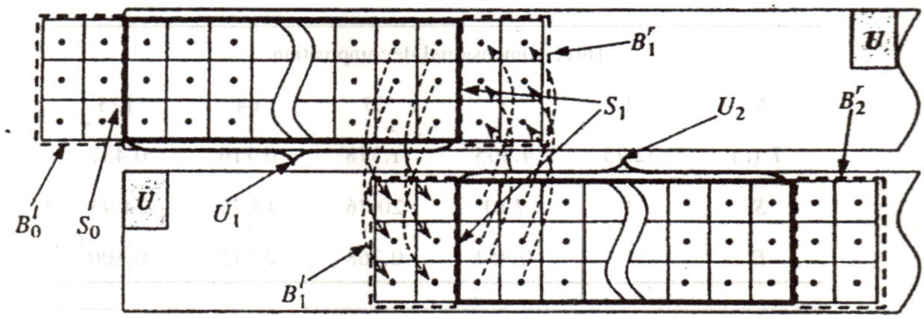

Fig.2.16. Data exchange scheme.

is in the determination of the values of the calculation components at these outer points B_0^l and B_1^r: for the outer boundary of the divided region S_0 they are determined by some method, on the basis of the determination of the boundary condition; for the points, geometrically falling into some other region of decomposition U_2, the values of the calculated components are taken (copied or approximated, taking their values into account) from the process responsible for the portion of the data.

Here, we require firstly the synchronisation of the processes realised in different processes. Secondly, it is necessary to exchange the data, processed in different processes – either directly through the common shear memory, or by some method of inter-process exchange of the data, for example, by sending messages. Figure 2.16 shows by the dotted arrows the corresponding exchanges between the processes. This scheme is not the only one possible. For example, it is impossible not to have doubled cells, and exchange the values of the flows to the surfaces S_1, S_2, etc. This scheme was selected because it enables calculation of each cell to be carried out in the same manner (in a homogeneous manner). In addition to this, it should be mentioned that the number of fictitious cells can be varied depending on the scheme (usually, the number corresponds to the order of the scheme).

The main processor time is of course outside the calculation of components in the cells. Nevertheless, with an increase in the number of decomposition regions, a significant role is played by the time used for the exchanges between the computing nodes. The number of exchanges for each type of decomposition and the ratio of the time, used for the exchange, to the entire processing time of the programme will be evaluated.

Let it be that the number of cells of the region $C = K^3$ (the region – a cube with a rib of length K), and in this case K is divisible by the number of computing nodes (processors) N, and $N = M^2 = L^3$. Consequently, the region can be divided by $(N - 1)$ parallel planes into N identical layers in one-dimensional decomposition, $2(M - 1)$ planes ($(M - 1)$ parallel to the plane XY and $(M - 1)$ – to the plane exit) into M^2 regions for two-dimensional decomposition, $3(L - 1)$ planes into L^3 regions for three-dimensional decomposition. Each constructed region contains the same number of cells C/N and, consequently, the time used for computation by the calculation scheme coincides for all three types of decomposition. However, the number of exchanges N_{comm} is proportional to the total area of the surface of the planes of the section (with the proportionality coefficient $2qK^2$, where q is equal to the number of calculated components multiplied by the number of fictitious cells along the same direction), differs and is equal to $2q\sqrt[3]{C^2}(N-1)$ for unidimensional decomposition, to $4q\sqrt[3]{C^2}\left(\sqrt{N}-1\right)$ for two-dimensional decomposition $6q\sqrt[3]{C^2}\left(\sqrt[3]{N}-1\right)$ (the exchanges into both sides are taken into account).

Evaluation of efficiency

The time of a single iteration of the program T consists of the time required for calculations of the calculated components in all cells T_{calc} and the time used for inter-process exchanges T_{comm}. It is assumed that the calculation time in a single cell is equal to τ_{calc}, and the duration of a single exchange is τ_{comm}. Consequently, the duration of operation in a single processor computer $T_1 = C\tau_{cals}$, and in a multiprocessor with N processors $T_N = \dfrac{C\tau_{calc}}{N} + N_{comm}\tau_{comm}$, where N_{comm} is the number of exchanges which depends on the type of decomposition. For the quantitative evaluation of the efficiency of parallelization we shall use the following [238] characterics:

Speedup

$$S(N,C) = \frac{T_N}{T_1},$$

Efficiency

$$E(N,C)=\frac{S}{N}=\frac{T_N}{N\cdot T_1}.$$

In the present case

$$S(N,C)=\frac{N}{1+\frac{N_{\text{comm}}}{C}\cdot k},\quad E(N,C)=\frac{1}{1+\frac{N_{\text{comm}}}{C}\cdot k},$$

where $k=\frac{\tau_{\text{comm}}}{\tau_{\text{calc}}}$ is the ratio of the time of a single exchange to the duration of a single iteration in a single cell. We use S_1, E_1, S_2, E_2, S_3, E_3 to denote the speedup and the efficiency for one-dimensional, two-dimensional and three-dimensional decomposition, respectively. Substituting the values of N_{comm}, we obtain

$$S_1(N,C)=\frac{N}{1+2qk\frac{(N-1)}{\sqrt[3]{C}}},\quad E_1(N,C)=\frac{1}{1+2qk\frac{(N-1)}{\sqrt[3]{C}}},$$

$$S_2(N,C)=\frac{N}{1+4qk\frac{\left(\sqrt{N}-1\right)}{\sqrt[3]{C}}},\quad E_2(N,C)=\frac{1}{1+4qk\frac{\left(\sqrt{N}-1\right)}{\sqrt[3]{C}}}$$

$$S_3(N,C)=\frac{N}{1+6qk\frac{\left(\sqrt[3]{N}-1\right)}{\sqrt[3]{C}}},\quad E_3(N,C)=\frac{1}{1+6qk\frac{\left(\sqrt[3]{N}-1\right)}{\sqrt[3]{C}}}$$

The presence of non-zero k causes that it is principally not possible to achieve a linear speedup with respect to the number of processes N at a constant size of the grid C. For example, for unidimensional decomposition, speedup S_1 is generally always limited by some constant (depending on C). However, in reality, since k may be sufficiently small, in the sections of the values N and C, used in practice, it is possible to ensure a large increase in speedup which is efficiently approximated by the linear dependence, even for unidimensional decomposition.

Since it is not possible to increase without limits the number of decomposition regions (and, consequently, the number of the processors), at a constant size of the grid, it is rational to discuss the dependence of the speedup on the number of processors N with an increase in the size of the grid $C(N)$ with increasing N. For example, for a linear increase of the size of the grid it is possible to ensure the asymptotically linear speedup for three-dimensional decomposition (in contrast to uni- and two-dimensional decomposition).

Figure 2.17 shows the graphs (in different scales) of the dependence of speedup and efficiency on the number of processors for $k = 1/100$, $q = 2 \times 6$, for a constant size of the grid of $C = 100 \times 100 \times 100$.

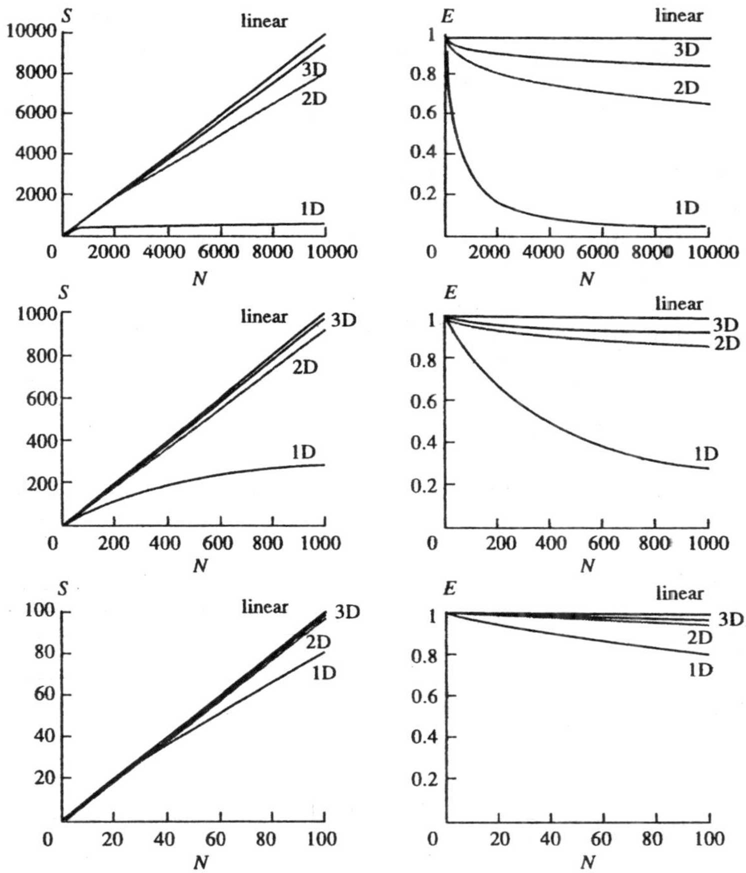

Fig.2.17. Graphs of the dependence of speedup S and the scaling coefficient E on the number of processors N for $k = 1/100$, $q = 2 * 6$, at a constant size of the grid $C = 100^3$.

The graphs show that for the values of $N \leq 100$, used most frequently in practice, the two- and three-dimensional decompositions show an almost 100% linear speedup, and the unidimensional decomposition lags behind by no more than 20%. For $N \approx 1000$ (this value is used less frequently but is still used in practice), the speedup of the unidimensional decomposition amounts to 70% of the theoretical possible value, and of the two-dimensional composition to 10%. For $N \approx 10\,000$ (used only seldom in practice), the speedup of the undimensional decomposition approaches its limit $S_1^{as} = \frac{\sqrt[3]{C}}{2qk}$, that of the two-dimensional composition decreases by 20% in relation to the theoretically possible value, and of the three-dimensional decomposition by 5%. It should be mentioned that in all likelihood (at the present time), the actual multiprocessors system consisting of 10 000 computing nodes has a cluster architecture in which the exchange time is considerably greater and the nodes can be characterised by different productivity and, consequently, more advanced models are required for evaluating the productivity of such a scheme.

We shall also present graphs (Fig.2.18) of the dependence of speedup and efficiency on the number of processors for the same values of the parameters as in the previous graphs, and the linear increase of the size of the grid $C(N) = cN$, $c = 125$.

In the present case, for three-dimensional decomposition, efficiency E_3 is even limited at the bottom by the value $\frac{\sqrt[3]{c}}{\sqrt[3]{c}+6qk}$.

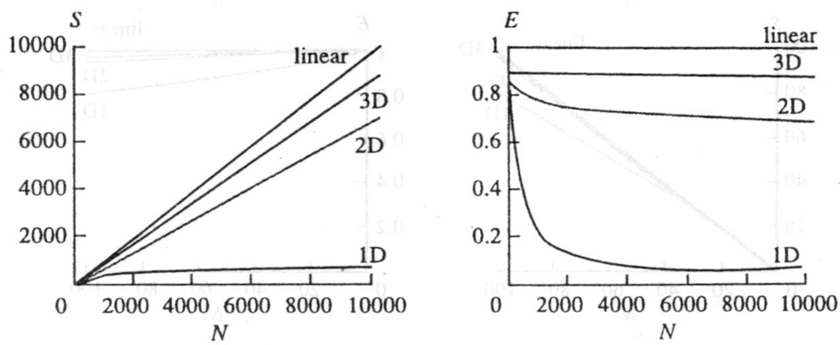

Fig.2.18. Graphs of the dependence of speedup S and the scaling coefficient E on the number of processors N at the same values of parameters as in Fig.2.17, and for a linear increase of the size of the grid $C(N) = cN$, $c = 125$.

Programs

Within the framework of investigations into numerical modelling [137], carried out at the Institute for Computer Aided Design of the Russian Academy of Sciences, a package of functions was developed for realising a sequential programme, developed for a single processor computer, with minimum losses for the modification of the programming code. The packet uses a standard exchange interface of MPI messages [230], so that it can be used in different platforms, and the control of different operating systems, without modifying the program code, written in the language *C*. The programming model SPMD (Single Program, Multiple Data) is also used which employs simultaneously several copies of the same program parallel in different processors. The advantage of this model is that it is necessary to maintain only one version of the programming code which, in addition to this, is independent of the number of processors and is specified as a parameter. It is possible to start the resulting programme, consisting of a single process (being a master). It is also possible to start a programme on a cluster of machines with different operating systems and MPI realisations. The package was tested on platforms Intel, Param, MVC-1000, with operating systems Windows, Solaris, Linux. The following MPI realisations were used: WMPI, MPICH.

When using the package, the external structure of the programme does not change, and in addition to this, the majority of the programming code remains sequential as previously. Only the separated number of areas in the programme changes. In these areas it is necessary to expand the logics of operation of a normal programme in comparison with the logics of operation of a single process of a parallel programme (with its part of the distributed data):

- Initiation (function initparallel()) and completion (function finishparallel()) of the parallel part of the programme in the initialisation process and completion of the entire programme;
- Expansion of the part of the net template (the scattergridarray() and exchange gridboundary()) functions in the process of initialisation of the programme;
- Expansion of part of the calculated data (functions scatterdataarray()) in the process of initialisation of the programme;
- Gathering of the calculated distributed data (function

gatherdataarray()) in the process of outputting the results of computation;

- Selection of the minimum (function getglobalminimum()) step with respect to time amongst the steps determined from the Courant condition with respect to the part of the calculated data;
- Summation (function getglobalsum()) in the process of counting the integral values of the loads of the retention;
- Exchange of doubled overlapping parts of the calculated data (function exchangedataboundary()) in the process of setting the boundary conditions.

The programme carries out decomposition with respect to these spatial variables leading to the volume of communications of the order $\sqrt[3]{N}$, instead of N in decomposition with respect to single spatial variables. For the exchange of data stored in the program memory with a specific template (not in a sequence) and also for abstracting from the features of the template and the exchange data we used the so-called user types of data MPR [230]. These types of data, describing the template, are determined in the localised block in the process of initialisation of the programme and can be easily exchanged. In the exchange functions, the type of data is used as a parameter and makes it possible to realise the logics of exchange in an abstract manner, for example, exchanges in different directions are carried out by the same programme code. For application in different types of communications – master-slave, processes – neighbours and networks of processes – several types of data have been determined:

- the type describing collective calculation variables (used only on the master-processor);
- the type describing local calculated variables;
- the types describing the overlapping parts of the calculated data (in respect to three different directions x, y, z).

A standard blocking sample regime is used for the master-slave communications. The master-process on its own distributes data not through sending the blocking sample but by means of direct copying of the data. For communications between the processes-neighbours it is necessary to use optimised combined operations of sending and receiving a message produced for several platforms using the SendRecv non-blocking operation and for some others using Send and Recv alternation.

Calculation characteristics and the results of numerical modelling

The programme was tested on a problem of lift of a surface thermic in a stratified atmosphere. This problem has been examined many times in the literature [7,74], and similar results have been obtained independently. We shall present the results obtained using the parallel programme described.

We shall examine a problem of the lift of a thermic – a heated cloud of gas having the form of a hemisphere, radius R = 1360 m. The temperature in the centre of the cloud T_0 = 7085 K decreases exponentially to the boundary of the cloud $T = T_0 \exp\left(-B\frac{r^2}{R^2}\right)$, $B=1$.

The atmosphere is modelled by a linear decrease of temperature to the level of the stratosphere at 11 km, the pressure on the surface of the earth is 1 atm, the temperature on the surface of the earth 288 K. Figure 2.19 shows the distribution of the cloud of the impurity, formed into a ring, at times t = 0, 1, 2, 3, 4, 5 min.

The calculation characteristics of operation of the real programme will be described on a testing problem. Measurements were taken in an MBC-1000-M supercomputer with 768 processors. A series of measurements were taken for unidimensional, two-dimensional and three-dimensional decompositions. The number of processes in all three cases was approximately the same. Figure 2.20 gives speedup and efficiency, determined from the time of a single iteration, on a grid with a size of 120^3.

The data in the graphs have the following form (Fig.2.20).

The graphs show that the effect of the type of decomposition on the speedup and efficiency is already strong at N > 25.

2.6 MATHEMATICAL MODELLING OF AN EXPLOSION OF A SUPERNOVA IN A PARALLEL COMPUTER

The problem of an explosion of a supernova has not as yet been solved (see, for example, a review by A. Mezzacappa [200]). Calculations were carried out mainly for unidimensional models and this has resulted in physical restrictions in the models of supernova explosions. Recently, in studies of a group of the Institute of Applied Mathematics of the Russian Academy of Sciences, the authors proposed model of the explosion of a supernova based on the development of large-scale convective instability [117,122],

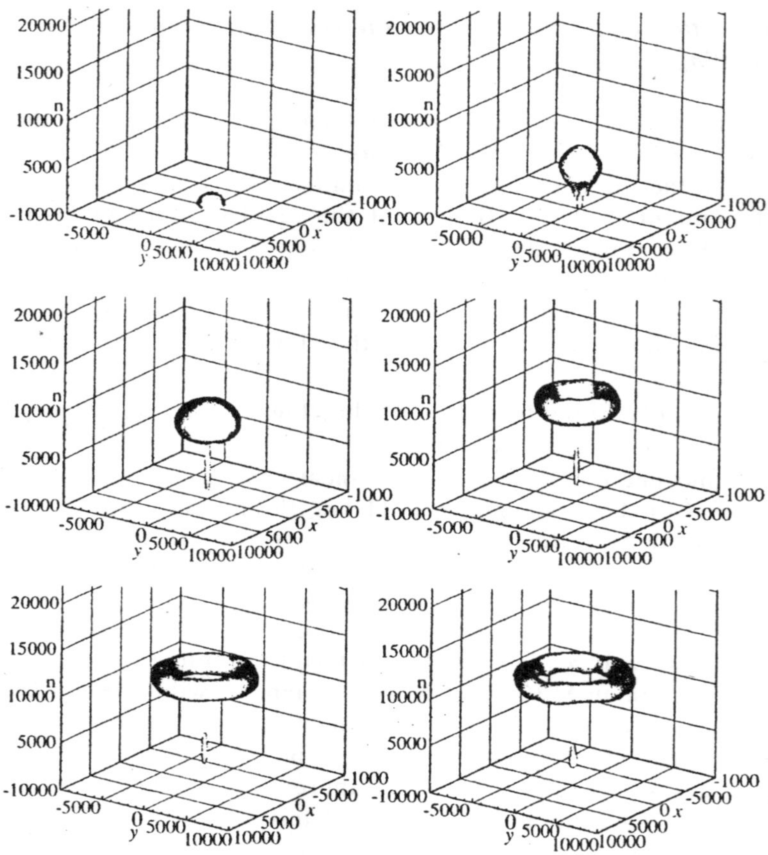

Fig.2.19. Distribution of an impurity cloud (the isosurface of impurity concentration is presented) rolled up into a ring, at times t = 0, 1, 2, 3, 4, and 5 min.

based on the concepts proposed in [59,150].

This programme requires a numerical solution of three-dimensional equations of hydrodynamics with a good spatial resolution. As shown in [177], even for an approximate description of the process, it is necessary to use a difference grid with a resolution of 50×50×50 only for the central region. This grid is insufficient for more matched description of the processes taking place. However, even under these conditions, the problem cannot be solved because of the more-or-less acceptable computing time, not talking about the need for checking several variants of the calculations. In addition, the calculation conditions have become

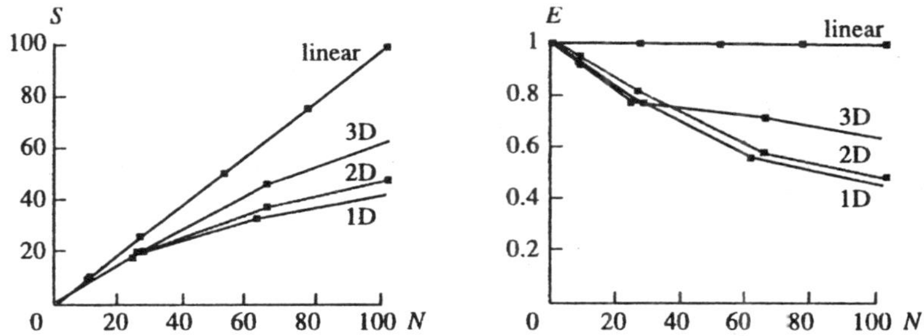

Fig.2.20. Speedup S and scaling coefficient E, determined by calculations of the problem of the liftup of a thermal in an MBS-1000 supercomputer on a grid with a size of 120^3.

more exacting in the case of expansion of the computing grid to the range of low densities (for example, to densities of $\rho \approx 10^8$ g/cm^3).

To solve this problem, it is necessary to use parallel computers. In this section, we present the preliminary results of calculations using the technology of parallel computations. We also describe the methods developed for parallel computations.

Description of the algorithm with embedded grids

In the numerical modelling of convection, we solved three-dimensional equations of hydrodynamics with gravitation. Degenerate relativistic electrons and non-relativistic nucleons provide a contribution to the equations of state. The initial equations were written in the divergent form:

$$\frac{\partial \mathbf{U}}{\partial t} + \nabla \cdot \mathbf{F} = \mathbf{S}.$$

where $U = (\rho, \rho u, \rho v, \rho w, E)$ is the vector of conservative variables, $\mathbf{F}$ is the vector of fluxes, $\mathbf{S} = (0, \rho g_x, \rho g_y, \rho g_z, \rho \mathbf{ug})$. The calculation region was covered by three levels of the embedded rectangular grids with a homogeneous step in space (Fig.2.21). In numerical calculations on each embedded grid we used the explicit conservative TVD difference scheme of the Godunov type with a second order with respect to space and time:

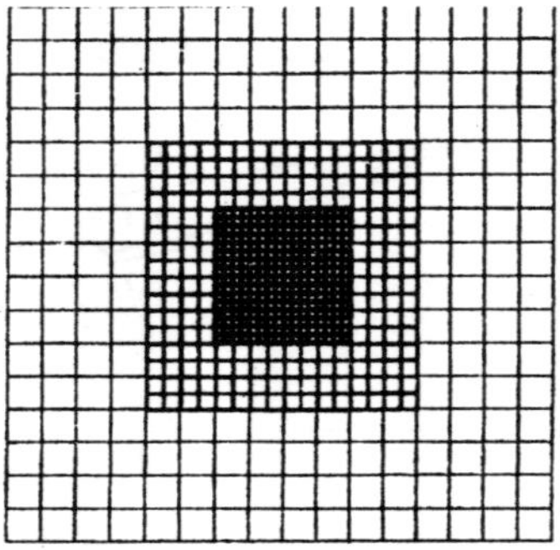

Fig.2.21. Embedded grids.

$$\left(U_{i,j,k}^{n+1}-U_{i,j,k}^{n}\right)/\Delta t=$$
$$=\left(F_{i+1/2,j,k}^{n}-F_{i-1/2,j,k}^{n}\right)/\Delta x-\left(G_{i,j+1/2,k}^{n}-G_{i,j-1/2,k}^{n}\right)/\Delta y-$$
$$-\left(H_{i,j,k+1/2}^{n}-H_{i,j,k-1/2}^{n}\right)/\Delta z+S_{i,j,k}^{n}.$$

The flows through the boundaries of the grid F, G, H were calculated using Roe's scheme and the local-characteristic approach. For example, in one direction:

$$F_{i+1/2,j,k}=\left(F_{i,j,k}+F_{i+1,j,k}+R_{i+1/2}W_{i+1/2}\right)/2,$$

where $R_{i+1/2}$ is the right eigenvector of the Jacobi matrix $A = \partial F/\partial U$, and $W_{i+1/2}$ is the vector of numerical dissipation. The second order with respect to time was ensured by using the Runge–Kutta method, retaining the TVD property of the difference scheme. The stability of numerical computation was guaranteed using a single Courant number equal to 0.1 in our calculations for all embedded grids. In this case, one step in respect of time on a course net corresponded to two steps on a finer net. The boundary values of the calculated quantities for each embedded grid, required for continuous computing, were determined by monotonic linear interpolation of the values of the quantities from a coarser grid. To determine the values of the cells of the coarse grid, covered by the cells of a finer grid, we used a simple averaging procedure. To maintain the conservative nature of the calculations, the fluxes through the cells of the coarse grid, bordering with the cells of the finer grid, were replaced by the fluxes computed on the embedded grid.

At the initial moment of time, the initial perturbation of the entropy of the Gaussian type was specified for the centre of the calculation region:

$$S = S_0 \left(S' - S_0\right) \exp\left(-\left|\mathbf{r} - \mathbf{r}_0\right| / \left|b\right|\right), \quad S' = 3S_0, \quad r_0 = 0.$$

The boundary conditions were specified as historic, i.e. the initial values of all quantities were retained in respect of time at the boundary.

Description of the parallel algorithm

The most important moment in the design of the parallel algorithm is the selection of the scheme of distribution of the data between the processes of the parallel programme. We selected a distribution of the data assuming that the programme, realising the algorithm, is applied on a uniform parallel computing system and that only one process of the parallel programme will be realised in each processor (therefore, they will not be described in this book).

The processes of the parallel programme are logically examined as a three-dimensional net with three co-ordinates I, J and K. To ensure the optimum distribution of the load between the processes, it was decided to distribute all the nodes of all grids uniformly in the lattice of the processes (Fig.2.22). In this case, a parallelepiped of the nodes of the grid of the same size was distributed on each node of the lattice of the processes. Figure 2.22 shows the projection of the distribution of a 12×12×12 grid in the nodes of the lattice of the processes with a size of 4×3×1. A 3×4×12 parallelepiped is distributed on each process.

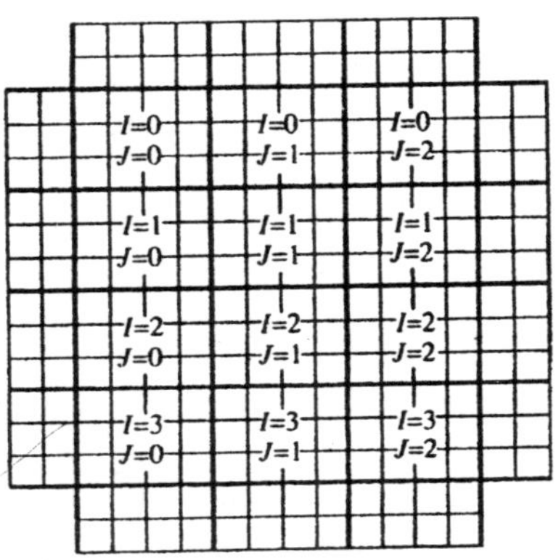

Fig.2.22. Distribution of the lattice in the processors.

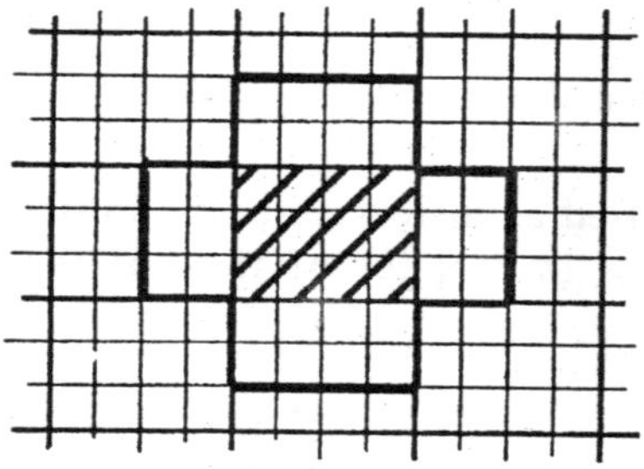

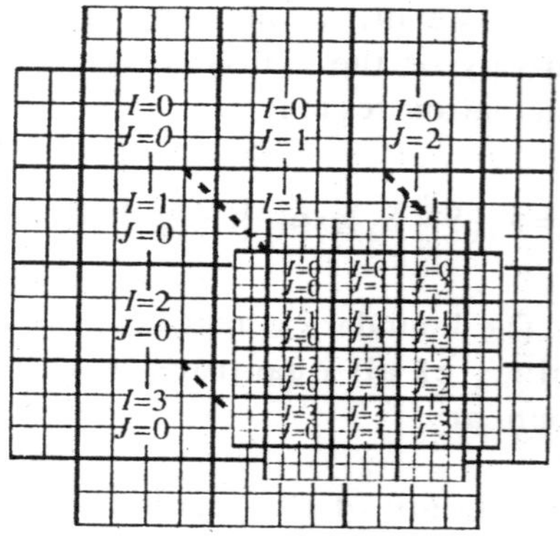

Fig.2.23. Data for a single processor.
Fig.2.24. (right) Relationship between lattices and computing space.

It should be mentioned that the algorithm of solving the problems based on the five-point template with respect to each coordinate, i.e. for the calculations in every node of the net it is necessary to know the values of the variables into adjacent cells of the grid with respect to all six directions. Figure 2.23 shows in greater detail one of the nodes of the lattice of the processes. The broken lines indicate the nodes of the grid distributed on the given process, and a thick line indicates the nodes of the grid containing the data required for calculations in the given process. Figures 2.23 and 2.24 show that the boundary cells (not included in the computing region of the processor) are either included in the composition of the boundary layer of the entire grid or relate to the computing region of the adjacent processor. In the second case, the data of these cells will be regularly renewed in the computation process, carried out in parallel by all processors, and the new values should be sent to all who require these data for work. Thus, communications are essential in any two adjacent processors (one of the co-ordinates differs by unity). To increase the operating speed of the programme, it would be necessary to organise transfer (if allowed by communication equipment) in such a manner that they take place in parallel, i.e. the communication time during the calculation period depends only on the maximum volume of a single message between any two processors, and there is no explicit dependence of the number of processors. On the basis of these considerations, the communications were organised as follows: the passage of the net on each dimension was carried out in sequence and in two iterations.

In the first change, information was exchanged by the processors with the front neighbour with even numbers, in the second iteration the processors with the odd numbers.

As already mentioned, each grid was projected separately onto the entire computing space (Fig.2.24). This approach enabled the most uniform distribution of the computing load between the processors but this was accompanied by the formation of additional transfers in two functions of the programme. One of them (Interpolation) is responsible for the determination of the boundary conditions on the embedded grid (the data in the cells with the boundary layer around the grid). The values were calculated on the basis of the given cells, containing the boundaries of the embedded grid, and the adjacent cells. The computed values must be attributed to the corresponding variables of the grid of the cells of the boundary layer and this requires sending these values.

The second function (Overlaid) is responsible for averaging. In each iteration, finding a more detailed solution on the embedded grid, it is necessary to improve the accuracy of the results in the corresponding physical region on the grid-carrier, i.e. we average out the values of four (eight in the three-dimensional case) cells of the embedded grid, belonging to one cell of the grid-carrier, re-determine it, and for this purpose the average value must often be sent to the second processor (Fig.2.24). A similar procedure should be carried out with the entire embedded net but this results in a large volume and number of endings. The values of the cells of the given region of the net-carrier are not important because in every step they will be re-determined by averaging. The important values are only those in the two extreme cells, obtained by averaging, because they are required in computing the variable cells, bordering with the region covered by the embedded grid. Taking this into account, when writing the parallel programme it was decided, in the averaging function, to restrict ourselves to calculations and sending only the given outer cell.

There is another basic function of the programme-function of determination of the step in respect of time. For this purpose, it is necessary to analyse the values in all cells of the net and on the basis of analysis find the minimum value of the step which is then used for further computations. A specific value is obtained in every processor. These values must be collected in a single processor, select the minimum value and sent out again. In this case, it is not possible to organise the parallel exchange of data, but when using the sendings in respect of the binary tree it is possible to shorten greatly the communication time of the given function – it will be proportional to $\log_2 P$.

Justification of the selection of the instrumental means for realisation of the parallel algorithm

The principal problem in the realisation of the complex parallel algorithm is the problem of the selection of the instrumental means of realisation. These means can be divided into two groups – with explicit and implicit parallelism. The first (for example MPI [198] and mpC [184]) required the programmer to carry out parallelization of the data and computations. For the second (high-level languages of parallel programming, for example HPF [198] and Fortran DVM [73] carry out this automatically often using additional information, supplied by the programmer. At first sight, the second approach is preferred but practice shows that the parallelization is a very complicated task and even the most advanced HPF compilers generate programmes whose speed is several times slower than that of their MPI analogues [218].

At the present time, the MPI is most popular in programming systems with distributed memory (and this includes the majority of Russian high-productivity parallel computing systems). However, differing low level means and development using this system of the realised task requires a very high qualification of the programmer and is time consuming.

The Institute of System Programming of the Russian Academy of Sciences have developed a language of parallel programming mpC which, being a high level means, enables programmes to be written that are comparable in efficiency (a difference of 5–10%) with the most advanced packages of applied programmes, developed by means of MPI on uniform computing systems and greatly exceed them in respect of productivity on non-uniform computing systems [181]. The language mpC is a two-step expansion of the C language. In the first step, language C has been expanded by introducing vector operations. This is the expansion – C[], which is equivalent, in respect of the computing power, to Fortran 90 language. In the second step, the language C[] was expanded by the concept of the computing space which the programmer can control in the same manner as when controlling the memory in language C. Language mpC is characterised by a high expressive power. For example, the application of vector operation in programming the examined algorithm has made it possible to reduce the size of the programming code by approximately a factor of three and greatly shorten the realisation time.

Analysis of the efficiency of the algorithm

One of the main characteristics of the parallel algorithm, solving the problem of the characteristic size n on P processors, is its efficiency determined from the following equations [161]:

$$E(n,P)=\frac{T_{\text{seq}}(n)}{P\cdot T(n,P)},$$

where $T_{\text{seq}}(n)$ is the time of solution of the problem using a sequential algorithm; $T(n, P)$ is the realisation time of the problem by means of the parallel algorithm on P processors.

If the efficiency depends on n and P only through their ratio, the parallel algorithm is referred to as scaling. In other words, the efficiency of application of the scaling algorithm remains constant with the proportional variation of the dimension of the problem (in the present case $n = N^3$, where N is the linear size of the cubic net) and the number of processors P.

We shall estimate the efficiency of our parallel algorithm within the framework of several assumptions. Initially, we shall assume that the communication equipment sustains the parallel transfer of data. When evaluating the sending time we shall regard this time as directly proportional to the volume of the transferred information; this is accurate only if the length of the reports is relatively large. For simple evaluation, we shall also assume that the processes form a cubic lattice and that a cubic piece of the grid is projected onto each processor. The programme consists of four main functions. We shall examine each function.

The first is a function directly occupied by the calculation of the values of the required variables of the problem in the next step in respect of time. Since the values are determined in every cell, the duration of calculations by this function is proportional to the number of cells in the cross-hatched region in Fig.2.23 (with a correction for the three-dimensional form of the problem), i.e.

$$T_s^{ex}=\alpha_1\cdot\frac{N^3}{P} \quad \text{(sector } N\times N\times N \text{ is counted on } P \text{ processors).}$$

After carrying the calculations, it is necessary to send the restored values to the adjacent processor. Examination of these sendings of the algorithm presented in the description shows that the volume of the transferred information is proportional to the area of the surface of the cube of the computing region of the processor. Thus, including all numerical coefficients in a proportionality constant, we obtain the following equation for the sending time of this function

$$T_s^{\text{com}} = C \cdot \left(\frac{N}{P^{1/3}} \right)^2 = C \cdot \frac{N^2}{P^{2/3}}.$$

The second function (Interpolation) is the determination of boundary conditions on the embedded grid. Since we determine the values of the variables in the cells of the boundary layer and their number is proportional to the area of the layer (in the three-dimensional case), then

$$T_I^{\text{ex}} = \beta_1 \cdot \frac{N^2}{P^{2/3}}.$$

The upper estimate for the sending time in the given function is:

$$T_I^{\text{com}} = D_1 \cdot \frac{N^2}{P^{2/3}}.$$

The third function is the averaging function (Overlaid). As in the Interpolation function, we examine only the outer cells and consequently, we obtain:

$$T_O^{\text{ex}} = \beta_2 \cdot \frac{N^2}{P^{2/3}} \quad \text{and} \quad T_O^{\text{com}} = D_2 \cdot \frac{N^2}{P^{2/3}}.$$

The last function is the function of determination of the step in respect of time. Since all cells are analysed, then as in the first function

$$T_T^{ex} = \alpha_2 \cdot \frac{N^3}{P},$$

and the communication time has the form

$$T_T^{com} = A_2 \cdot \log_2 P.$$

We assume that the time of execution of the parallel programme is equal to the sum of the computation and communication times: $T(n, P) = T^{ex}(n, P) + T^{com}(n, P)$.

The total computation time is

$$T^{ex}(n, P) = T_S^{ex} + T_T^{ex} + T_I^{ex} + T_I^{ex} + T_O^{ex} =$$
$$= (\alpha_1 + \alpha_2) \cdot \frac{N^3}{P} + (\beta_1 + \beta_2) \cdot \frac{N^2}{P^{2/3}} = \begin{Bmatrix} \alpha = \alpha_1 + \alpha_2 \\ \beta = \beta_1 + \beta_2 \end{Bmatrix} = \alpha \cdot \frac{N^3}{P} + \beta \cdot \frac{N^2}{P^{2/3}}.$$

The total communication time is

$$T^{com}(n, P) = T_S^{com} + T_T^{com} + T_I^{com} + T_O^{com} = (C + D_1 + D_2) \cdot \frac{N^2}{P^{2/3}} + A \cdot \log_2 P.$$

The total volume of the computations is the same for sequential and parallel algorithms. Consequently

$$T_{seq}(n) = T^{ex}(n, P)\Big|_{P=1} = \alpha \cdot N^3 + \beta \cdot N^2.$$

In our problem $\beta < \alpha$ and $N \gg 1$, and consequently $T_{seq}(n) \approx \alpha \cdot N^3$.

Therefore, the expression for the efficiency of the parallel algorithm has the form

$$E(n,P) \approx \frac{\alpha \cdot N^3}{P\left(\alpha \cdot \frac{N^3}{P} + \beta \frac{N^2}{P^{2/3}} + (C + D_1 + D_2) \cdot \frac{N^2}{P^{2/3}} + A \cdot \log_2 P\right)} =$$

$$= \frac{\alpha \cdot N^3}{P\left(\alpha \cdot \frac{N^3}{P} + (\beta + C + D_1 + D_2) \cdot \frac{N^2}{P^{2/3}} + A \cdot \log_2 P\right)} =$$

$$= \{\lambda = \beta + C + D_1 + D_2\} =$$

$$= \frac{N^3}{P} \cdot \frac{\alpha}{\alpha \cdot \frac{N^3}{P} + \lambda \cdot \left(\frac{N^3}{P}\right)^{2/3} + A \cdot \log_2 P} =$$

$$= \frac{\alpha}{\alpha + \lambda \cdot \left(\frac{P}{N^3}\right)^{1/3} + A \cdot \frac{P}{N^3} \cdot \log_2 P}.$$

We examine the problem for scalability. For this purpose, as mentioned previously, it is necessary to examine the behaviour of the efficiency with the proportional variation of the dimension of the problem N^3 and the number of processors P. We set $\frac{P}{N^3} = \mu = \text{const}$. Consequently, the expression for efficiency has the form:

$$E(n,P) = \frac{\alpha}{\alpha + \beta \cdot \mu^{1/3} + A \cdot \mu \cdot \log_2 P}.$$

It may be seen that, generally speaking, the problem is not scalable. However, it should be mentioned that $\log_2 P$ is a very 'sluggish' function at high P: $\frac{d(\log_2 P)}{dP} \approx \frac{1}{P} \xrightarrow{P \to \infty} 0$. Correspondingly, the problem may be regarded as scalable at high P. This conclusion is in good agreement with the results obtained by the authors in analysis of the realisation time of the programme. In particular, we measure the time of realisation of the problem on the condition that $\frac{N^3}{P} = \text{const}$. The previous considerations show that this time is determined by the following equation:

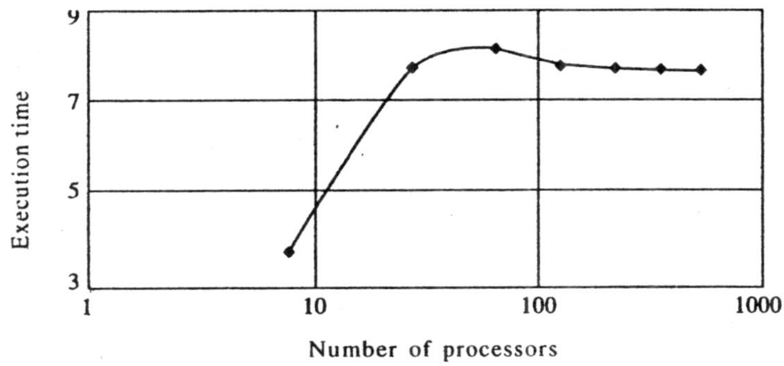

Fig.2.25. Dependence of the execution time on the number of processors of MBC-1000M supercomputer (number of cells per processor N^3/P = 4096).

$$T(n,P)=\frac{T_{seq}(n)}{P\cdot E(n,P)}=\frac{\alpha\cdot N^3}{P}\cdot\frac{\alpha+\beta\cdot\left(\frac{P}{N^3}\right)^{1/3}+A\cdot\frac{P}{N^3}\log_2 P}{\alpha}=$$

$$=\left\{\frac{N^3}{P}=\nu=\text{const}\right\}=\alpha\cdot\nu+\beta\cdot\nu^{2/3}+A\cdot\log_2 P.$$

The results of the calculations, carried out in MPC-1000M are presented in Fig.2.25. Here $\frac{N^3}{P}=4096$ or, in other words, 4096 net cells per processor. The graph shows the logarithmic dependence. This is in agreement with our estimates and it may also be seen that

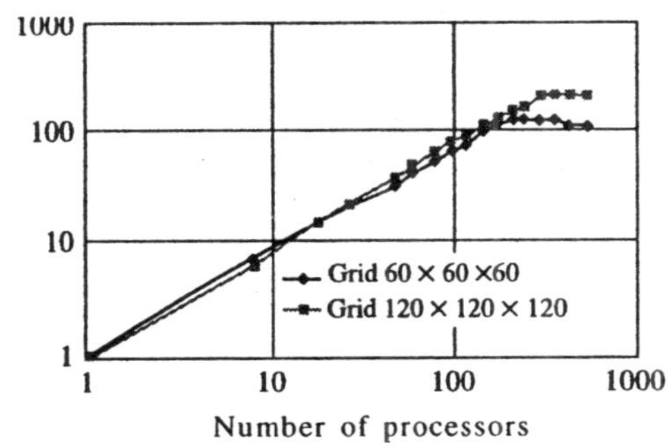

Fig.2.26. Dependence of the speedup on the number of processors of MBC-1000M supercomputer for two grids: 60^3 and 120^3.

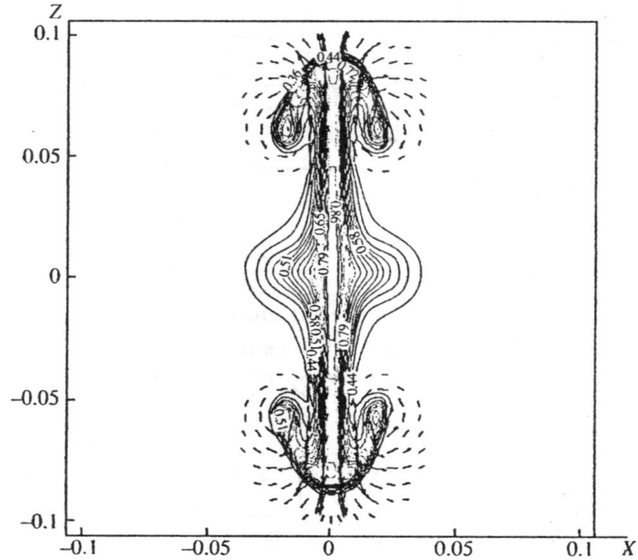

Fig.2.27. Distribution of the density of matter and the field of velocities in explosion of a supernova at t = 0.00432 s. determined in computing experiments on 100 processors using embedded grid technology

this realisation time is practically constant, i.e., the algorithm and realisation are scalable at:

$$\begin{cases} P \gg 1, \\ \left(\dfrac{N}{P^{1/3}}^{2}\right) = \dfrac{N^2}{P^{2/3}} \gg 1. \end{cases}$$

It should be mentioned that the second restriction corresponds to the condition of applicability of our estimates (the processors exchange information in large packets).

On the other hand, it should be mentioned that at low P, the efficiency tends to unity. This is confirmed by the graph (Fig.2.26) of the dependence of the speedup of the programme with the number of processors P on which the programme is realised at the dimension of the problem of N = 60 and N = 120.

These methods were used to carry out preliminary calculations of an explosion of a supernova. Figure 2.27 shows the distribution of the ejected matter in the form of two jets and the appropriate field of velocity of the matter in the process of development of large scale convection. The results correspond to the moment of time t = 0.00432s. This calculation requires ten minutes of computing in the supercomputer of the Russian Academy of Sciences using 100 parallel processors.

Chapter 3

Selected results

3.1. CALCULATIONS OF COHERENT STRUCTURES IN A NEAR WAKE BEHIND A SOLID FOR LIMITING MOTION REGIMES

We shall present several results of calculating the movement of ordered large scale turbulent structures.

As an example of calculating the real turbulent wake behind a moving solid, we examine the problem of the flow of a compressible gas around a circular cylinder at supercritical speeds (calculation schemes, methodological computations and some of the results obtained in this region were described in [10,19]). It is interesting to note that in numerical modelling of the flow of a subsonic supercritical flow of a compressible gas around a circular cylinder ($M_\infty \geq M_\infty^{cr} = 0.4$, $\gamma = 1.4$) there is a distinctive possibility (both in the case of viscous and non-viscous gas) of existence of two detachment regimes: symmetric stationary and non-symmetric periodicity. There is agreement with the experimental data [12,13,81]. We shall pay more attention to these results obtained by A.V. Babakov ([9,10,21] *et al.*).

Figure 1.8 shows the experimental [236] patterns of the flow of a compressible gas around a circular cylinder at different Mach numbers M_∞.The Reynolds numbers in the examined experiments are in the range 1.1×10^5–1.35×10^6 and correspond mainly to the turbulent regimes of motion. These experimental data indicate that the large scale vortex structures (reflecting mainly the wave process) exist on the background of fine-scale turbulence. It is not possible to take the latter adequately into account. However, as reported in [12], 'investigations carried out in the last 20 years in the area of turbulence have resulted in a growing acceptance of the fact that the processes of transfer in the majority of turbulent shear flows are determined by the large scale vortex movements' and 'the direct link between large- and small-scale turbulence does not have any controlling effect on the main characteristics of the macrostructures'.

Taking into account that at high Reynolds numbers the convective transfer in a considerably greater part of the flow region prevails over viscous transfer (with the exception of the regions of the type of boundary layers), and examining the problem of modelling flow 'as a whole', we shall use as a basis the complete non-stationary model in the form integral laws of conservation (see equation (1.14) in Chapter 1) for a non-viscous compressible gas (Euler model). The realisation of the model of the viscous gas presents no principle difficulties. However, as already mentioned, the adequate application of the model of the viscous gas at high Reynolds numbers from most stationary flows with large scale structures in currently available computers is, in our view, difficult, not only because of technical reasons. In addition, the problem with practical description capacity for the limiting developed turbulent flows remains open (it is necessary to take into account all scales of the phenomenon, up to molecular).

Figure 3.1 shows stationary solutions for the flows of a compressible gas around the cylinder on subsonic and sonic regimes. For the numbers $M_\infty < M_\infty^{cr}$ the numerical solution is continuous (the molecular mechanism of detachment and falling jumps are not present) – this type of flow corresponds to the potential one. At $M_\infty > M_\infty^{cr}$, there are local supersonic zones with a closing jump which results in detachment of the flow (this is in complete agreement with the experimenting [236]) and, consequently, the bottom part is characterised by the formation of the zones of reciprocal–circulation flow (the solutions, in particular, were obtained on the condition of the flat symmetry of the flow). The stationary nature of the presented solution is not disrupted even in integration for long times in the entire region around the cylinder without formulation of the symmetric conditions. Evidently, this is explained by the symmetry of both the computation algorithm and the initial and boundary conditions. In all likelihood, the errors of the purely scheme and computing origins are not capable of causing an instability of the stationary solution which in the natural conditions is easily disrupted by natural perturbations.

In this case, it is necessary to examine the problem of the possibility of existence of stationary detachments in natural phenomena and full-scale experiments. In our view, the situation in this case is very similar to the current fashionable direction of

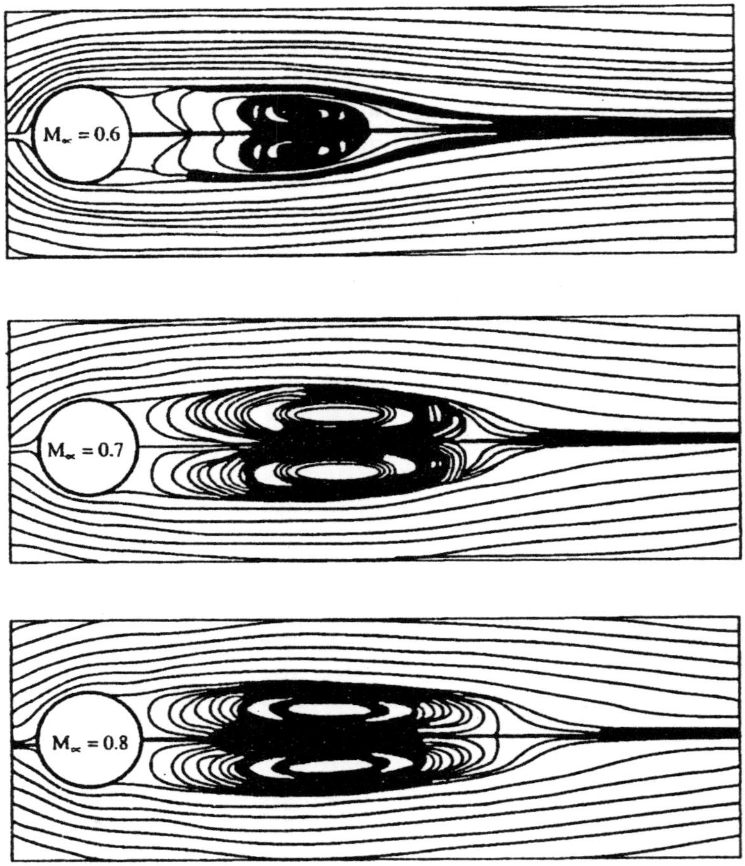

Fig.3.1. Stationary (metastable) regimes at different M_∞. Instantaneous current lines in the coordinate system connected with the body.

mathematical description of the loss of stability associated with strange attractors where 'the stationary regime remains theoretically possible, but becomes unstable and, therefore, is not observed closely' [120]. In the examined phenomena, the stationary flow can be, in all likelihood, interpreted as a metastable state unstable in hydrodynamic sense, not realised in the natural conditions and being very difficult to achieve in the experiments.

Experimental investigations [12, 13], carried out in a low-noise aerodynamic pipe with an open part in the range of Re_∞ numbers from 2×10^3 to 4×10^4, in the conditions ensuring the absorption of acoustic perturbations (evidently, this was of controlling importance), in movement of the model into the working part, the 'aeolian tone', characteristic of the stationary regime, disappeared. Both in these experiments and in numerical calculations [11] there was a large (approximately by a factor of 2) increase in the coefficient of the wave resistance of the cylinder in transition from the stationary regime to a non-stationary one (the mean value is considered for the latter).

In the natural phenomena and in experimental investigations, the heterogeneities of the flows and design special features lead to the realisation of a stable periodic regime with a developed trail [81,123,124,222]. In numerical modelling, these non-symmetric perturbations do not occur (if they are not specially introduced).

In order to obtain a non-stationary regime in a numerical experiment, the stationary state acted by means of short-term or instantaneous perturbations, and the latter were of both physical and purely computing nature. One of the types of these perturbations modelled in principle a weak supersonic blow into the bottom part of the cylinder in the sector $160° \leq \varphi \leq 180°$, acting for a short (in comparison with the characteristic duration of the phenomenon) period of time $\Delta t = 0.5$. The second type of perturbations consisted of 'volitional' rotation of the vectors of the speeds of the flow of the stationary solution through the angle of 10° in the sector $120° \leq \varphi \leq 240°$ of the bottom part of the flow. The third type of perturbation was expressed by the simulation of the screen of the form of the cylindrical arc $150° \leq \varphi \leq 180°$, situated at a distance of one gauge behind a cylinder and acting on the flow during a period of time ($\Delta t = 1$).

It is important to mention that, in all cases, irrespective of the type of introduced non-symmetric perturbation, the numerical solution led to the same strictly periodic non-stationary regime. The degree of perturbation of the flow affected only the time to establishment of a new regime which was 20–100 relative units. The resultant non-stationary periodic solution was again subjected to these perturbations, but after a specific period of time it again returned to the established periodic state, and so on. The 'independence' of the examined flow from the initial data of the type of perturbation and grids indicates in this case that there is a sufficiently large reserve of the stability of the investigated flow (within the framework of the model used). After establishing a stable periodic regime, a decrease in the time step (for example, by a factor of 10) and further integration for five periods also did not result in any significant changes of the flow characteristics. This indicates that the resultant numerical results are not a manifestation of the internal scheme effects. In the given series of methodological calculations [9,10,21] the outer boundary of the integration range was at a distance of 25–100 radii of the cylinder, and the number of volumes-cells in the region was varied from 1200 to 5400. Some of the results of these investigations are presented in Table 3.1.[1]

Table 3.1

Grid	40×30, 3.25	60×60, 3.100	60×90.3, 100 τ = 0.005	60×90.3, 100 τ = 0.0005
max $\|C_y\|$	0.76	0.71	0.72	0.725
$\overline{C}_x$	1.126	1.01	1.09	1.09
Sh	0.232	0.232	0.244	0.241

Calculations [10] show that the minimum distance of the outer boundary of the integration region should be 25 gauges. Further numerical modelling of the vortex trail was carried out for M_∞ values in the range 0.4–0.95 on a grid (60×90.3,100). The investigated regime is characterised by descent of the vortex sheet from the surface of the cylinder in its bottom region which during its further evolution forms large scale vortex structures. The position of the detachment point on the surface of the body in the supercritical regime (for a circular cylinder $M_\infty \approx 0.4$) is determined mainly by the compression shock closing the local non-stationary supersonic zone. Practically in the entire period (with the exception, possibly, of relatively short periods of time), only one point $\varphi = \varphi_\omega$ with a zero tangential component of the speed is found on the surface of the cylinder. This point moves continuously on the surface of the cylinder in the sector $180°-\varphi_0<\varphi<180° +\varphi_0$, slows down in the upper and lower positions and passes quite rapidly in the vicinity of the point $\varphi = 180°$. Figure 3.2 shows the time dependence of the coefficient of the lifting force of the cylinder in the non-stationary regime for different values of M_∞.

The experiment [123,124,222,236] shows that in the examined range of the Mach number, the Strouhal number Sh = $2R/(TV_\infty)$,

[1] Notations of the grid parameters are as follows: ($l \times m$, R_0, R_l), where l and m are the number of divisions in respect of angular and radial coordinates; $r = R_0$ is the surface of the solid, $r = R_l$ is the outer boundary

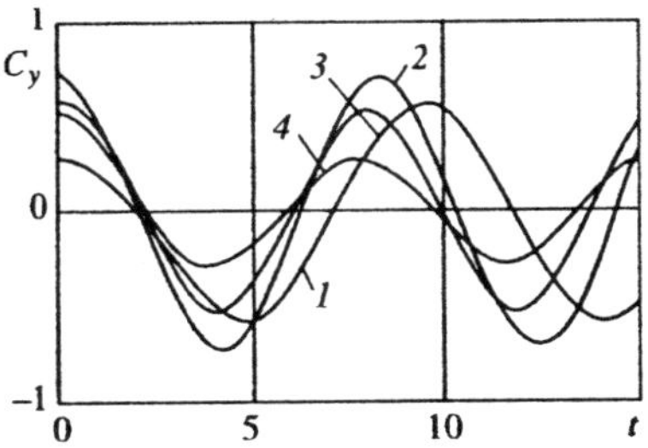

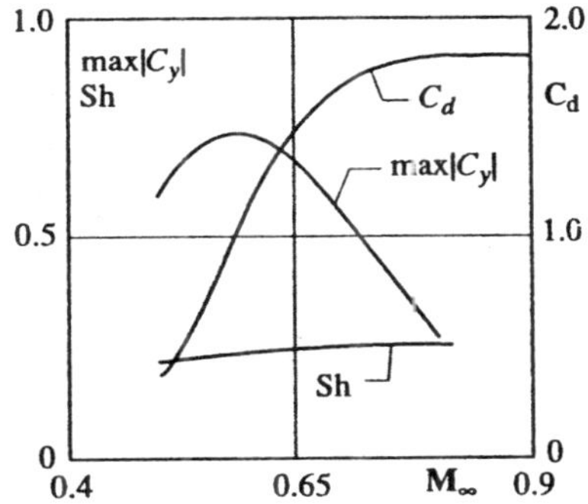

Fig.3.2. Nonstationary regime. Behaviour of the coefficient of lifting force with time: 1) $M_\infty = 0.5$; 2) $M_\infty = 0.63$; 3) $M_\infty = 0.7$; 4) $M_\infty = 0.8$.
Fig.3.3. (right). Nonstationary regime. Effect of Mach number (M_∞) on aerodynamic characteristics: C_x, max $|C_y|$ and Strouhal number Sh.

where T is the period of the phenomenon, is in the range 0.18–0.24. Figure 3.3 shows the numerical dependences on M_∞ of the values of Sh, the maximum value of the modulus of the coefficient of the lifting force max $|C_y|$ and the mean (for the period) value of the coefficient of wave resistance $\overline{C}_x$. For the maximum $|C_y|$ there is an extreme value at $M_\infty \approx 0.6$.

In experimental investigations, special attention has again been given to the visualisation of the flow as a means of examining its structure [71]. In numerical modelling, the 'visualisation' of the flow may be carried out using different procedures: in the form of instantaneous flow lines in the co-ordinate systems moving with different speeds in relation to the flow or in the form of isolines of vorticity, density, pressure, etc. In this case, the vortex structures existing in the flow are manifested in different manners. For example, Figure 3.4 and 3.5 show the flow patterns in the form of instantaneous flow lines for $M_\infty = 0.8$ ($\Delta t = 0.25T$, T is the period). In Figure 3.4, the instantaneous flow lines are depicted in the system of the co-ordinates linked with the incident flow. This corresponds to the photograph of tests on a ballistic track (developed vortex structures, forming a vortex track, are clearly visible). Figure 3.5 corresponds to the instantaneous flow lines, depicted in the co-ordinate systems linked with the solid – an analogue of the photographs of the traces of impurity particles in an aerodynamic pipe. Here, the vortex structures are manifested in the form of wave-shaped motions of the media.

Fig.3.4. Nonstationary wake behind a cylinder: $M_\infty = 0.8$. Instantaneous lines of current in the coordinate system connected with the incident flow.

Fig.3.5. Nonstationary wake behind a cylinder: $M_\infty = 0.8$. Instantaneous lines of current in the coordinate system connected with the body.

The patterns of the instantaneous flow lines depends strongly on the speed of the co-ordinate system in which they are depicted. The instantaneous flow lines are a peculiar visualisation of the flow combined with a considerable amount of quantitative information on them. However, there are considerable conceptual difficulties in interpreting the relationship of the non-stationary patterns of the flow lines with the actual flow patterns [71]. The patterns, invariant in relation to the co-ordinate system, can be represented in, for example, the form of vorticity and density isolines. Figures 3.6–3.8 show the isolines of vorticity for M_∞ equal to 0.5; 0.6 and 0.7, respectively. They show clearly the evolution of the wake behind the solid with increasing Mach number M_∞.The isolines of the density correspond, as regards the structure, to the visualisation of the experimental examination in the form of shadow photographs (Fig.3.9). Visualisation of the flow makes it possible to determine the geometrical dimensions of the vortex structures, the distance between them and the speed of their motion. It can be seen that the numerical experiment provides a very large amount of information.

We shall now compare the results in greater detail with the experimental data [236]. For example, Figure 3.10 shows film frames of the flows for $M_\infty = 0.64$ 'pasted in' patterns of the flow obtained by the numerical procedure ($M_\infty = 0.6$) and selected, if possible, in the phase and reduced to the same scale. Comparison shows that the positions of the vortex structures and the geometrical characteristics of the flow are in relatively good agreement. This is indicated by both visual and quantitative comparison of the calculated flow fields [9,10] with the experiments [236] (Re = 1.35×10^6), the values of the aerodynamic coefficients, the positions of the points of separation and the nature of the behaviour of the pressure along the surface of the cylinder [19]. For example, at $M_\infty = 0.54$, the experiment [81] gives Sh = 0.18, $C_x^H = 0.9$ (C_x^C = 0.34), and from the results of calculations in [9,10] Sh = 0.178, $C_x^H = 0.9$ ($C_x^C = 0.45$). Here Sh = D/TV_∞ is the Strouhal's number, C_x^H, C_x^C are the coefficients of resistance for non-stationary cases, respectively.

To evaluate the reliability of the results, calculations were also carried out for these problems with the introduction of a semi-empirical ($K - \Theta$) model of turbulence, where K is turbulent energy, and Θ is the rate of dissipation [213]. The model makes it possible to improve the accuracy of the integral characteristics and evaluate the effect of small-scale 'subgrid' structures (which

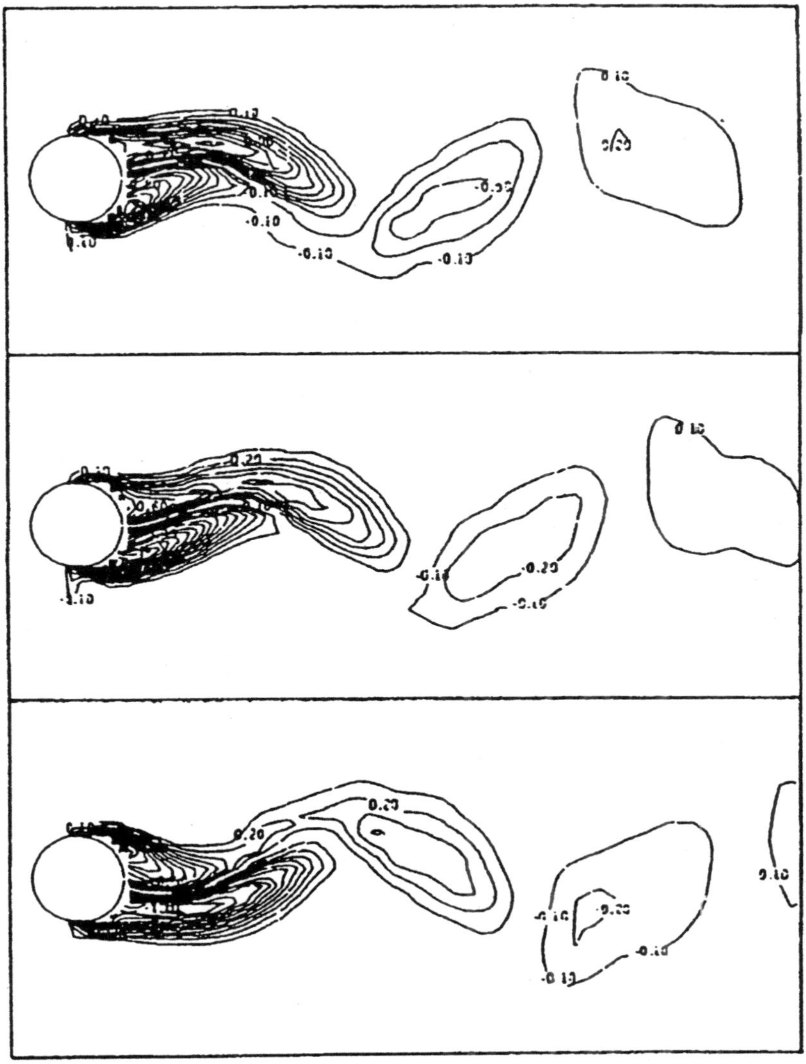

Fig.3.6. Isolines of vorticity in flow of a compressed gas around a cylinder: $M_\infty = 0.5$; $\Delta t = 2.5$.

cannot be resolved directly) on the large-scale phenomena and obtain certain energy characteristics of fine-scale turbulence.

The structure, the characteristics of the wake and the visual pattern of the flow change only slightly (Fig.3.11). This also relates to the drag factor of the body C_x, whose value is controlled by the

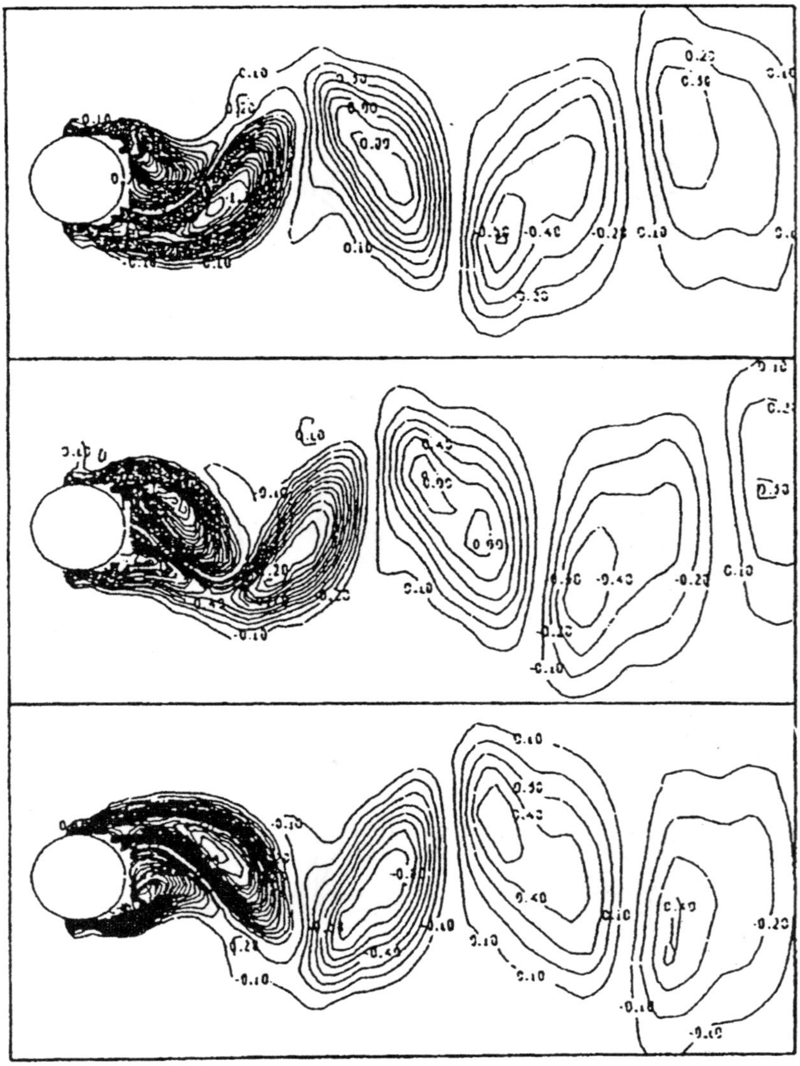

Fig.3.7. Isolines of vorticity in flow of a compressed gas around a cylinder: $M_\infty = 0.6$; $\Delta t = 2.5$.

wave properties of the phenomenon. The application of the (K – Θ) model (with turbulent viscosity) results in a change of the Sh number in the range 10%, but as expected, decreases the amplitude of pulsations of the parameters of the flow in the vicinity of the

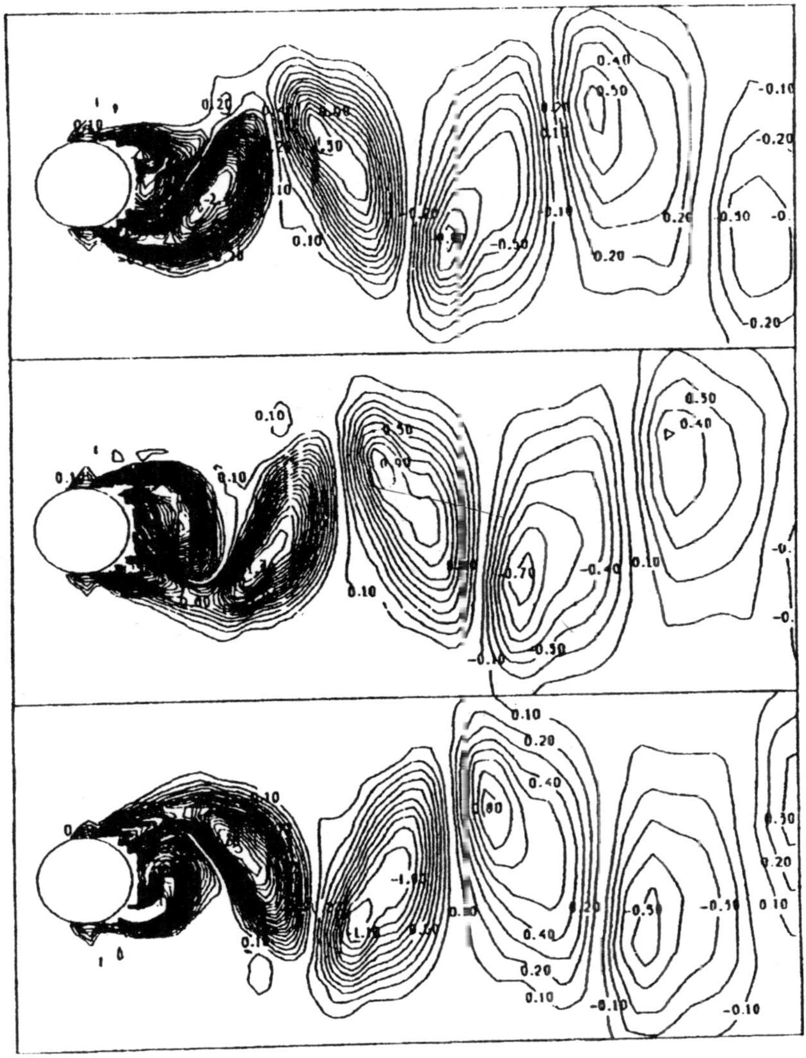

Fig.3.8. Isolines of vorticity in flow of a compressed gas around a cylinder: $M_\infty = 0.7$; $\Delta t = 2.5$.

body because of the appearance of the turbulent boundary layer. The flow in the vicinity of the body becomes more stationary and this results in a small decrease of the lifting force coefficient C_y.

The small quantitative variation of the flow parameters in the

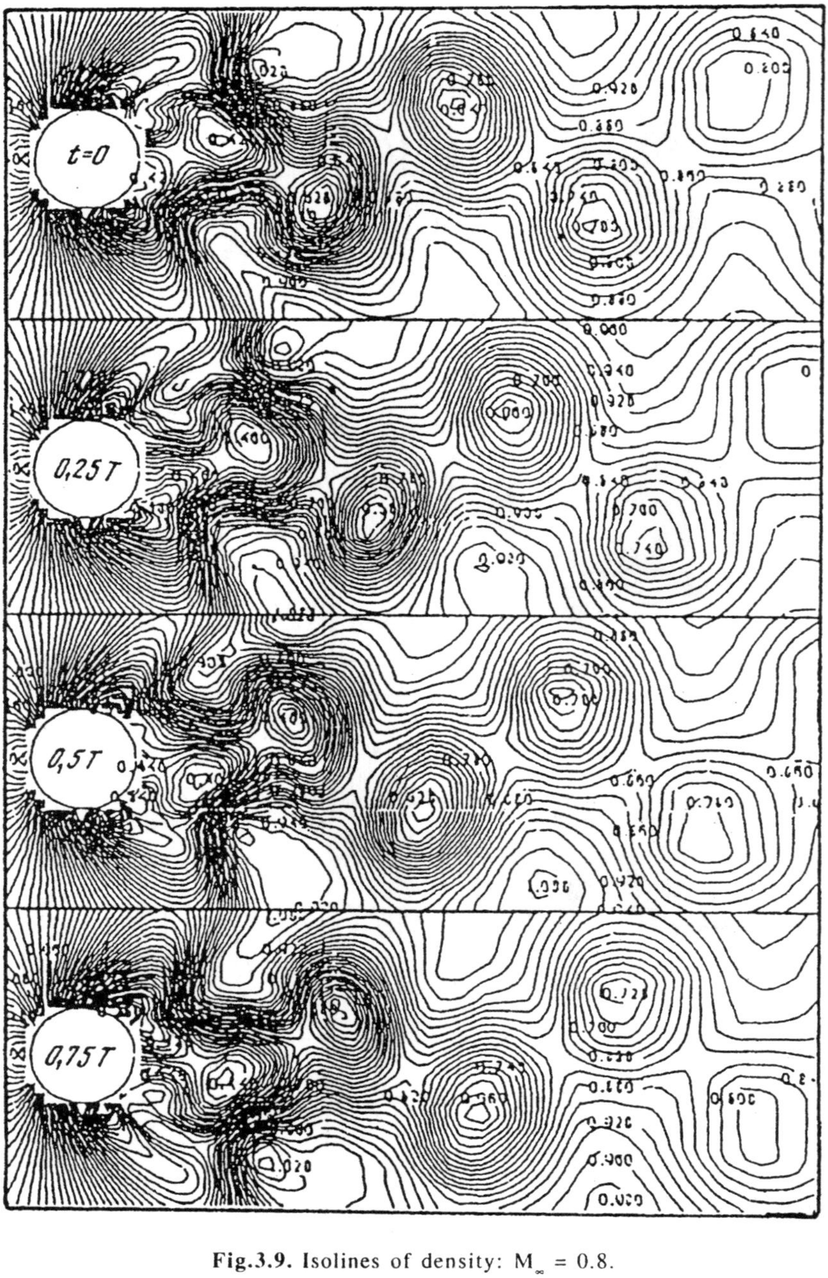

Fig.3.9. Isolines of density: $M_\infty = 0.8$.

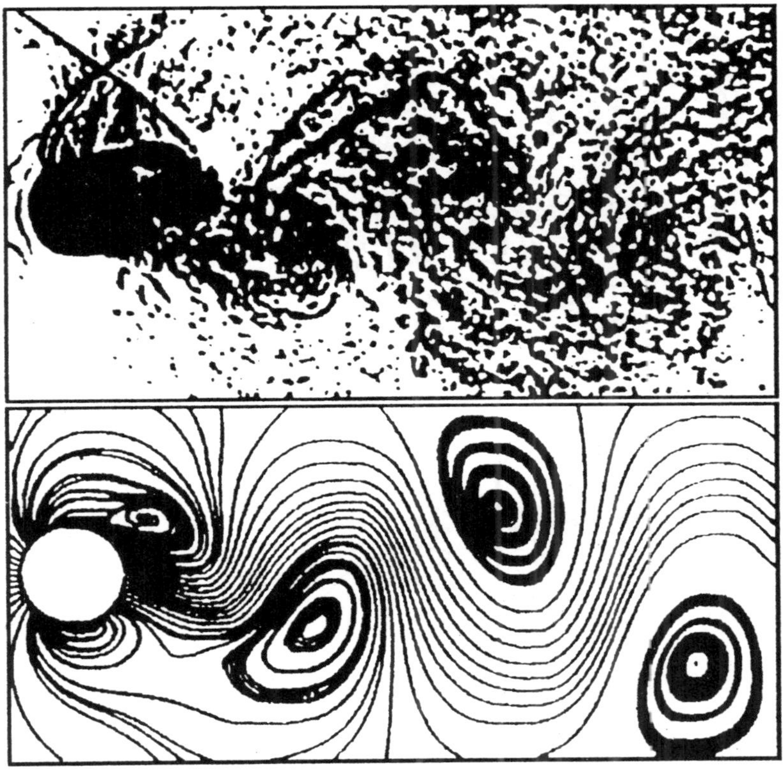

a

b

Fig.3.10. Turbulent wake behind a cylinder. Comparison of experimental (a) [236] and calculated (b) data.

near wake (with the exception of the immediate vicinity of the body) indicates the weak dependence of the properties of the large scale structures on the subgrid approximation (the nature of dissipation, the presence of the turbulence model). By direct resolution it is possible to 'grab' the main part of the energy spectrum, situated in the low frequency and inertia sub-regions on the scale of the wave numbers. This can be carried out on 'real' calculation grids and this also confirms the hypothesis on the separation of large- and small-scale phenomena in maximally developed turbulence.

The numerical experiment methods also made it possible to model the trans-sonic regime, for example at $M_\infty = 0.9$ (Fig.3.12) where the vortex structures are formed behind the tail jump and not at the

Fig. 3.11. Nonstationary wake behind a cylinder ($M_\infty = 0.6$): a) instantaneous lines of current, non-viscous model; b) instantaneous lines of current with introduction of the K–Θ model; c) isolines of turbulent energy with introduction of the K–Θ model.

a

b

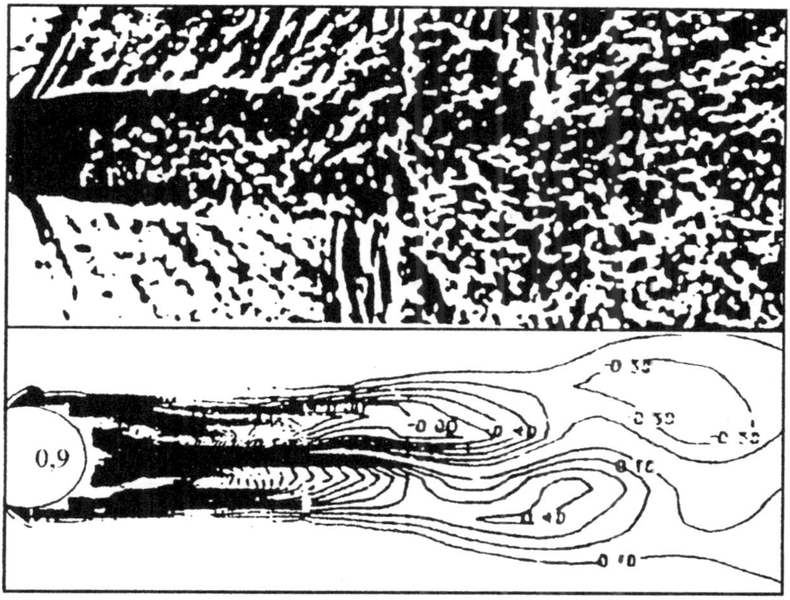

Fig.3.12. Isolines of vorticity in trans-sonic flow around a cylinder ($M_\infty = 0.7$); a) experiment [236]; b) calculations.

surface of the cylinder. In this case, the flow around the cylinder is almost completely stationary (the vortex structures are detected at a distance of more than 3 gauges behind the cylinder), and this results in a decrease of max $|C_y|$ with M_∞ approaching the sonic value (Fig.3.3). The non-stationary nature of the flow in the wake results in particular in the situation in which the local supersonic zones 'breath': Figure 3.13 shows isolines of local values of the Mach number for $M_\infty = 0.7$ for different moments of time. The value of $M = 1$ is denoted by the thick line, restricting the local supersonic zone.

It is interesting to examine the behaviour of the flow parameters with time. For example, Fig.3.14 shows, for M_∞ of 0.6 and 0.8, the time dependence of pressure at different points of the cylinder surface. Pressure pulsations are characterised by a high amplitude, and for $M_\infty = 0.6$ the maximum amplitude is found in the range $\varphi \approx 110°$, and for $M_\infty = 0.8$ in the range $\varphi \approx 135°$. The oscillations of these functions in the flow are strictly periodic, but in addition to the main frequency, harmonics with high frequencies are also detected. Fourier analysis makes it possible to determine the

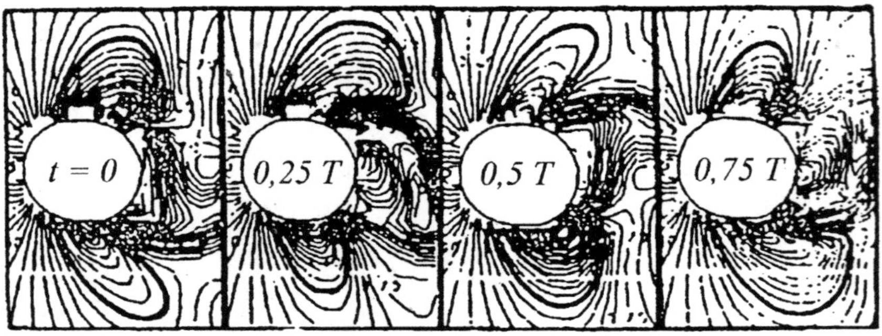

Fig.3.13. Non-stationary local supersonic zones for different moments, $M_\infty = 0.7$.

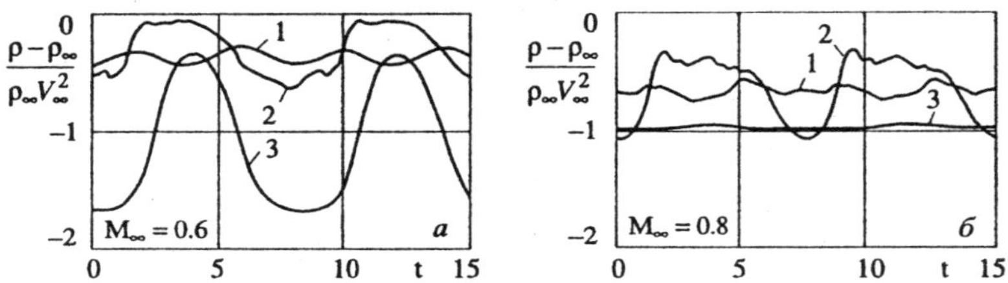

Fig.3.14. Pulsations of pressure on the surface of the cylinder: a) $M_\infty = 0.6$, b) $M_\infty = 0.8$ (1 – φ = 180°, 2 – φ = 135°, 3 – φ = 111°; angle φ is counted from the front critical point.

amplitude of these harmonics. These facts indicate that in the vicinity of the large scale 'coherent' structures there are also pulsation fluctuations of different scales, whose average characteristics should be determined.

The strict periodic nature of the numerical solution with time makes it possible, having a sufficient number of 'frames' of the flow for a single cycle, to separate from the solution the mean value of the functions and their pulsation components at all points of the integration region:

$$f'(x_i,t) = f(x_i,t) - \bar{f}(x_i),$$

$$\bar{f}(x_i,t) = \frac{1}{T}\int_{t_0}^{t_0+T} f(x_i,t)dt.$$

This processing of numerical information makes it possible not only to obtain mean and pulsation components but also determine moments for different functions:

$$\overline{f_i(x,t)\cdot f(x_i,t)} = \frac{1}{T}\int_{t_0}^{t_0+T} f_1'(x_i,t)\cdot f_2'(x_i,t)dt.$$

Some results of the appropriate analysis of the flows for M_∞ equal to 0.5; 0.6; 0.7 and 0.8 are presented in Figs.3.15 and 3.16.

Figure 3.15 shows the isolines $\overline{\vartheta'_{x_i}\vartheta'_{x_2}}$ – analogue of the Reynolds stresses, and Fig.3.16 – isolines of 'pulsation energy' $\sum\overline{\vartheta_i'^2}$. For the 'pulsation energy' in the near wake there is one local maximum situated behind the cylinder at a distance of $r \approx 3$ which is almost completely independent of the Mach number. Two extrema (max and min), positioned symmetrically, are found for the 'Reynolds stresses' in the near wake. In addition to this, the examined functions have local extreme values on the surface of the cylinder at the points close to the position of the jumps closing the local supersonic zones. In the same regions, at relatively long periods of time, the vortex cover moves from the surface of the solid.

We shall present results of another interesting numerical experiment in which the motion of the cylinder ('apparatus') with $M_\infty = 0.5$ was accompanied by the appearance, in the wake, of a screen (a 'parachute') in the form of an arc with $150° \leq \varphi \leq 210°$, situated at a distance of seven radii behind the cylinder. After approximately 100 periods, the flow was changed. In the vicinity of the body, the oscillations in the flow weakened and the flow became almost stationary, but more intensive (than prior to the opening of the parachute) large-scale vortex structures (Fig.3.17) formed behind the screen. In this case, the drag coefficient of the cylinder–screen system increased by an order of magnitude in comparison with a single cylinder.

Of considerable interest is the calculation of spatial-nonstationary problems. Examination of the characteristics of these problems is regarded as a basis for the systems of automated design of complex technical structure. It is well known that because of the 'non-adequate mathematics', many technical projects using systems for automatic control of processes have not been realised in practice.

For example, the prediction of the field of speeds behind a ship or calculation of the characteristics of an oscillating wing with a finite dimension range in the liquid flow are very important practical

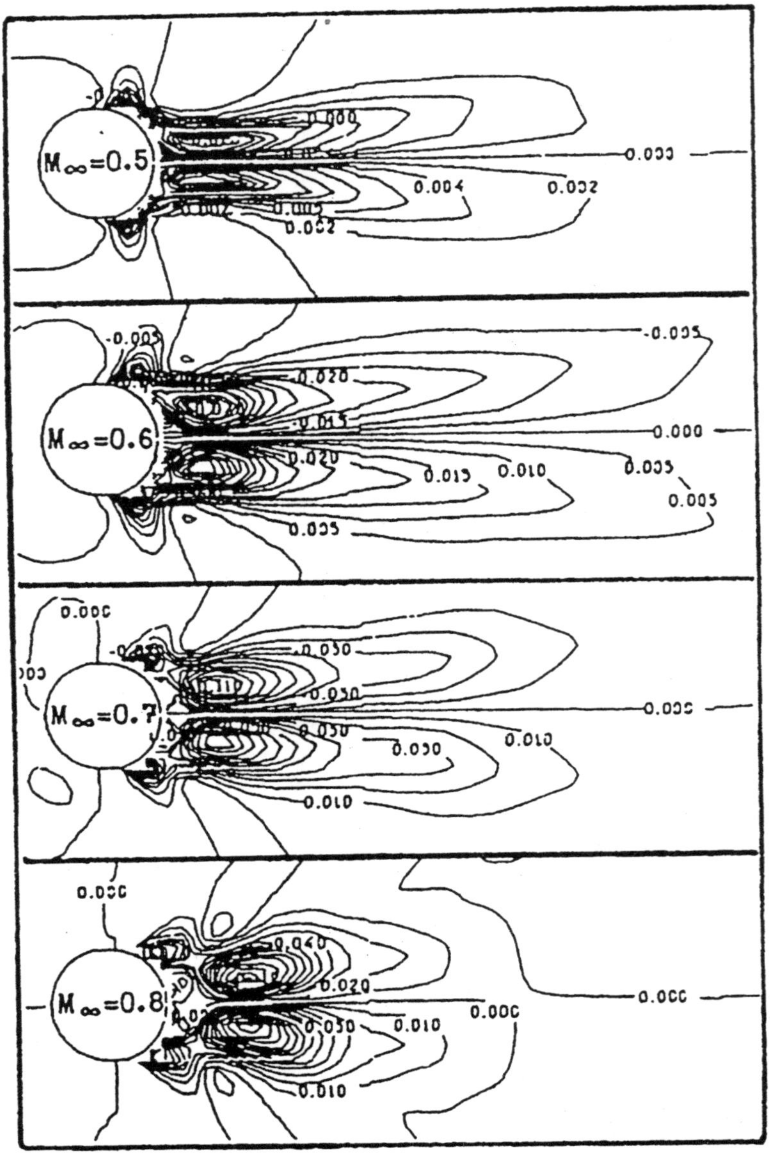

Fig.3.15. Isolines of an analogue of Reynolds stresses $\overline{\vartheta'_{x_1}\vartheta'_{x_2}}$

problems, especially in the case of non-stationary motion regime. Figure 3.18 shows the results of calculation of the spatial (3*D*) flow around turn of a sheet ($\alpha(t) \geq 0$) in the flow of an

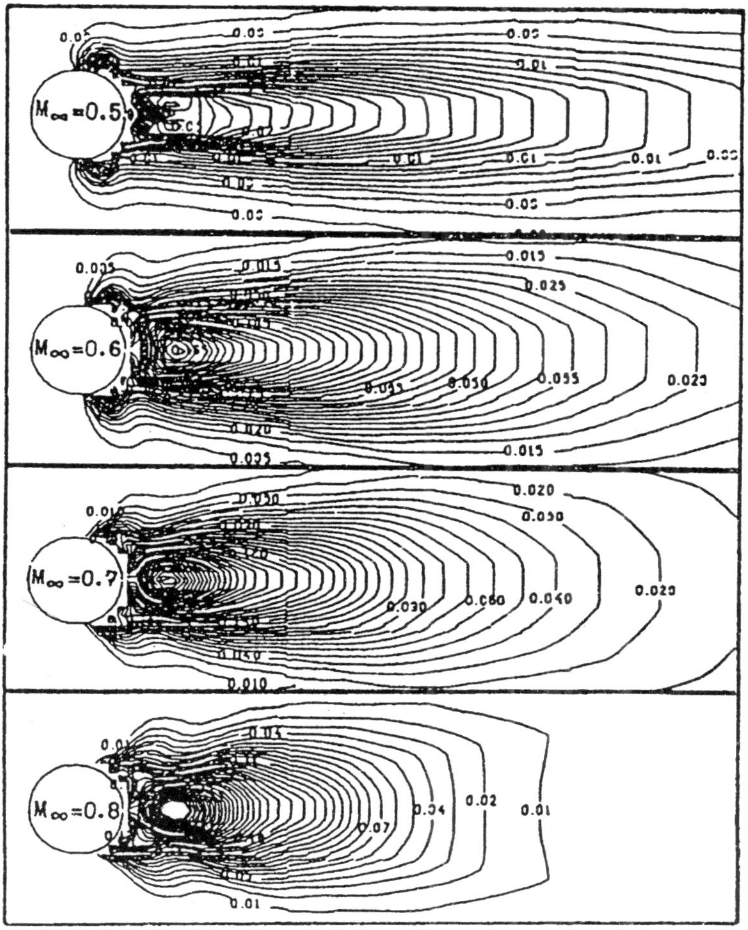

Fig. 3.16. Isolines of 'pulsation energy' $\sum\overline{\vartheta_i'^2}$

incompressible fluid, calculated by V.V. Rykov [108]. Figure 3.18a shows the formulation of the problem, Fig.3.19b, c the pattern of the fields of speeds at different sections, y = const. The vortices formed during rotation of the sheet, interact with the vortex wisps, converging from the side edges, forming a complex pattern of the flow in the turbulent wake behind the sheet.

Finally, Figs.3.19–3.21 illustrates some data obtained by A.V. Babakov in examining the dynamic properties of an 'automobile' in the spatial configuration with a compressible gas flowing around it. Figure 3.19 shows a fragment of the generated

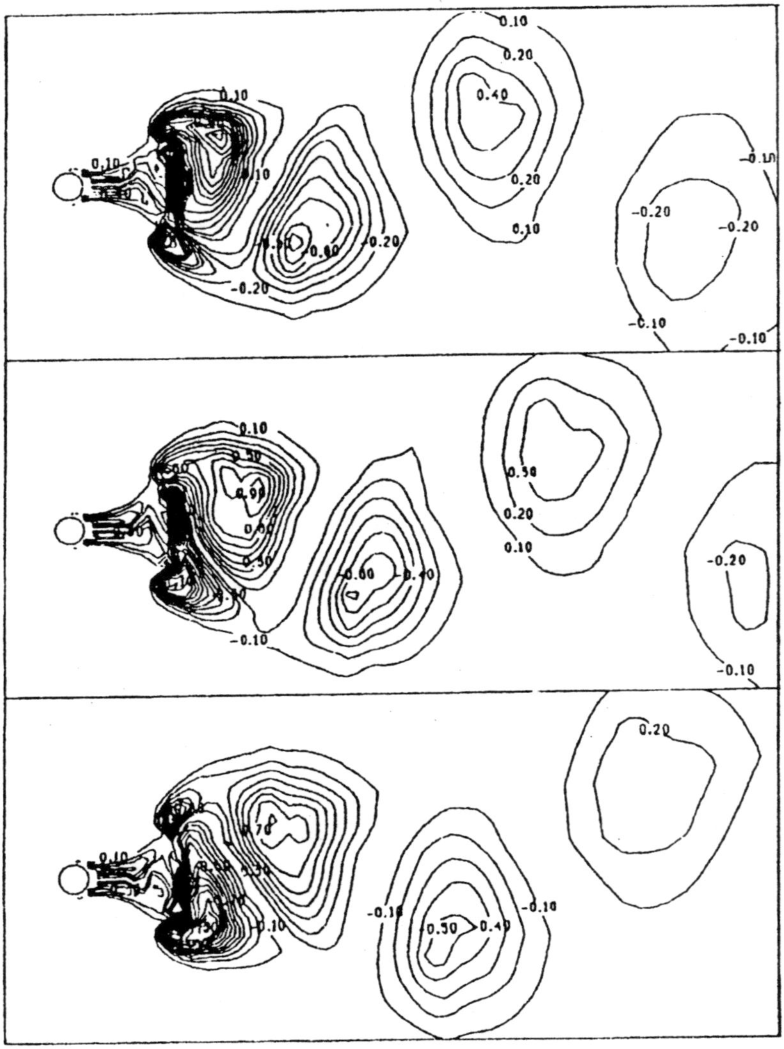

Fig.3.17. Isolines of vorticity in movement of a solid with a 'parachute' (M_∞ = 0.5).

calculation grid, Figs.3.20 and 3.21 shows the patterns of the field of speeds and isolines of pressure in the immediate vicinity of the automobile. These results require a very detailed calculation grid (~10^6 nodes). A.V. Babakov carried out these calculations in a high-productivity multiprocessor computing system.

Thus, the problems of this class (macrostructure of detachment

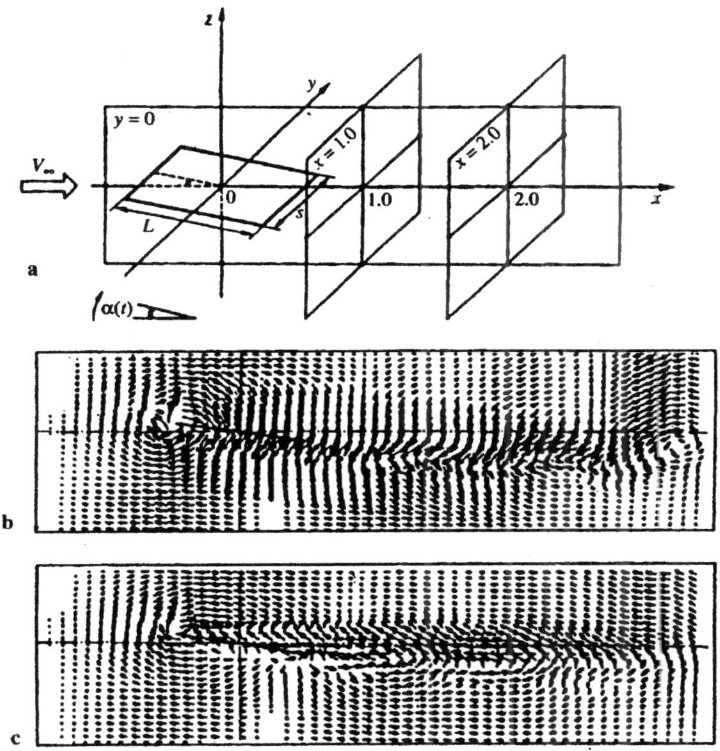

Fig.3.18. Rotation of a sheet of a fluid flow (3*D*): a) formulation of the problem; b–c) vectors of velocity on the logarithmic scale in sections y = const (b – $y = 0$, c – $y = 0.25$).

flows of the compressed gas) are investigated on the basis of the models of a non-viscous medium. The non-stationary Euler equations (in the form of integral laws of conservation in the difference representation with 'rational' averaging) ensure a correct description of the mechanism, being evidently the main mechanism in separation of this type for limiting flow conditions. As shown by estimates, the scheme viscosity of the 'difference Euler' in a detachment zone is relatively low (and can be 'controlled') thus ensuring the sufficiently adequate description of the flow pattern in relation to the actual phenomenon. Comparison of the experimental data shows that there is not only good qualitative but also quantitative agreement between the results of physical and numerical experiments.

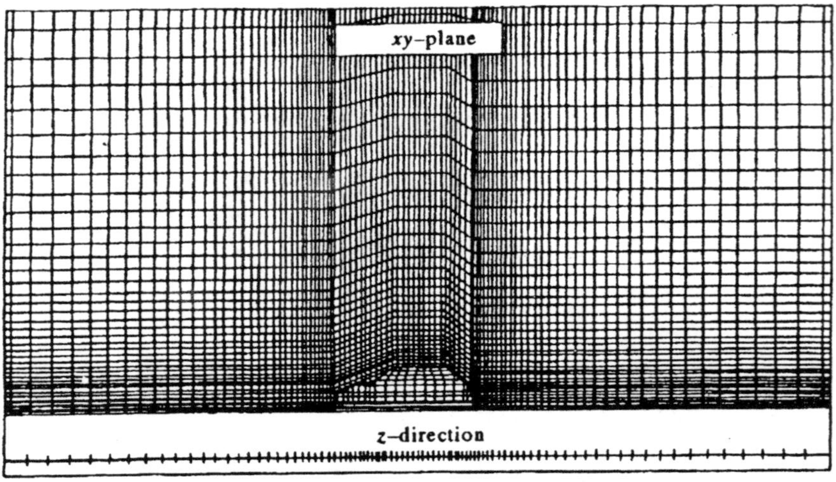

Fig.3.19. Spatial problem of flow around an automobile (fragment of the generated calculation grid).

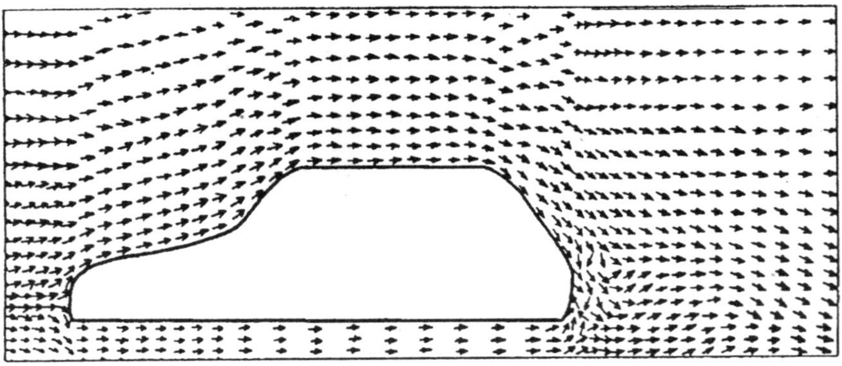

Fig.3.20. Field of velocities.

Thus, the average characteristics of the ordered large scale structures for the limiting regimes of detachment have been examined using dissipative linear models, obtained on the basis of non-stationary equations of the ideal medium. This in complete agreement with the methodology proposed by S.M. Belotserkovskii and M.I. Nisht [31] in examining the detachment flows of the ideal compressible fluid.

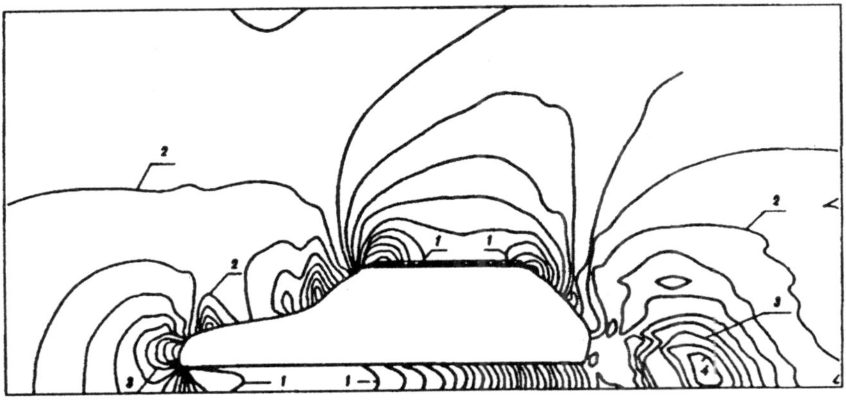

Fig. 3.21. Pressure isolines (C_p = const).

3.2. NUMERICAL MODELLING OF STOCHASTICAL COMPONENTS OF TURBULENCE (TURBULENT BACKGROUND)

1. The numerical modelling of the 'stochastic' (small-scale) component of turbulence in the centre of the wake may be carried out by a statistical method of the kinetic level. If the previous results illustrated modelling of large-scale macrostructures of the flow, then the statistical characteristics of the 'turbulent' background, forming the centre of the wake (Fig.1.11) should be examined using the Monte-Carlo procedures, constructing imitation models for appropriate kinetic equations. We shall present here the approaches developed by V.E. Yanitskii [24,33,131,132] with special reference to the problem of the calculation of the breakdown of a turbulent spot, examined as the cross-section of a wake.

Here, as in the dynamics of the rarefied gas (DRG), the problem is solved on the level of the distribution function but it is now the distribution of the instantaneous values of hydrodynamic velocity V of the 'liquid' particle. In turbulence, the ideology of constructing the computing process on the basis of the splitting method for statistical approaches remains unchanged (in comparison with the DRG problem). The particle is also characterised by the position and speed, but now it is already a model of the liquid particle. The difficulties in developing such a model are associated with the non-stationary nature of this phenomenon and with the absence in turbulence of the universial kinetic equation similar to the Boltzmann equation in DRG. In principle, modelling can be carried out for different types of kinetic equations. One of these attempts, following [131], will be demonstrated later. The main aim is to maintain, when solving the non-stationary problem of turbulence, the main principle of the method of direct statistical modelling [33] writing them however, in different terms (Table 3.2).

2. The statistical method of the particles, described in [33], can be strictly justified as the numerical Monte-Carlo method for solving the problems, formulated on the basis of the Boltzmann equation. One of the attractive qualities of this approach is its constructive nature. Each procedure is easy to interpret physically. Naturally, attempts are made to use similar approaches for modelling turbulence (according to V.E. Yanitskii [131]).

An analogy between molar mixing in the turbulent flow and molecular transfer in gases was already used by G. Bussinesq and L. Prandtl for deriving the well-known equations of turbulent friction. The Bussinesq equation

$$\tau = \rho \nu_T \frac{\partial u}{\partial y}, \quad \nu_T = \langle l\nu \rangle;$$

where the random quantities l and ν are the length of mixing and the pulsation speed of liquid particles (moles).

The Prandtl equation

$$\tau = \rho \nu_T \frac{\partial u}{\partial y}, \quad \nu_T = L^2 \left| \frac{\partial u}{\partial y} \right|.$$

Table 3.2

Object	DRG	Turbulence
Particle	Model of the atom:	Model of liquid volume:
Distribution function	$\begin{cases} \mathbf{r} - \text{position} \\ \mathbf{c} - \text{gas velocity} \end{cases}$	$\mathbf{x}$ – position $\mathbf{V}$ – instantaneous velocity
Moments and macroparameters	Single-particle: $\begin{cases} f = f(t, \mathbf{r}, \mathbf{c}) \\ \int f\, d\mathbf{c} = \text{– density} \end{cases}$	Single-point $f = f(t, \mathbf{x}, \mathbf{V})$ $\int f\, d\mathbf{V} = 1$ – normalisation
	Velocities: $\rho^{-1} \int \mathbf{c} f\, d\mathbf{c} = \mathbf{u}$ – gas velocity $(\mathbf{c} - \mathbf{u})$ – thermal velocity	Velocities: $\int \mathbf{V} f\, d\mathbf{V} = \mathbf{u}$ – mean flow velocity $(\mathbf{V} - \mathbf{u})$ – pulsation velocity

Developing this analogy, it is natural to assume that the chaotic motion of liquid moles in the turbulent flow is governed by some kinetic equations for the density $f(t,\mathbf{x},\mathbf{V})$ of distribution of the probabilities of the instantaneous values of the speed v of the mole at a given moment of time t at point x. The problems of describing turbulence by kinetic equations have been investigated in studies of many authors. Unfortunately, the turbulence theory contains no universal closed kinetic equation for the single-point density $f(t,\mathbf{x},\mathbf{V})$ of the distribution of probabilities similar to the Boltzmann equation in the dynamics of rarefied gases. The turbulence theory remains the science of semi-empirical models also on the kinetic level of description. As a basis for constructing the simulation scheme, it is much more efficient to use mathematical models in which the elementary physical relationships of the phenomenon are not 'buried] in a large number of additional assumptions. The basis qualitatively reflecting the analogy of molar mixing in the liquids and molecular mixing in gases is, for our purposes, the following relatively simple relaxation kinetic equation [101,194]:

$$\frac{\partial f}{\partial t}+\mathbf{V}\frac{\partial f}{\partial x}+\frac{\partial}{\partial \mathbf{V}}\left[\left(-\frac{\mathbf{v}}{2\tau_1}+\mathbf{F}\right)f\right]=\frac{f_M-f}{\tau_2}. \tag{3.1}$$

where $\mathbf{v} = \mathbf{V} - \mathbf{u}$ is the pulsation speed and $\mathbf{u} = \langle \mathbf{V} \rangle$ is the mean speed of the flow. The function

$$f_M=\left(\frac{3}{4\pi q^2}\right)^{3/2}\exp\left(-\frac{3\upsilon^2}{2q^2}\right) \tag{3.2}$$

is the density of probabilities of the normal law of distribution of pulsation of the speed; $q^2=\langle \mathrm{v}_x^2+\mathrm{v}_y^2+\mathrm{v}_z^2\rangle$ is the doubled mean value of the specific kinetic energy of these pulsations. (The density of turbulent energy is $E = q^2/2$.) Vector field $\mathbf{F}$ is the field of mass forces acting on the liquid particle. For an incompressible liquid $p = 1$, and the force $\mathbf{F} = \mathbf{g} - \nabla P$, where P is the mean pressure in the turbulised fluid, $\mathbf{g}$ is the vector of free-fall acceleration.

The relaxation parameters τ_i are expressed formally by means of the turbulence scales L_i and the specific energy of turbulence q^2 in such a manner that $\tau_i = L_i/q$. The turbulence scales L_i are not functionals of the single-point distribution f, they are determined by the two-point correlation function. In the semi-empirical theories of turbulence, the laws of variation of L_i are selected from difference considerations, for example, they are determined by the geometry of the problem or are determined from approximate equations, linking evolution L_i with the variation of the mean characteristics of the turbulent medium. We shall assume that the relationship $L_i(q)$ is given. In a number of cases, it may actually be determined. For example, in a turbulent flow behind a grid, L_i and q are linked by the Loitsyanskii invariant $L_i^{5/2}q$ = const. In the given approach, we assume the exponential dependences L_i and q:

$$L_1^{-1} \sim q^{(2\gamma_1 - 1)} \quad \text{and} \quad L_2^{-1} \sim 1^{(2\gamma_2 + 1)}. \tag{3.3}$$

It should be mentioned that at $\gamma_1 = 0.7$ from (3.3) we obtain the Loitsyanskii invariant for the dissipation scale L_1. The main empirical constants of the model are the exponents γ_1 and γ_2 and the initial values of the integral scales L_1^0 and L_2^0.

If this assumes that the force field **F** is the given function of t and x, then equation (3.1) differs from the BGC model (P. Bhatnagar, E. Gross, M. Crook) used widely in the dynamics of rarefied gases, by the additional term $-\frac{\partial}{\partial \mathbf{V}}\left(\frac{\mathbf{v}}{2\tau_1}f\right)$. In order to understand the meaning of this term, we shall examine Cauchy's problem for the distribution function that is independent of the spatial variables:

$$\frac{\partial f}{\partial t} - \frac{\partial}{\partial \mathbf{V}}\left(\frac{\mathbf{v}}{2\tau_1}f\right) = \frac{f_M - f}{\tau_2}, \tag{3.4}$$

$$f(t = +0, \mathbf{V}) = f_0(\mathbf{V}). \tag{3.5}$$

Multiplying (3.1) and (3.2) by **V** and $\mathbf{v}^2$ in succession and integrating each time in respect of the entire space of speeds $\{\mathbf{V}\}$, we obtain equations for $\mathbf{u}(t)$ and $q^2(t)$:

$$\frac{\partial \mathbf{u}}{\partial t} = 0, \quad \frac{\partial q^2}{\partial t} = -\frac{q^2}{\tau_1}, \tag{3.6}$$

$$u(t = +0) = u_0, \quad q^2(t = +0) = q_0^2 \tag{3.7}$$

These equalities show that in the system described by equation (3.4), the pulse **u** is maintained and the energy q^2 is a monotonically decreasing function of time. Dependence (3.3) determines the following type of the function $\tau_1(q^2)=\tau_1^0\left(q_0^2/q^2\right)^{\gamma_1}$. Consequently, equation (3.6) has the following form:

$$\frac{\partial \tilde{q}^2}{\partial t}+\frac{\partial \tilde{q}^{2(1+\gamma_1)}}{\tau_1^0}=0, \quad \tilde{q}=q/q_0.$$

The equation is solved analytically and, consequently, at $\gamma_1 > 0$ we obtain:

$$\tilde{q}^2=\frac{1}{\left(1+\gamma_1 t/\tau_1^0\right)^{1/\gamma_1}}.$$

This dependence approximates the decrease of the turbulent energy in wakes (see paragraph 3).

The meaning of the relaxation member in the right-hand part of (3.1) is also easily examined in the spatial-homogeneous case. Without dissipation and force **F** we obtain the conventional Bhatnagar–Gross–Crook equation:

$$\frac{\partial f}{\partial t}=\frac{f_M-f}{\tau_2},$$

whose solution has the following form:

$$f(t,\mathbf{V})=e^{-t/\tau_2}f_0(\mathbf{V})+(1-e^{t/\tau_2})f_M(\mathbf{V}), \tag{3.8}$$

here $q^2 = q_0^2$ = const, $\tau_2(q^2) = \tau_2^0$ = const. Here, equation (3.8) shows that τ_2 characterises the time during which any initial distribution $f_0(\mathbf{V})$ evolves to the normal law $f_M(\mathbf{V})$.

3. The experiments with the application of simulation as the method of numerical modelling of turbulence were carried out by V.E. Yanitskii on the example of solving a two-dimensional nonstationary problem of the 'turbulent spot' [19,24,33,131,132]. At the moment of time $t = 0$ in a still liquid with $u \equiv 0$ the initial distribution $q^2(0,Y,Z)$ of the energy of turbulent pulsations develops in some fashion. The problem is to determine the spatial distribution of the

turbulence characteristics of the arbitrary moment of time t. This problem formed from the examination of the pulse-free wake behind a generator of pulsations with diameter D, distributed in the flow, whose velocity is equal to U_0. In the experiments, this axisymmetric wake, extended along the OX axis, was realised and investigated in [206]. The pulse-free nature, i.e. the absence in the mean section of mass transfer along the axis of the wake OX makes it possible to ignore, at a sufficiently large distance from the source, the heterogeneity of the profile of the mean speed and friction stresses. In this region, the spatial distribution of energy q^2 is self-similar, the development of the turbulent wake takes place as if it was initiated by a point source of pulsations, situated in the accompanying flow with the speed U_0, and when calculating the fields it can be accepted that $t = X/U_0$ [206]. At the distances $X \geq 4D$ or in terms of the appropriate time variable $t = X/U_0$, at $t \geq t_0 = 4D/v_0$, the measured function $q^2(t,Y,Z)$ assumes the Gaussian form:

$$q^2(t,r) = q^2(t,Y,Z) = q_m^2(t)\exp\left(-0.69\,r^2 - r_q^2(t)\right), \tag{3.9}$$

where $r = (Y^2 + Z^2)^{1/2}$, $q^2(t, r_{1/2}) = q^2_m(t)/2$.

Energy $q^2_m(t)$ at the centre of the spot and its characteristic radius $r_q(t)$ are approximated by the following dependences [115]:

$$q_m^2(t) = q_0^2(r)\left(1+\frac{t-t_0}{\tau_0}\right)^{-n_1}, \quad r_q(t) = r_0\left(1+\frac{t-t_0}{\tau_0}\right)^{n_2}; \tag{3.10}$$

where $n_1 \approx 1.58$, $n_2 \approx 0.35$, $\tau_0 = 2D/U_0$ for $t_0 \leq t \leq 50D/U_0$ (for $50D/U_0) < t \ll 130D/U_0$ the exponents n_1 and n_2 are equal to 1.80 and 0.25, respectively).

The mathematical formulation of the problem of the turbulent spot on the kinetic level of description may be described as follows. We solve the Cauchy problem for the kinetic equation (3.1) in the unlimited 5-dimensional phase space $\{Y, Z, V_x, V_y, V_z\}$. The initial distribution function $f_0(Y, Z, V)$ is represented by the density of the normal law for the distribution of the probabilities of the instantaneous values of hydrodynamic velocity V, i.e.

$$f_0(r,V) = \left(\frac{1}{4\pi q_0^2(r)}\right)^{3/2} \exp\left(-\frac{3v^2}{2q_0^2(r)}\right),$$

where $q^2_0(r) = q^2_0 \exp(-0.69r^2 / r^2_0)$.

The diagram of modelling was constructed on the basis of the principles, described in the previous paragraph, as well as in the dynamics of rarefied gases. Each particle is a model of the small element of the turbulised liquid and is characterised by its position $\mathbf{r}(t)$ and the instantaneous hydrodynamic speed $\mathbf{B}(t)$. The calculation region, characterised by the evolution of the system from N such particles, has the form of a rectangle with the dimensions $F_y \times F_z = 2 \times 2$ in the first square of the co-ordinate plane, which assumes the mirror symmetry of the problem in relation to the axes OY and OZ. The external boundaries $Y = F_y$ and $Z = F_z$ are assumed to be open for the particles leaving the calculation region. We use the uniform grid with a total number of cells of 20×20 and the dimensions of each cell $h_z - h_y = 0.1$. At the initial moment of time $t = 0$, the distribution of the particles in the calculation region was uniform, and the mean number of the particles in each cell was assumed to be 10. The speeds $\mathbf{v}_i$ of the particles were 'raffled off' in accordance with the density of the normal law of distribution of the probabilities in which $r = r_i$ is the distance of the i-th particle from the origin of the co-ordinates.

Kinetic equation (3.1) includes the following physical processes:

- convective transfer $\mathbf{V}\frac{\partial f}{\partial x} + \mathbf{F}\frac{\partial f}{\partial \mathbf{V}}$;
- dissipation of turbulent energy $-\frac{1}{2\tau_1}\frac{\partial}{\partial \mathbf{V}}(\mathbf{v}f)$;
- redistribution of pulsations in respect of the degrees $\frac{f_M - f}{\tau_2}$.

In accordance with these considerations, the modelling scheme of the evolution of the model in a short time period Δt, equal to 0.085 in the calculations, is the sequence of three stages of splitting, described previously in [19,131].

Investigations were carried out in order to investigate the principal possibility of simulating turbulence by the statistical method of particles in cells. Taking this into account, the authors assumed that in the given stage of investigations it is permissible to make additional simplifications of the formulation of the problems which are not principle from the viewpoint of main problems.

The degeneration of the turbulent spot in the homogeneous (non-stratifies) surrounding the medium, takes place as a result of dissipation of the energy of pulsations and of its diffusion. To a certain approximation, the effect of the gradient of the mean pressure on these processes can ignored, and because of the absence of stratification it may assumed that $g = 0$. Thus, force F_i, acting on each mole, is equal to zero. At zero values of the initial field of speeds $u(t = +0, r) = 0$ the condition P = const leads with time to the appearance of the mean speed u_r of the turbulised liquid in the radial direction. Starting from some moment, the derivative

$\frac{\partial q^2}{\partial t}$ becomes comparable with the convective member $u_r \frac{\partial q^2}{\partial r}$ and,

consequently, the field of the mean pressure should be taken into account in this case. The initial values of q_0^2 and r_0, equal to 0.0104 and 0.612, respectively, were selected in such a manner that in the time period $t \leq 20$ there were simultaneous small effects of the zero ∇P and finite dimensions of the calculation region.

The moment $t = 0$ was related to point $x_0 = 4D$ in experiment [206] which determines $\tau_0 = 2D/U_0$, $n_1 = 1.58$, $n_2 = 0.35$ in approximation (3.10) of physical experiments. The initial kinetic model has four empirical constants – initial values of the integral scales L_1^0 and L_2^0 and exponents γ_1 and γ_2 in the exponential dependences $L_i(q)$. Special analysis [115] shows that these constants are determined unambiguously by the initial data, if the criteria of agreement with the experiments represented by the agreement of the calculated function $q_m^2(t)$ and $r_q(t)$ with the approximation (3.10), assuming in this case self-similarity of the profile $q_m^2(t,r)$. In the calculations it was assumed that $\gamma_1 = 0.63$, $\gamma_2 = 0.80$, $L_1^0 = L_2^0 = 0.14$ $(= 0.23r_0)$. This corresponds to the following values of the relaxation parameters: $\tau_1^0 = \tau_2^0 = 1.31$. In the dynamics of rarefied gases, the flows with identical values of the parameters L and τ relate to the 'transition' regime, i.e. it is examined on the kinetic level of description.

Comparison with the experimental data obtained by E. Naudasher [206] is shown in Figs.3.22 and 3.23. In the calculations, the distribution of energy $E = q^2/2$ in space assumes the self-similar form with time. However, this form differs from the profile by

E. Naudasher (Fig.3.23) and in the range $r/r_q \le 1.5$ is approximated efficiently by Bessel's function $J_0\,(1.52r/r_q)$. Calculations with the following initial distributions of the energy of pulsations in respect of radius were verified:

$$q_0^2(r) = q_0^2 J_0(1.52r/r_0), \quad q_0^2(r) = \begin{cases} q_0^2 = \text{const}, & r \le r_0, \\ 0, & r > r_0. \end{cases}$$

They confirm the same nature of the steady self-similarity, as presented in Fig.3.23. The identical discrepancy between the results of calculation and experiments was recorded by many investigators solving the given problem numerically on the basis of different turbulent models. In [115] it was shown that the presence of 'a tail' on the experimental curve $q^2(t,r)$ and, correspondingly, the absence of a specific edge at the spot are explained by the intermittency

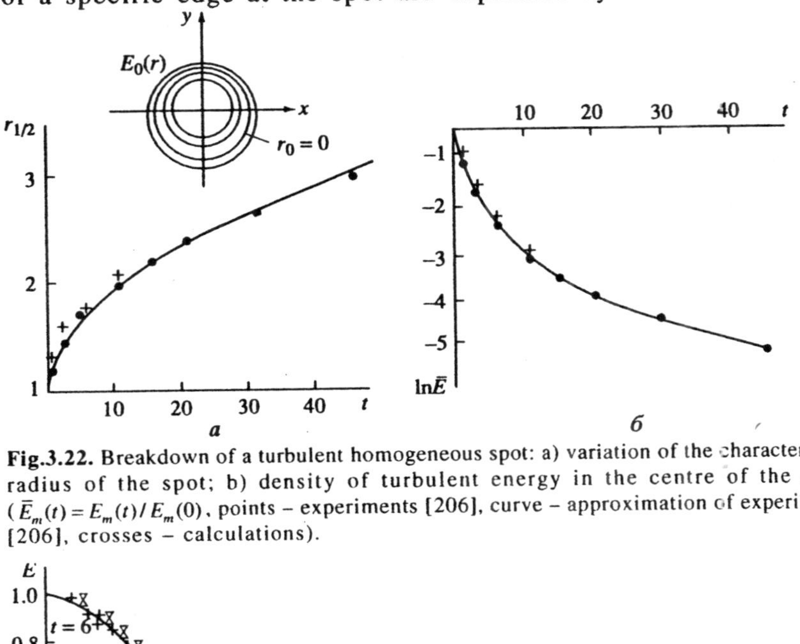

Fig.3.22. Breakdown of a turbulent homogeneous spot: a) variation of the characteristic radius of the spot; b) density of turbulent energy in the centre of the spot ($\bar{E}_m(t) = E_m(t)/E_m(0)$, points – experiments [206], curve – approximation of experiment [206], crosses – calculations).

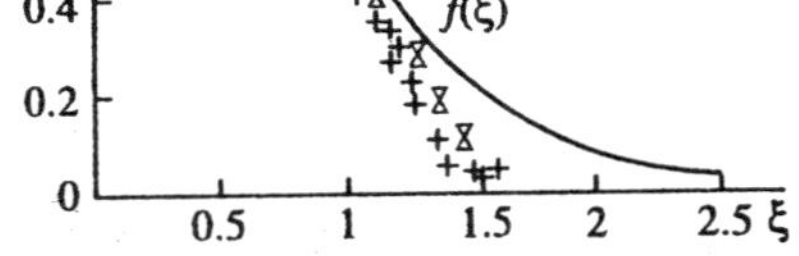

Fig.3.23. Distribution of turbulent energy in the radius of a homogeneous spot: curve – Nadausher's experiments [206]; ⧗ – calculated data for t = 4, + – for t = 6, × – for t = 11.

effect – irregular intermittency of the laminar and turbulent phases. Thus, the above mentioned discrepancy shows the inadequacy in the physics of the phenomenon common for many numerical models – it is the ignorance of the intermittence effect.

4. One of the advantages of the kinetic approach in turbulence is the possibility of obtaining the function of distribution of the pulsations of the speed and, consequently, verify the hypotheses associated with the function. A hypothesis on the normal law of distribution of the probabilities of speed pulsations is used widely. One of the best known hypotheses is the hypothesis proposed by Millionshchikov, which assumes that the cumulants of the fourth and higher orders are equal to zero. It makes it possible to express the moments of the distribution function of high orders to the moments of low orders: the first (mean speed), second (Reynolds friction stresses) and third (energy fluxes). These relationships may be used to express the closed system of Reynolds equations. To verify the proposed hypothesis in the experiments, we measure the exponents of asymmetry and the excess which are the cumulants of the third and fourth order of the distribution of the speed of pulsations **v**. A natural generalisation of this approach is the analysis of the characteristic function of distribution of pulsations carried out by V.E. Yanitskii for a turbulent spot [131,133].

Let us assume that $\text{v}_r = (V_r - u_r)/\sqrt{D(V_r)}$. The characteristic function, corresponding to the distribution is, according to the definition in [80]:

$$\chi_{t.r}(m) = \int e^{im\text{v}_r} f(t,r,\text{v}_r)d\text{v}_r.$$

It can be represented in the exponential form

$$\chi_{t.r}(m) = e^{\rho(m)+i\theta(m)}.$$

The functions $\rho(m)$ and $\theta(m)$ are the functions of the cumulants κ_n of the even and odd orders, respectively

$$\rho(m) = \frac{m^2}{2} + \sum_{k=2}^{\infty}(-1)^k \kappa_{2k}\frac{m^{2k}}{(2k)!}, \qquad \kappa_2 \equiv D(\text{v}_2) = 1;$$

$$\theta(m) = -\alpha\frac{m^3}{6} + \sum_{k=2}^{\infty}(-1)^k \kappa_{2k+1}\frac{m^{2k+1}}{(2k+1)}, \qquad \kappa_3 \equiv \alpha;$$

Here α is the asymmetry and κ_4 is the excess of distribution of the pulsations of the radial speed v_r

The hypothesis on normality corresponds to the following equalities:

$$\theta(m) \equiv 0 \quad \text{and} \quad \rho(m) \equiv -m^2/2,$$

or which is the same

$$U(m) \equiv \mathrm{Re}_{\chi}(m) = e^{-m^2/2} \quad \text{and} \quad V(m) = \mathrm{Im}_{\chi}(m) \equiv 0.$$

The Millionshchikov hypothesis corresponds to the assumption that $\rho(m) = -m^2/2$ and $\theta(m) = \alpha m^3/6$, consequently,

$$U(m) = e^{-m^2/2}\cos(\alpha m^3/6), \quad V(m) = e^{-m^2/2}\sin(\alpha m^3/6). \tag{3.11}$$

The statistical method of the particles makes it possible to evaluate directly $U(m)$ and $V(m)$ as the mean values of the quantities $\cos(mv_r)$ and $\sin(mv_r)$. Using the equations of the consistent estimates, we obtain, from the results of modelling, the approximate values $\widehat{U}(m)$ and $\widehat{V}(m)$; the rms scatter (standard deviation) of these quantities is denoted by S_U and S_V, respectively. If the mean number of the particles in the region where measurements are taken is N_*, and the number of realisation in averaging in respect of the ensemble is L, then:

$$S_U^2 = \frac{D\cos(mv_r)}{N_* L}, \quad S_V^2 = \frac{D\sin(mv_r)}{N_* L}.$$

In the statistical modelling of the turbulent spot we obtain the characteristic function $\chi_{t,r}(m)$ of the distribution of the radial pulsations v_r are the moments of time t = 0, 1, 3, 6, 11. Measurements were taken in ring-shaped regions $\left\{\underline{x}r_q(t) \le r \le \bar{x}r_q(t)\right\}$, shown in Fig.3.24. The areas of the regions 1–4 were selected in such a manner that on average they contain the same number of particles N_*. Table 3.3 shows their boundaries.

Regions 1 and 2 form the 'core of the spot', 3 and 4 are in the

Table 3.3

N of region	1	2	3	4	5	6
x	0.00	0.40	0.92	1.00	1.17	1.23
$\bar{x}$	0.40	0.57	1.00	1.08	1.23	1.30

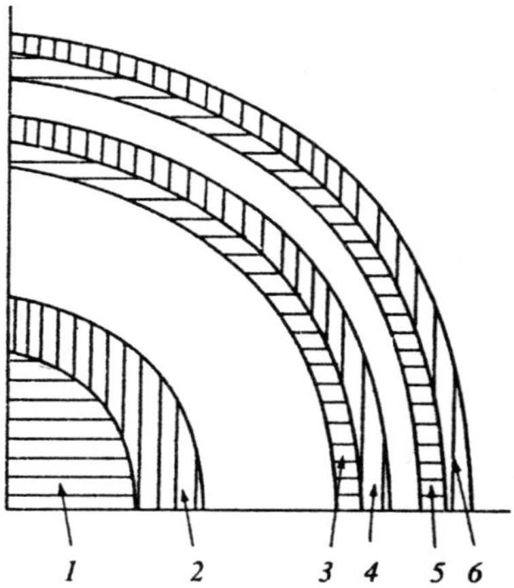

Fig.3.24. Determination of the characteristic function of a turbulent spot $\chi_{t,s}(m)$ (boundaries of ring-shaped regions).

immediate vicinity of the conventional boundary of the spot $r = r_q$, 5 and 6 characterise the 'external region'. Figure 3.25 shows the results for the moment $t = 6$. The solid lines show the values of $\widehat{U}(m)$ and $\widehat{V}(m)$ calculated in simulation in the given regions (numeration of the curves in Fig.3.25 corresponds to the numeration of the regions in Fig.3.24). The circles indicate the accurate values of $U(m)$ and $V(m)$ for the normal distribution. The vertical sections give the band of the statistical scatter $\pm 3S$. It may easily be seen that only the curves 1 fit this band. This means that the hypothesis on the normal law for the distribution of the pulsation speeds can be accepted only for the centre of the spot.

The straight crosses in Fig.3.25, almost coincide with the curves 2, denote the values of $U(m)$ and $V(m)$ calculated from equation (3.11), corresponding to the Millionshchikov hypothesis at $\alpha = 0.25$. The inclined crosses, coinciding with the curve 3 at $m \leq 1.4$, also denote $U(m)$ and $V(m)$ according to Millionshchikov, but at $\alpha = 0.65$. It is evident that at $m \geq 2$ the curve 3 for $V(m)$ leaves the band $\pm 3S$ of the confidence range for the Millionshchikov hypothesis. The curves 4 match the Millionshchikov hypothesis only at $m < 1$, the curves 5 and 6 are not in qualitative agreement with the curve.

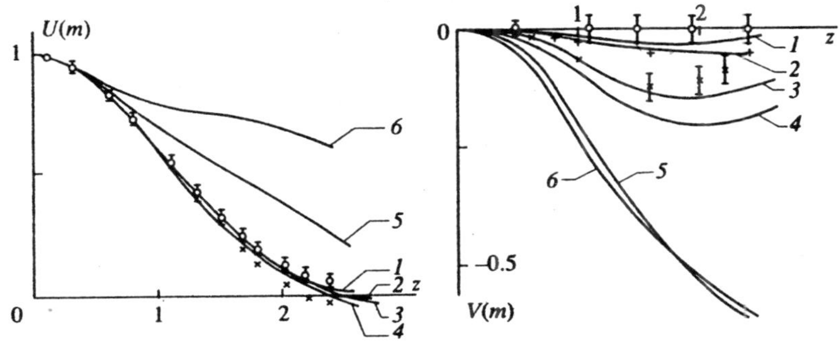

Fig.3.25. Values of functions $U(m)$ and $V(m)$ for time $t = 6$ (numeration of curves correspond to numeration of regions in Fig. 3.24).

The following conclusion can be made on the basis of this analysis. The distribution of pulsations of the speed is isotropic only in the centre of the spot and is governed by the normal law. With increase of the distance from the centre to the periphery, the anisotropy becomes greater, and in this case the Millionshchikov hypothesis may be accepted in the core of the spot, and the deviations from the hypothesis start in transition through the conventional boundary $r \sim r_{q^*} < r_{1/2}(t)$. In the external region, the distribution of the pulsations cannot be matched with this hypothesis.

Possibly, by taking into account the intermittency phenomena, the region of applicability of the Millionshchikov hypothesis is expanded [115].

5. The method of direct statistical modelling makes it possible to obtain in principle the single-point function of distribution of pulsations or its moments. Figure 3.26 shows the distributions of the exponents of asymmetry α_r and excess β_r obtained at different moments of time (in respect of the relative radius $\xi = r/r_{1/2}(t)$). The component of asymmetry $\alpha_r(\xi) = \mu_3/\mu_2^{3/2}$ characterise the third moment of the function of distribution (the flow of turbulent energy) and the excess exponent $\beta_r(\xi) = \mu_4/(3\mu^2_2-1)$ corresponds to the fourth moment of the distribution function. The boundary of the spot is clearly visible here.

Calculations were also carried out for a stratified (in respect of density) spot [132]. Figure 3.27 shows on the logarithmic scale in the form independent of the stratification (by introduction of the frequency Froude number Fr_D), comparison of the results obtained by V.E. Yanitskii and N.N. Slavyanov with the experiment [192] in

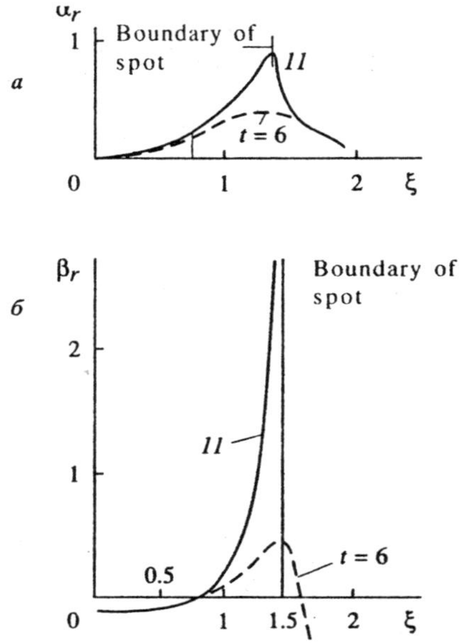

Fig.3.26. Parameter of asymmetry α_r (a) and excess β_r (b) for a homogeneous turbulent spot.

respect of the speed of attenuation of the axial pulsation of speed (here $\mathrm{Fr}_D = \dfrac{U}{ND}$ is the Froude frequency, $\mathrm{Re}_D = \dfrac{UD}{\nu}$ is the Reynolds number, $N = \dfrac{1}{2\pi}\sqrt{-\dfrac{g}{\rho_0}\dfrac{\partial p}{\partial z}}$ is the Vaissala frequency $\lambda = \dfrac{\mathrm{Fr}_D^{3/4}\sqrt{\overline{u}_{\max}^{-2}}}{U_0}$, $\eta = N(t-t_0)$).

Figure 3.28 shows the dependence on time (related to the Vaisalla period T) of quantity $\theta(t)$ equal to the relative potential energy of $\pi_m(t)$ in the centre of the spot:

$$\theta(t) = \frac{\pi_m(t)}{E_z^m(t) + \pi_m(t)},$$

where $E_z^m(t) = E_z(t,\ y = 0,\ z = 0)$ is the kinetic energy of equal pulsations in the centre of the spot, $\pi_m(t) = \pi_m\ (t, y = 0,\ z = 0)$.

If the first maximum is taken as the conventional start of the collapse of the spot, this will correspond to the moment of time (from the start of development t_0 of the spot) $t^* = t-t_0 \approx 0.27T$, which is close to the experimental estimate $T/3$ [93].

It should be mentioned that at the initial moment the spot was assumed to be non-mixed (mixing occurs only as a result of

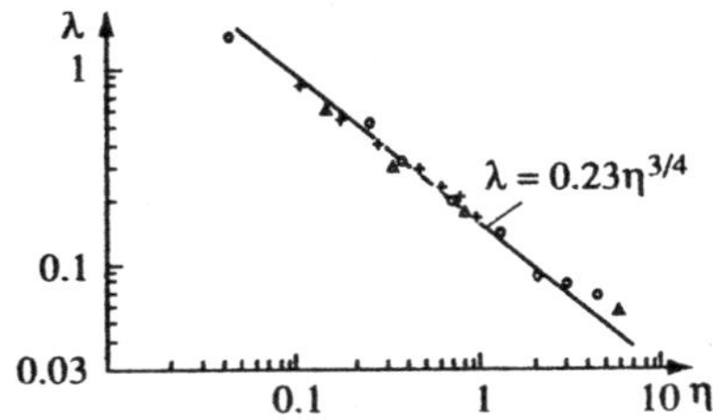

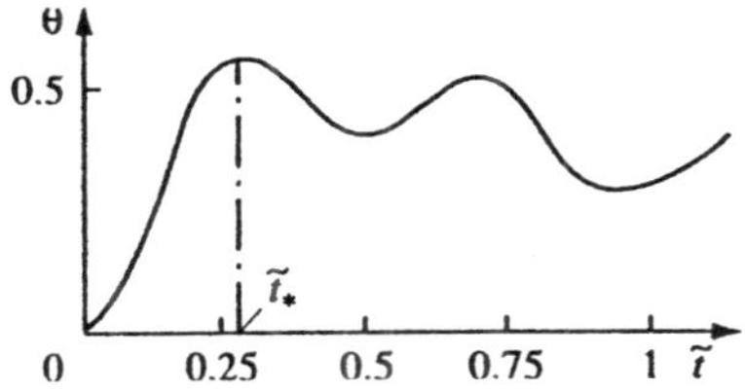

Fig.3.27. Attenuation of axial pulsation of velocity in variables (λ, η) for a stratified turbulent spot: Δ – $Fr_D = 23$, $Re_D = 2 \cdot 10^4$; o – $Fr_D = 31$, $Re_D = 3 \cdot 10^4$ – experiments by Lin and Pao [192]; + – direct statistical modelling.

Fig.3.28. Relative potential energy of mixing in a stratified turbulent spot (start of collapse of the spot at $\tilde{t}_* = 0.27$ in calculations and $t_* \approx 0.33$ in Merrit's expertiments [93]).

turbulence). Quantity θ is not suitable as the degree of mixing because it gives too high an upper estimate. If we accept the definition of the mixing exponent, according to [93], the numerical results give, for this exponent, the value equal to 0.10÷0.15 at the moment of the start of the collapse. This corresponds to the experimental data [192]. Thus, the physical and numerical experiments give a low estimate of the degree of mixing determined by turbulence only.

Figure 3.29 shows the distribution, on the axis of the stratified spot of the horizontal component of the flow of turbulent energy, normalised as the asymmetry of the distribution function of pulsations

$$\alpha(t, y/y_{1/2}) = Q_y(t, y/y_{1/2})/E_m^{3/2}(t),$$
$$Q_y = \left\langle v\left(u^2 + v^2 + w^2\right)\right\rangle.$$

It may be seen that the maximum value of the asymmetry of the function of distribution of pulsations, determined by this procedure, is reached approximately at half the Vaissala period.

6. The statistical method of the particles is designed for solving problems formulated on the basis of kinetic equations. The version of the method of particles in the cells, used in the present work, is oriented to kinetic equations describing the medium as a set of particles (molecules or liquid moles), whose scale of free (without interaction) motion is comparable with the macroscopic scale of the flow. The statistical method of the particles makes it possible to obtain the distribution function of the probabilities of pulsations or

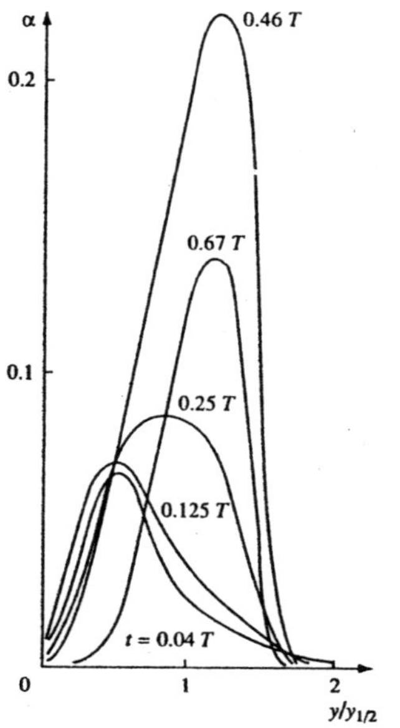

Fig.3.29. Distribution along the axis of the stratified spot of the horizontal component of the flux of turbulent energy (asymmetry of the distribution function of pulsations).

derivatives or any moments of the distribution. In this sense, the approach, and also the kinetic equation used as its basis, provides more information than the approaches based on the truncated chain of the Reynolds equations.

The kinetic methods corresponds to an infinite chain of evolution equations for momenta. The general principal of the equations obtained in this manner from the kinetic equation is well known and is based on the integration of the kinetic equation with different weight functions $\Psi(\mathbf{V})$. The first three equations of the infinite chain of the equations of the moments, equivalent to (3.1), have the following forms [131]:

$$\frac{\partial}{\partial x_i} u_i = 0; \tag{3.12}$$

$$\frac{\partial u_i}{\partial t} + u_j \frac{\partial u}{\partial x_j} + \frac{\partial}{\partial x_j}\left(\left\langle v_i v_j \right\rangle + P\delta_{ij}\right) = 0; \tag{3.13}$$

$$\frac{\partial \left\langle v_i v_j \right\rangle}{\partial t} + u_k \frac{\partial}{\partial x_k}\left\langle v_i v_j \right\rangle + \left\langle v_i v_k \right\rangle \frac{\partial u_i}{\partial x_k} + \left\langle v_i v_k \right\rangle \frac{\partial u_j}{\partial x_k} +$$
$$+ \frac{\partial}{\partial x_k}\left\langle v_i v_j v_k \right\rangle = -\frac{1}{\tau_1}\left\langle v_i v_j \right\rangle + \frac{1}{\tau_2}\left(\frac{2E}{3} - \left\langle v_i v_j \right\rangle\right). \tag{3.14}$$

This standard method may be used to obtain equations for the moments of the third, fourth and higher orders. Finally, for the kinetic equation there is also a problem of 'closure' but its solution may be simpler because it does not require additional hypotheses for breaking the infinite chain of the appropriate moment equations. We believe that this is also the advantage of this approach. There is another important aspect of the problem. The simulation, in particular the statistical method of the particles for solving the kinetic equations, is direct modelling of the mechanism of turbulent exchange. Investigations, carried out by this method, make it possible to understand these mechanisms better.

The main obstacle for the extensive application of the statistical simulation by the method of particles is the absence of universal kinetic curve of turbulence. There are also difficulties of another type mentioned by V.E. Yanitskii [131]. We shall discuss them.

a. The given method has inherited many characteristic features of the appropriate method of dynamics of rarefied gases and this restricts its possibility to kinetic equations identical to the equations for the rarefied gas. They (and the appropriate formulations of the problems) do not reflect the specific features of turbulence to a sufficient degree. For example, the equations (3.12) does not contain the members of the type.

$$\frac{\partial}{\partial x_i}\langle p v_i \rangle.$$

Thus, kinetic equations, which were used as the basis, ignores the effect of pulsations of pressure on the diffusion of turbulence energy. The previously mentioned numerical results show that by selecting the empirical constant it is possible 'compensate' the influence of the defect on the time and spatial characteristics of the energy $q^2(t,r)$. However, the problem of the influence of this defect on the distribution functions of pulsations of the speed has not yet been answered.

b. Solving the problem of the pulse-free wake (of the problem of the turbulent spot), we ignore the quantity ∇P. Therefore, in the calculations of time t we restrict ourselves to the relatively low value of $t < 20$. The correct formation of the problem should include, in addition to the kinetic equation, the condition of non-compressibility div $\mathbf{u} = 0$. The application of this condition for the equation of pulse transfer (13) give the well known equation for the mean pressure:

$$\nabla^2 P = -\frac{\partial^2}{\partial x_i \partial x_j}\left(u_i u_j + \left\langle v_i v_j \right\rangle\right).$$

Thus, the simulation scheme should include the additional procedure solving the Poisson equation. It is still necessary to explain how to combine the available numerical methods of solution of this equation with the scheme of the statistical method of the particles in cells.

c. Because of the differences between the numerical and experimental results for the radial distribution of the energy of the spot, previously we have mentioned the mixing effect. Analysis carried out in [115] shows that to take this effect into account, it is necessary to stop using the local dependences of the integral scale L_i, functions of energy q^2 in the given point. At least, the calculated field of the energy $q^2(t,r)$ will be in agreement with the experiment if it assumed that L_i are exponential functions of energy only in the centre of the spot $q^2_m(t)$ with the same exponents γ_1 = 0.63 and γ_2 = 0.80. However, this method on its own may prove to be insufficient for accurate calculation of more special features of the distribution function f in the vicinity of the boundary of the spot. The contribution of the pressure gradient to modelling of mixing is also unclear. Statistical methods make it possible to solve the problem of anisotropy and 'non-equilibrium' turbulence when the distribution function greatly differs from the normal law. The main obstacle for extensive application for the method is the absence of universal kinetic equations of turbulence.

The development of these approaches is being continued. It is hoped that successes will be achieved in solving more complicated problems of turbulence when large scale processes are calculated directly from the schemes of splitting for dynamic equation (for example, using the method of large particles of flows), and local fine-scale fluctuations are modelled by a statistical method [16, 19].

3.3. STATISTICAL MODELLING OF FLOWS OF FREE TURBULENCE IN A LONG-RANGE WAKE

1. S.A. Ivanov and V.E. Yanitskii [61] developed recently very original schemes of direct statistical modelling of molar transfer in a turbulent flow of a long-range wake, based on the realisation of Brownian trajectories. We shall discuss several assumptions of this approach, following [61].

The proposed method is based on the concept of L. Prandtl of turbulent mixing. The flow is determined by the movement of small volumes of the liquid, referred to as moles, each of which has its own speed. The speed field changes as a result of turbulent mixing of the moles with different speeds. Generally speaking, any characteristics of the flow is the averaging of identical characteristics of moles, its components.

The basic problem used for the development of the method was represented by the planar flow of the liquid in a long-range wake behind a cylindrical solid. For the mean speed of the flow there is the approximate analytical solution proposed by Görtler. It uses the approximation of the turbulent section by a function proportional to the derivative of the longitudinal speed with the accuracy of the constant ν^*. The derivative is taken along the axis y, normal to the axis x, coinciding with the direction of the wake. This solution at a relatively large distance from the solid provides satisfactory agreement with the experiment, and can be used as a reference solution.

Returning to direct modelling, it should be noted that the speed of the moles in the given planar problem can be represented efficiently by a means of the longitudinal component u along the axis of the wake and the transverse (or vertical) component v, normal to this axis. The vertical speed v is of a purely pulsational nature because the flow in the long-range wake is such that its mean value $\bar{v}$ is equal to zero (here and in the rest of the book the mean value of the quantity is denoted by the line above). Thus, moving in the transverse direction, each mole behaves as a Brownian particle. In this case, the moles retain (according to Prandtl's assumptions) the x component of the pulse, i.e. they did not change in the volume, they have a constant speed along the entire length of the examination range. All moles have the same size. The position of the mole is characterised by the coordinate of its centre. At the initial moment, the moles are continuously distributed along the y axis. Each mole is related to the longitudinal speed which is such that the profile of the longitudinal speed of the flow forms in the mean. In addition, for the determination of the characteristics of the flow it is sufficient to examine the evolution of this (generally infinite) chain of the moles. During the time period Δt all the moles are dispersed along the x axis by the step determined by the speed of the incident flow $\Delta x = U_\infty \Delta t$. In this case, the coordinates of the moles on the y axis change in accordance with the law of movement of the Brownian particles.

The diffusion coefficient ν_* is a parameter of the movement. It determines the intensity of displacement of the moles and of their mixing. It should be mentioned that all moles are equivalent to each other at any moment of modelling and move on the y axis in an isotropic fashion so that their uniform distribution on the axis is not disrupted.

As a result of Brownian motion at the end of the examination range at any point of y_* axis there will be in the general case a mole which differs from which was situated at this point of the start of observation, consequently, the mean speed of the moles, i.e. the speed of the flow of the given point, changes. In order to calculate the mean speed of the point at y_*, it us sufficient to average-out the speeds of the moles whose coordinates are close to this point. The model makes it also possible to calculate the quantity such as the friction stress. It is well known that the friction stress is a flow of the appropriate pulse. Therefore, in modelling it is sufficient to fix all the moles, intersecting point y_*, during the time Δt, taking their pulses into account in this case. Each mole carries a pulse equal to the product of its longitudinal speed by its size and by the density of the liquid. The sum of these pulses with the sign taken into account (plus, if the mole moves in the direction of increasing y, and minus if it moves in the opposite direction) is the general transfer of the pulse to the point y_* during time Δt.

The numerical experiments in a computer, carried out using the scheme described previously, showed complete agreement on the modelling profile of the mean speed with the Görtler profile and, consequently, the fact that this profile is similar to the experimental profile (Fig.3.30). In addition to this, the viscosity obtained in the course of modelling, also coincides with the value of viscosity according to Görtler. The only free parameter of the model is the constant ν_*. The selection of this parameter determines the value of variation of the mean speed of the axis of the wake during the step Δt. Therefore, ν_* is selected from the condition of coincidence of the modelling speed with the results of physical experiments. The resultant value $\nu_* = 0.0225$ is in agreement with the identical Görtler constant.

Thus, this methods gives the results completely corresponding to the Görtler model of the long-range wake for the mean speed and for the friction stress. However, this is not only the Monte Carlo method for the Görtler equation. Using in the calculations only the initial profile of the mean speed, we obtain almost complete information on the flow, including viscosity, which is included in the

Görtler equation *a priori*. The agreement of the results of the model with the Görtler solution should understood only as an efficient physical interpretation for the equation taken from empirical considerations. At the same time, this agreement reflects the fact that the modelling viscosity does not correspond to the experimental data. For example, according to the measurements presented in [5], the profile of viscosity in the long-range wake in self-similar variables does not change and has the form shown in Fig.3.31. Consequently, it is necessary to define additionally the model in such a manner as to overcome this discrepancy.

Direct modelling is an attempt to construct physically sustainable model and, consequently, it is natural to return to empirical physical considerations. Primarily, it is necessary to introduce the boundaries of the region of turbulence. It is well known that the examined lee flow consists of two parts: a strongly turbulised core and laminar liquid surroundings. The separation of the regions is characterised by the parameter of intermittency γ, reflecting the multi-scale nature of turbulence. According to Landau [85], we accept the existence of two scales – a small scale, characterising pulsations inside the turbulent part of the flow, and a large scale, associated with the oscillations of the shape of the restricted region, occupied by the turbulised liquid. The model described previously is of the

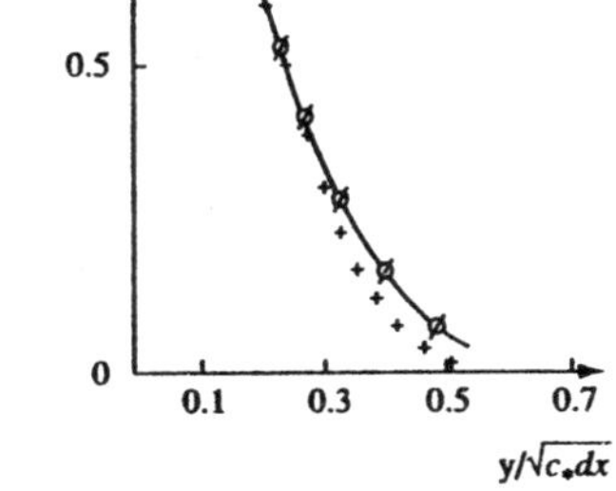

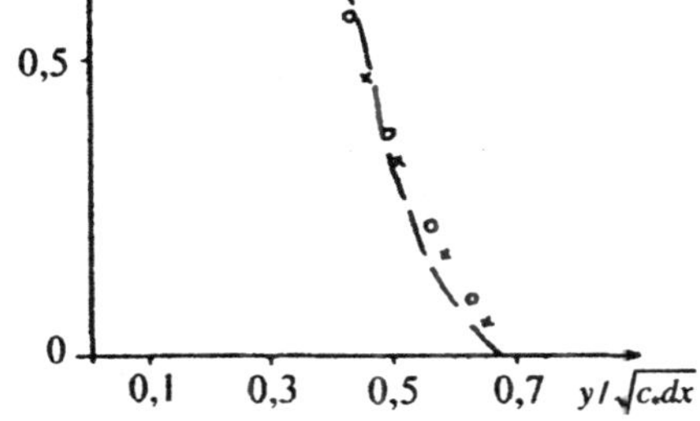

Fig.3.30. Long-range wake. Profile of mean velocity: solid line – Görtler model, ⌀ – model with intermittency not take into account, × – Ivanov–Yanitskii model [61] with mixing.

Fig.3.31 (right). Long-range wake. Profile of viscosity: solid line – Görtler model, ⌀ – model with intermittency not taken into account, × – Ivanov–Yanitskii model [61] with intermittency, o – experiment [5], broken function – function γ^2 (t, y).

single-scale type in this sense and examines the flow of the liquid as a perturbation in the entire space. The restriction of the turbulent region by a means of the boundary of separation of the flow into two logical parts corresponds to the behaviour of the second scale of turbulence. To determine this boundary, we can use the well known experimental expression for the intermittency coefficient. The division of the flow into regions also changes the nature of movement of the moles. In this case, the moles which have penetrated into the turbulent zone, i.e. the moles whose coordinates are smaller than the coordinates of the boundary, move, like Brownian particles, in the direction normal to the axis of the wake. If the mole reaches a boundary during its movement, reflects from it as a Brownian particle. The moles, form in the laminar part of the flow, i.e. the moles whose coordinates extend outside the boundary of turbulence, do not have vertical speeds in general. However, it should be mentioned that the boundary change with time, and the mole, situated in the laminar zone at this moment, may be with time captured by the turbulent region.

In the second method of modelling the turbulent mixing, the characteristics of the flow greatly differ from those obtained in the case of the modelling with intermittency not taken into account. In particular, modelling viscosity is completely proportional to the experimental (Fig.3.31). Thus, on the basis of natural references, constructing a physically understandable model, we obtain results similar to the experimental data.

2. We examine a problem of statistical modelling of turbulence in the core of the wake. It is assumed that a flow of an incompressible liquid arrives at a cylindrical body in the direction normal to its generating line with a speed U_∞ and the axis of the wake coincides with the x axis. The Reynolds equations for the mean speeds are reduced [126] to the form

$$\bar{u}(x,y)\frac{\partial \bar{u}(x,y)}{\partial x} = -\frac{\partial \overline{u'v'}}{\partial y}, \quad \bar{v}(y) = 0.$$

where the pulsation speeds $u' = u - \bar{u}$, $v' = v - \bar{v}$, and the line above the quantities indicates the averaging symbol. We introduce the defect of the instantaneous speed:

$$u_* = U_\infty - u.$$

For the mean value, the Reynolds equation transformed to the form:

$$U_\infty \frac{\partial \bar{u}_*}{\partial x} = \frac{\partial \overline{u'_* v'}}{\partial y}.$$

It is well known that the profile of the mean defect of the instantaneous speed in the long-range wake, when the flow is self-similar, is described satisfactorily by the Görtler model. The model assumes the following approximation of the friction stress:

$$\tau = \rho \overline{u'_* v'} = -\rho \nu_* \frac{\partial \bar{u}_*}{\partial y}, \quad \nu_* = \text{const}. \tag{3.15}$$

The main equation of this model has the form

$$U_\infty \frac{\partial \bar{u}_*}{\partial x} + \nu_* \frac{\partial^2 \bar{u}_*}{\partial y^2} = 0.$$

Its self-similar solution on the condition of conservation of the pulse is

$$\int_{-\infty}^{+\infty} \bar{u}_* dy = \frac{1}{2} C_W d U_\infty$$

(C_W is the coefficient of resistance of the solid, d is the diameter of the cylinder), has the form

$$\bar{u}_*(x, y) = 2U_\infty \left(\frac{\ln 2 C_{W^d}}{\pi x} \right)^{0.5} \exp\left(-16 \ln 2 \frac{y^2}{C_{W^{dx}}} \right). \tag{3.16}$$

In statistical modelling we shall examine the following Cauchy problem:

$$\begin{cases} \dfrac{\partial \bar{u}(t, y)}{\partial t} - \nu_* \dfrac{\partial^2 \bar{u}(t, y)}{\partial y^2} = 0, \\ u(0, y) = u_0(y). \end{cases} \tag{3.17}$$

$u_0(y)$ is represented by the Görtler solution (3.16) at the point $x = x_0$. Here x_0 is the coordinate on the axis of the wake from which modelling starts. t is represented by the expression $(x - x_0)/U_\infty$.

To solve the formulated problem, we propose the following stochastic model of evolution of N moles in respect of time. In our understanding, the mole is some volume of the liquid. It is characterised by its size Δy_0, the coordinate of its centre $\xi_i(t)$ and the natural longitudinal speed u_i. Modelling is carried out up to the moment T in discrete steps Δt. The space on the y axis is restricted by the value y_m so that we examine only the range $[-y_m, y_m]$. To facilitate calculation of the average values, a uniform grid with the step Δy: $y_i = y_m + \Delta y(j-1)$, $= j = \overline{1,J}$ is placed on the y axis.

At the initial moment of time the moles in the range $[-y_m, y_m]$ are distributed in the following manner: $\xi_i(0) = -y_m + i\Delta y_0 - \Delta y_0/2$, $i = \overline{1,N}$. With this initial distribution, the moles are not only distributed continuously on the y axis but also the number of moles in every cell of the grid is the same and equal to $N_0 = \Delta y/\Delta y_0$. In accordance with the coordinate, for each mole we determine its longitudinal speed: $u_i = u_0(\xi_i'(0))$. The mole retains this speed throughout the modelling stage.

The calculation of the evolution of the model in the small range t is used to realise N Brownian trajectories (the method of realisation of the model is presented in [61]). The friction stress, obtained in this model, corresponds to the Görtler approximation (3.15). The numerical calculations, carried out in accordance with the described model, have confirmed the agreement between the resultant profile of the mean speed of the flow and the viscosity with the identical profiles of the Görtler model (Fig.3.30).

3. We shall now take into account the intermittency effect. The described model gives a value of turbulent viscosity $nu_T = \overline{u'v'} \Big/ \dfrac{\partial \bar{u}}{\partial y} = \nu_*$ constant in the cross section of the wake. As shown previously, this does not correspond to the experimental data. Consequently, the model reflects inaccurately the turbulence mechanism. Above all, it does not take into account the intermittency effect, i.e. the division of the flow into the turbulent and laminar components.

In [61], S.A. Ivanov and V.E. Yanitskii proposed a heuristic model in which it is attempted to take this phenomenon into account. They examined a different method of modelling by introduction of a random boundary between the turbulised core of a wake and its laminar environment. According to experimental data [186], the intermittency coefficient, characterising this boundary, is determined from the equation

$$\gamma(x,y)=\int_{y}^{+\infty}\frac{1}{\sqrt{2\pi\sigma_\gamma^2(x)}}\exp\left(-\frac{\left(\xi-\sigma_\gamma(x)^2\right)}{2\sigma_\gamma^2(x)}\right)d\xi,\quad y>0. \tag{3.18}$$

here $\gamma(x, y)$ may be understood as the probability of the given point y being trapped by the turbulent part of the liquid; $\gamma(x,y) = P\{y<\xi_\gamma^+(x)\}$, where $\xi_\gamma^+(x)$ is the normally distributed random quantity with the probability density of $n(b_\gamma(x), \sigma^2_\gamma(x))$.

In the new model, the determination of the turbulence boundary in the upper half plane of the wake ($y > 0$) is reduced to determination of the random quantity ξ_γ^+. For $y < 0$, the flow is symmetric and the boundary is determined by quantity ξ_γ^-, distributed normally with the probability density $n(-b_\gamma(x), \sigma_\gamma^2(x))$.

On the whole, modelling is carried out in the same manner as in the previous model. Only the block of displacement of the models changes [61].

The difference between this model and the previous one is that only the moles situated in the zone of turbulence $[\xi_\gamma^-(t), \xi_\gamma^+(t)]$ move on the y axis and they cannot leave this zone. Because of the introduction of the boundary into the model, in addition to the constant ν_* there are also two empirical functions $b_\gamma(x)$, $\sigma_\gamma(x)$. For them we have the experimental dependence $\sigma_\gamma / b_\gamma \approx 0.24$ [186]. In the calculations it was assumed that $b_\gamma(x) \approx 2b_{1/2}(x)$, where $b_{1/2}(x)$ is the half width of the profile of the speed at half height, $b_{1/2} = 0.25\sqrt{x}$ [126].

We shall examine the results of numerical realisation of the model with the intermittency effects taken into account. The following characteristics were selected for the modelled flow: $U_\infty = 1$, $C_W = 1$, $d = 1$, $\rho = 1$ (here we accepted the following notations: C_W – the coefficient of resistance of the solid, d – the diameter of the cylinder, ρ – density, U_∞ – the velocity of the incident flow). Using the model, we calculated the profiles of velocity and viscosity. Computer calculations show that these profiles reached the self-similar condition for different parameters. The following values of parameters may be regarded as characteristic: $y_m = 8.1$; $\Delta y = 0.2$; $\Delta y_0 = 0.004$; $x_0 = 50C_{Wd}$; $\Delta t = 5$; $N_S = 2500$.

Quantity ν_*, which is the main parameter of both models (both with and without the random boundary), was determined in the numerical experiments from the condition of coincidence of the change of the velocity of the flow of the axis of the wake with the experimental data. It was shown that, in this case, it was necessary to take into account the Görtler value $\nu_* = 0.0225$.

The profile of viscosity, reduced to the unit scale, is shown in Fig.3.31. The graph also shows the viscosity in the Görtler model and the values of viscosity determined on the basis of experimental data [5]. The values of viscosity on the axis of the wake are equal to 0.0225 for the Görtler model and for the model with intermittency. For the experimental profile, this value is half the above value and approximately equal to 0.011. It should be mentioned that the calculated viscosity is self-similar throughout the entire modelling stage, regardless of the evolution of the profile, the velocity from the Görtler model at $t = 0$ to a steeper profile at high t (Fig.3.31). Its values are completely proportional to the experimental values with a factor of 2. It should be mentioned that the value of viscosity on the axis of the wake in our models was determined by parameter ν_*. This parameter characterises the value of change of the speed on the axis of the wake in relation to x. The parameters b_γ and σ_γ determine the 'boundaries' of the viscosity profile – the boundary separating the turbulent and laminar regions.

4. In the final analysis, we concluded that the model of statistical modelling of turbulence in the core of the wake uses one free parameter ν_* and gives the results coinciding with the well-known Görtler model, i.e. the speed of the flow, close to the experimental data, and the viscosity, equal to a constant.

Taking intermittency into account, parametric functions $b_\gamma(x)$ and $\sigma_\gamma(x)$, determining the boundary of the turbulent core of the flow, are introduced additionally into the model. This model gives the velocity which does not contradict the experimental results, and the viscosity proportional to the experimental data. However, the resultant absolute value of viscosity on the axis of the wake is higher than the experimental value. The work in this direction continues.

The values of turbulent viscosity, obtained here, can, generally speaking, be used in equations so that it is possible to construct an iteration process in order to improve the accuracy of the results.

3.4. FORMATION OF LARGE-SCALE STRUCTURES IN A GAP BETWEEN ROTATING CYLINDERS (RAYLEIGH–ZEL'DOVICH PROBLEM)

1. Recently, special attention has been paid to the numerical modelling of the process of stabilisation of the flow in different conditions both on the earth and in space [18,140,142]. For example, investigations of the rotational motion are of principal fundamental value for the physics of accretion discs [37] (see also chapter 4 of the present book). The problem of accretion of matter on compact sources is determined by the rate of loss of the angular moment in the accretion discs. Usually, it is assumed that the mechanism of the loss of the angular moment is associated with turbulent viscosity which exceeds molecular viscosity by many orders of magnitude. This results in the following problem: what is turbulent viscosity and what is its physical meaning for accretion discs. We believe that it is necessary to separate two physical processes. The first process is associated with the transfer (loss) of the angular moment. It is determined by large-scale turbulence structures. The second process is responsible for the transfer of the kinetic energy of turbulence into heat (dissipation). The problem of the relationship of these processes depends on many factors: the possible stationary nature of the process and the associated spectrum of turbulence, the geometrical special features of the flow, the possible effect of the magnetic field and other factors controlling the turbulence spectrum. Therefore, the problem of turbulence (including turbulent convection and turbulent mixing) becomes controlling for understanding the processes in astrophysical objects, as shown in [114]. The process of stabilisation of the flow is of the non-stationary nature and is extremely complicated to investigate. However, advances have been made in understanding this process in recent years because of the application of the methods of mathematical modelling [18].

It is well-known [18] that turbulence may be investigated by mathematical modelling or direct calculations of nonstationary movement of ordered and large-scale structures (solving the Euler equation) or by the application of specific type of hypotheses on the stochastic nature of the field of velocities and other physical quantities. In the second case, generally speaking, the physical reasons for the development of turbulence (supplying energy to chaotic motion) are not considered. We shall investigate the problem of formation of large structures. In this sense, the most representative picture is provided by the structure of the long-range wake during the flow around a cylinder consisting of ordered vortices, mainly large-scale ones. However, small-scale vortices are also presented at the same time. The main problem of the theory of turbulence is to explain the development of such a spectrum of scales in the turbulent flow. Of course, it should be mentioned that the majority of investigations have been concerned with specific tasks in a narrow range of the variation of the physical parameters so that the problem could not be examined in sufficient detail.

However, more and more studies have appeared recently (for example, [148]) casting doubts on the problem of isotropic free turbulence. It is claimed (for example, [15]) that turbulence is of the alternating nature, i.e. the regions of turbulence are replaced by regions of laminar flow. It has been claimed (see, for example, [15]) that turbulence is of the alternating nature, i.e. the regions of turbulence are replaced by the regions of laminar flow. Therefore, the author of [148] has suggested that it is more accurate to talk about the interaction in the flow (linear or nonlinear) and not about structures (laminar or turbulent). In this sense, the meaning of the Reynolds number, introduced for the characteristic of the problem, is difficult to understand because the Reynolds number starts to depend on the local structure of the flow. Therefore, instead of turbulent viscosity, it is necessary to introduce the concept of local dynamic viscosity and this has also been proposed in [148].

2. We started our investigations of the development of turbulence with a simple problem from the physical viewpoint: determination of the profile of velocity in the gap between the cylinders, starting with a highly nonequilibrium shear flow. The problem of the flow between cylinders has been studied for more than 100 years. The main results were obtained in the classic studies by Taylor (1923, 1936) where it was shown that convective cells appear at a Reynolds number of Re ~ 60 in the condition of incidence of the angular moment of the amount of motion outwards. At the same time, as for the plane-parallel flow, turbulence develops at Re ~ 2000. It was also shown [58] that the increase of the angular moment outwards results in stabilisation of the laminar flow (Re ~200 000 from the extrapolation of Taylor's results (1936)). Our investigations were directed to the examination of the nature of formation of the profile of velocity in the gap if in the initial stage the distribution of the shear flow was nonequilibrium. The aim of investigations was to examine the evolution of the structure of the flow (including vortices) on the model of the flow in the gap.

Our concept was that at high Reynolds numbers the inertia forces prevail over the viscous stresses and turbulisation of the flow start with the development of large scales – the formation of large vortices as a result of the shear flow (Kelvin–Helmholtz instability). In other words, the energy of chaotic mjotion in the turbulent flow is the energy of the shear flow itself!

We have considered the simplest case from the viewpoint of numerical modelling, i.e. the flow between two coaxial cylinders. Already in the studies by Rayleigh it was shown that the centrifugal force may stabilise the Taylor–Couette flow if the following condition is satisfied [201]:

$$\frac{d(Vr)^2}{dr} \leq 0. \tag{3.19}$$

The solution of the problem of the onset of turbulence is of great importance in astrophysical problems. For example, it is usually assumed that the accretion discs are characterised by the presence of conventional turbulent viscosity, because the Reynolds numbers are very high. Nevertheless, in a number of investigations these results are doubted and it is claimed that the flow of Kepler rotation may prove to be laminar. The main argument of these claims is justified by the stability of the differential rotation of the medium with the moment increasing with the radius. Zel'dovich carried out a number of analytical investigations of the flow between two rotating cylinders with different laws of the variation of moment in relation to radius. In particular, Ze'dovich [58] designed the dimensionless Taylor number Ty:

$$\mathrm{Ty} = \frac{\dfrac{d}{dr}\left(\omega r^2\right)^2}{r^5\left(\dfrac{d\omega}{dr}\right)^2}, \tag{3.20}$$

which can be used to obtain the condition of stabilisation of the flow homogeneous with respect to density:

$$\mathrm{Ty} < 0. \tag{3.21}$$

For example, if we consider the exponential dependence of the law of rotation $\omega \sim r^n$ which corresponds to the constant value of the Taylor number in the entire range Ty = 8 $(n+2)/n^2$, stabilisation starts at $n > -2$. In solid-state rotation Ty = ∞ and the motion is stable. The criterion, introduced by Zel'dovich, is more convenient for astrophysical conditions in the evaluation of transition to the turbulent flow. These investigations are important because the movement of a homogeneous liquid with a rotating external cylinder is identical with the gravitationally stabilised shear flow. Simple similar considerations and analogy with the Richardson number (the number determines the stabilisation of the laminar flow in the atmosphere with gravitation, stratified in respect of density) and the results in equation (3.20).

3. In the first chapter, we discussed the general concept of the direct numerical modelling of a large number of nonlinear spatially nonstationary problems of modern aerodynamics, hydrodynamics and gas dynamics – phenomena accompanied by the development of free shear turbulence and/or hydrodynamic instabilities [18,142]. The main concept of the proposed multiplan approach to the examination of turbulence and instabilities is the development of 'rational' numerical models, adequate to the investigated phenomenon, and greatly differs from the investigated approaches to the modelling of turbulence. Taking into account the structural representation of turbulent flows, it is convenient to use models closely related with the investigated interaction mechanism. For example, the large-scale transfer is set up on dynamic models for the ideal medium, the laminar–turbulent flows – taking into account the viscosity mechanism of interaction, and the stochastic process – on the kinetic level. This method of 'rational' modelling corresponds to the mechanisms of interaction in structural turbulence and makes it possible to reduce greatly the level of requirements on the resources of computers in comparison with other approaches.

The problem of turbulisation of the flow in the gap between cylinders is complicated because this is associated with the effect of different forces: the first is associated with inertia force $V\, dV/dr$, and the second force is determined by viscosity. In particular, the viscosity establishes the linear profile of the Couette plane-parallel flow between two moving sheets, if the Reynolds number is relatively small, although the problem of the value of the Reynolds number and the ratio of the forces require further examination. The point is that in the presence of a steep gradient of velocity and sufficiently high kinetic energy of the flow, the instabilities of the Kelvin–Helmholtz type can develop at the boundary of two flows with different velocities (tangential discontinuity). In this case, the main role is played by the dynamic member in comparison with the viscosity member. In the development of the Kelvin–Helmholtz instability, viscosity is not included in the evaluation of the rate of growth [84]. As shown in [18], these models can be investigated using Euler equations. The importance of the large-scale structures in turbulence was addressed for the first time in [158,234]. More detailed discussion of this problem and of this approach in the calculation of turbulent flows may be found in the first chapter and in books in [18,142]. It was shown that, basically, the energy of the turbulent flow is associated with large scales, i.e. inertia forces.

In this formulation of the problem, the inertia forces produce, as a result of the pair of forces, vortices (cyclones and anticyclones) which may result in the establishment of the profile of the flow in the gap. An important moment in this approach is the problem of break up of the vortices. It is important to determine the forces resulting in the disintegration (viscous or dynamic) and the characteristic duration of these processes. According to our assumptions, the high-frequency part of the spectrum is generated in the nonlinear interaction of the large-scale structures with each other and with the walls, i.e. molecular viscosity plays no important role here (for relatively high Reynolds numbers).

4. Following the concept proposed previously, we carried out the following numerical experiments. These experiments were conducted on two coaxial cylinders with the characteristic size of the gap $\Delta R/R = 0.227$. Investigations were carried out using the two-dimensional geometry in the polar system of coordinates (r,ϕ). The Euler equation was solved numerically. We used the quasi-monotonic grid-characteristic scheme of the second order of accuracy in respect of space and time [102] (see also equation 2.3 in this book).

Molecular viscosity is not taken into account but the numerical method includes implicitly some nonlinear dissipative mechanism which, generally speaking, depends on the step of the grid (decreases with decreasing step of the grid). However, it should be mentioned that for turbulent flows, the rate of disintegration of energy is determined by the rate of inertia transfer and is not sensitive to the value of the dissipative mechanism (molecular viscosity).

The ratio of the inertia and dissipative forces in the computing experiment may be evaluated from the experiment itself, measuring the rate of dissociation of kinetic energy θ_{num} and constructing the dimensionless number Re_{nam} from the similarity of the Reynolds number:

$$\mathrm{Re}_{num} \sim \frac{(\Delta u)^3}{\theta_{num} l},$$

where Δu is the characteristic variation of velocity on the characteristic scale l.

The characteristic dimensions of the finite-difference grids, used for the computing experiments in this study, were: 100 points in the direction of radius and 2500 in respect of the angle. The following variants, differing in the initial and boundary conditions, were investigated.

In the first variant, the inner cylinder was in the still state for the entire period of time ($V_0 = 0$), and the outer cylinder rotated in the anticlockwise direction with a constant velocity ($V_1 = 1$)[3]. At the initial moment, the flow had the following structure: in the middle of the gap, close to the inner cylinder, the liquid was in the still state. In the outer half of the cylinder, the angular speed corresponded to solid-state rotation, coinciding with the speed of the outer cylinder. At the boundary of a tangential discontinuity we specify a random, with low amplitude (less than 1% of velocity V_1), high-frequency (characteristic scale of the order of the size of the computing grid) perturbation of the radial component of velocity.

The boundary conditions on the inner and outer cylinders were established by means of an implicit correction of the angular component of velocity v_ϕ and internal energy (in the conservative manner) in the cells of the integration range, in the immediate vicinity of the cylinders:

$$v_\phi^{n+1} = \frac{\tilde{n} + \alpha v_{cyl}}{1+\alpha}, \quad \alpha = v_* \frac{2\tau}{h_r^2}, \tag{3.22}$$

where $\tilde{v}_\phi$ is the velocity obtained in the next step in respect of time of the solution of the Euler equations using the non-occurrence condition, v_{cyl} is the velocity of the cylinder, τ is the step in respect of time, h_r is the step of the grid in respect of the radial coordinate, v_* is the effective viscosity acting on the on near-boundary cells. If as a result of correction of velocity (3.22) the kinetic energy decreases, this indicated the dissipation of kinetic energy, i.e. the conversion of this energy into heat. If, on the other hand, the kinetic energy increases, the addition of kinetic energy related to the work of external forces sustaining the rotation of the cylinders.

Figure 3.32 shows the starting moment of development of the flow t = 0.05, 0.1, 0.2 (from top to bottom) when the Kelvin–Helmholtz shear instability rapidly develops. The left part of this figure shows the so-called conventional concentration, associated with the initial position of the particle in the upper (black colour) all lower (white) layers, and the vorticity is given on the right-hand side. The inverse increment of the development of the Kelvin–Helmholtz instability is proportional to $\Delta R/V_1$ [84]. This time is relatively short. It is equivalent to numeral 1/30 of the rotation of the outer cylinder. After this time, the growth of the vortices is 'constricted' by the size of the gap between the cylinders. This is followed by a decrease in the number of vortices and by intensification of the rotation in the course of attraction and merger of the vortices (the Zhukovski force).

Two vortices remain after completing the calculation of the given variant (t = 122, which equals approximately 20 revolutions). The instantaneous lines of the flow at this moment of time are shown in Fig. 3.33. The profile of the flow is established by means of the distribution of the angular moment in large vortices. Figure 3.34 shows the profile of the angular moment (markers), established in

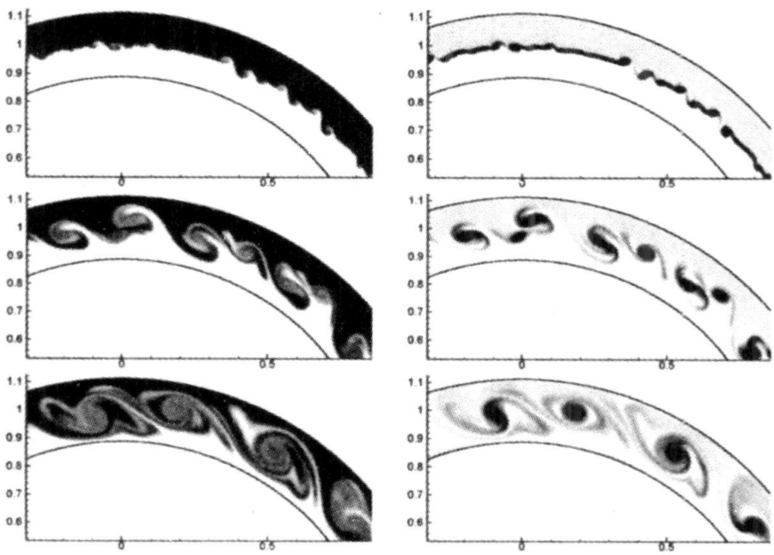

Fig.3.32. Structure of the flow for variant 1 (outer cylinder rotates in the anticlockwise direction; the inner cylinder is stationary) for starting times t = 0.005; 0.1; 0.2 (from top to bottom): left – vorticity.

the examined case, and the profile from [201] (solid line) when the profile forms by means of the viscosity mechanism. Also, as indicated by Taylor's and Zel'dovich theory, there is no turbulence after the establishment of the profile of the flow in the gap. However, the profiles, determined in the calculations, differ principally from the evaluation profile obtained in [15], giving the dependence for velocity $u = r^{6.5}$. This difference may be explained by low values of the Reynolds number in comparison with the critical value of the Reynolds number obtained in the experiment and in Zel'dovich's estimates [58].

Figure 3.35 shows the correlators of velocity (or more accurately, of the radial component of velocity u) at two different moments of time when the outer cylinder carried out a third of revolution (t = 2) and approximately 20 rotations (t = 122), respectively. Correlator K_u is determined as follows:

$$K_u(r,\varphi) = \int_0^{2\pi} u(r,\varphi+\xi)u(r,\xi)d\xi.$$

This figure shows clearly that the characteristic angular dimension (let us say, the distance from the minimum to the maximum of the correlator) corresponds to the size of the gap between the cylinders, i.e. the dimensions of the large structures are approximately equal and of the start of evolution (when they

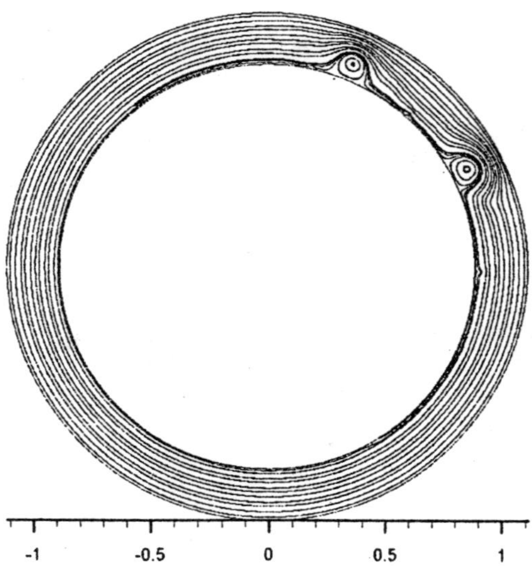

Fig.3.33. Instantaneous lines of current for variant 1 at t = 122.

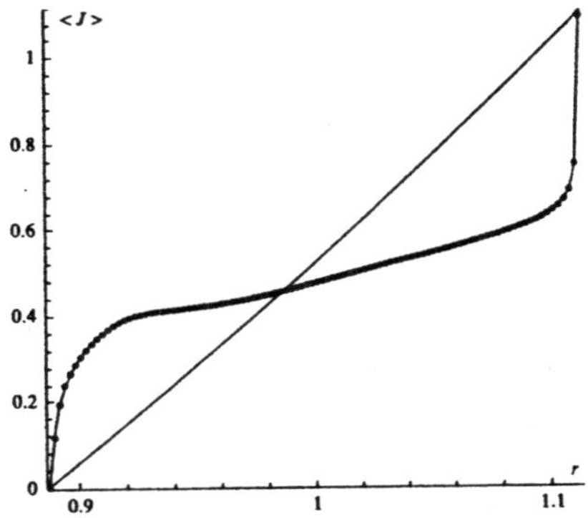

Fig.3.34. Average (in respect of angle) distribution of the angular moment for variant 1 (t = 122): markers – numerical modelling, solid line – viscous Taylor–Couette solution.

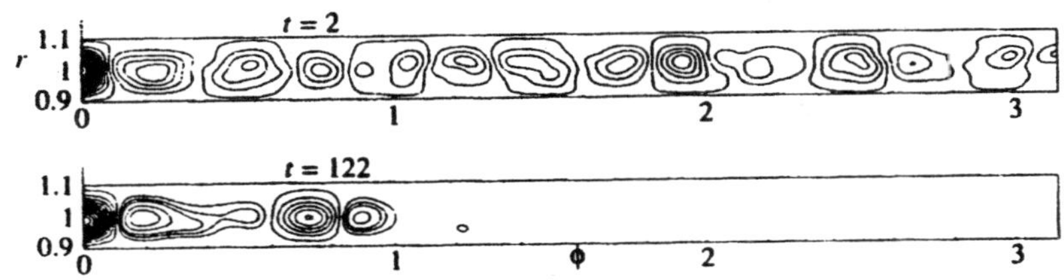

Fig.3.35. Correlators of velocity for variant 1 at times t = 2 and t = 122.

have already formed and their number is large) and also closer to the end of evolution (when their number is small).

Figure 3.36 shows the characteristic spectrum of the kinetic energy ε_n (in respect of the wave numbers ε_n), calculated from the equations

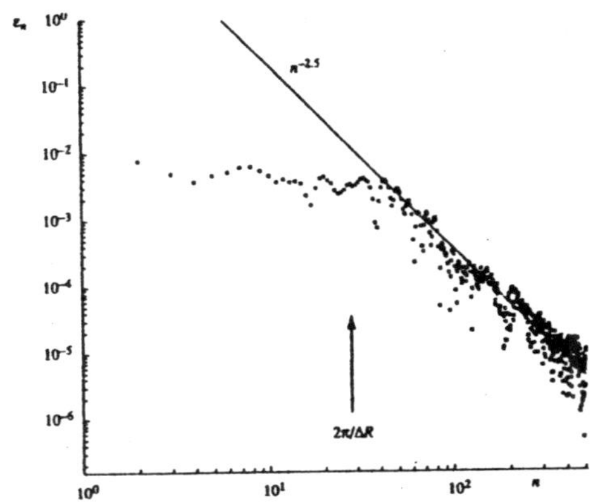

Fig.3.36. Spectrum of kinetic energy for variant 1 at time t = 122.

$$\varepsilon_n = \sqrt{a_n^2 + b_n^2}, \quad a_n = \pi^{-1}\int_0^{2\pi}\left(\frac{u^2}{2}+\frac{v^2}{2}\right)\cos n\varphi d\varphi,$$

$$b_n = \pi^{-1}\int_0^{2\pi}\left(\frac{u^2}{2}+\frac{v^2}{2}\right)\sin n\varphi d\varphi.$$

The spectrum was computed in the section in which the kinetic energy is maximum, i.e. approximately in the mean section $r \approx 1$. Two moments should be noted. Firstly, the fact that the 'tail' of the spectrum is not governed by the Kolmogorov–Obukhov law. Secondly, the given spectrum has a characteristic inflection point $n_1 = 2\pi/\Delta R$, determined by the width of the gap between the cylinders.

Two other variants are associated with the examination of the appearance of turbulence with the change of the boundary conditions in such a manner as to violated the condition (3.19) of the growth of the moment to the outside. After approximately 20 revolutions we arrested of the outer cylinder which should have resulted in a decrease of the angular moment in the external part of the gap. The results of the calculations show a decrease of the characteristic size of the vortex, and the entire vorticity is concentrated in the vicinity of the internal cylinder.

The system tends to the solid-state rotation and this is reflected in Fig. 3.37 showing the graph of variation of the angular moment, averaged out in respect of the angular coordinate, for the moment of time $t = 190$. Figure 3.38 shows the correlators of the radial speed component of the speed, also stresses the decrease of the characteristic size of the vortices and the pressing of the 46 to the inn internal cylinder.

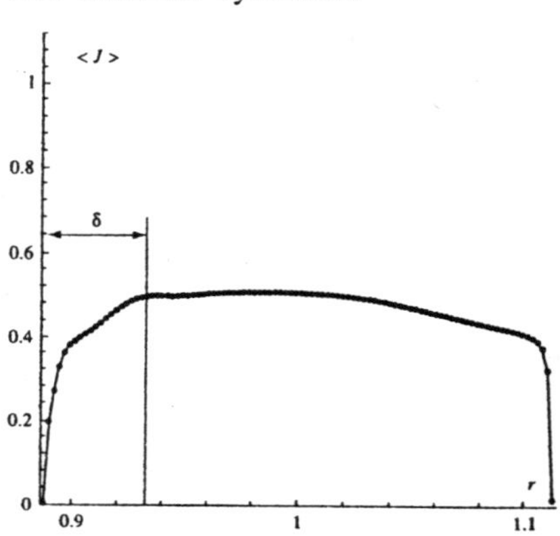

Fig.3.37. Averaged-out (with respect to angle) distribution of angular moment for variant 2 (after arresting the outer cylinder; inner cylinder is also stationary) at $t = 190$.

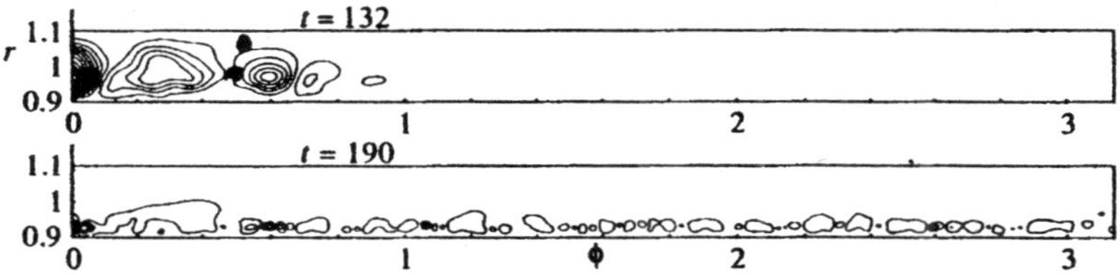

Fig.3.38. Correlators of velocity for variant 2 at times t = 132 and t = 190.

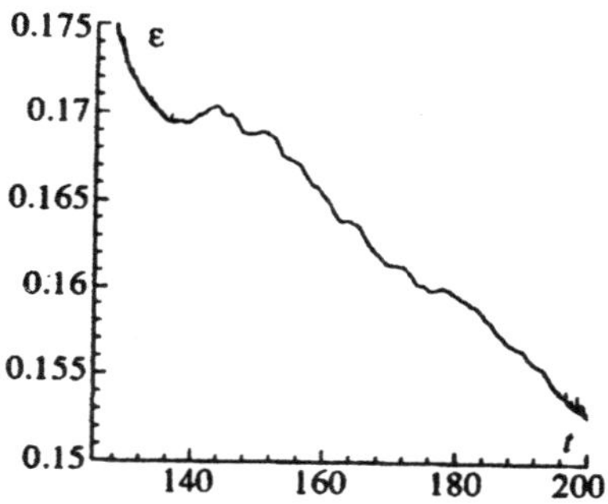

Fig.3.39. Determination of the mean velocity of dissipation: variation of kinetic energy with time (variant 2).

Taking into account the results of these numerical experiments, it is possible to evaluate the rate of dissipation of the kinetic energy and, consequently, the effective turbulent viscosity. Figure 3.39 shows the time dependence of the integral of kinetic energy. In accordance with equation (3.22), this corresponds to the 'numerical Reynolds' of the order of 1000.

Turbulence does not develop, as expected from the practically solid-state distribution of the angular moment (with the exception of boundary layers), obtained during the calculations.

In the last variant, we considered the classic case of the development of turbulence when the condition (3.19) or (3.21) is violated, i.e. the internal cylinder rotates whereas the external cylinder is stationary. Figures 3.40–3.43 shows a result of the numerical modelling relating to this case. Figure 3.40 shows the instantaneous line of current for the moment t = 840, Fig. 3.41 the velocity correlators for t = 700 and t = 840, and Fig. 3.42 shows the spectrum of kinetic energy for t = 840. The given distribution of energy in respect of the wave numbers shows that in contrast to the Kolmogorov–Obukhov classic spectrum, the high-frequency part of the spectrum contains a small amount of energy. This causes that the rate of transition of kinetic to thermal energy is rapidly reduced and this is also indicated by the small change of entropy in the calculations. This comment is also valid for the first

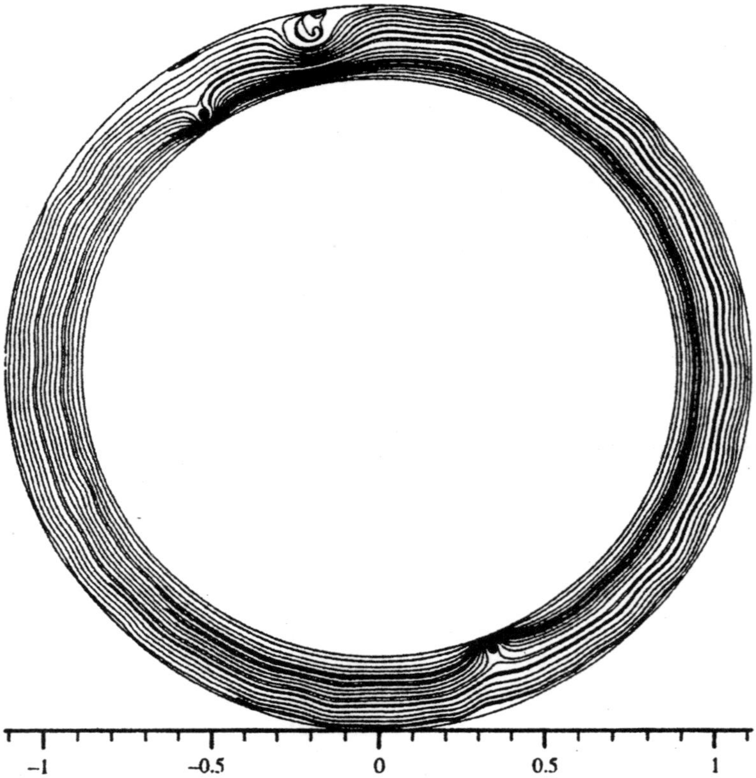

Fig.3.40. Instantaneous lines of current for variant 3 (the inner cylinder rotates in the anticlockwise direction; the outer cylinder is stationary) at moment t = 840.

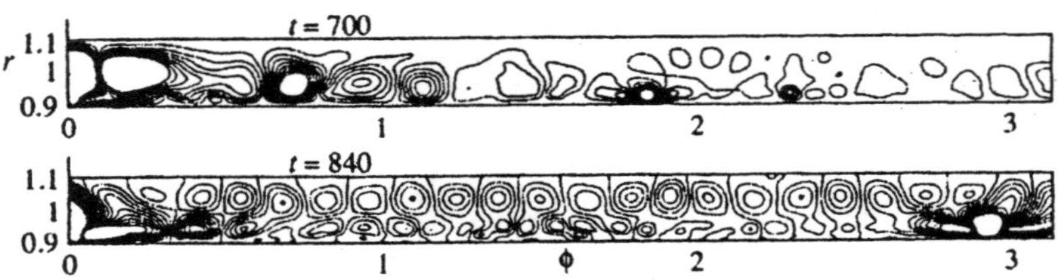

Fig.3.41. Correlators of velocity for variant 3 at times t = 700 and t = 840.

two variants.

Figure 3.43 shows the field of vorticity (velocity rotor) for the moment of time t = 840. The centre shows the enlarged fragment of the structure in the vicinity of a large vortex. Grey colour corresponds to low values of absolute vorticity, and lighter and dark

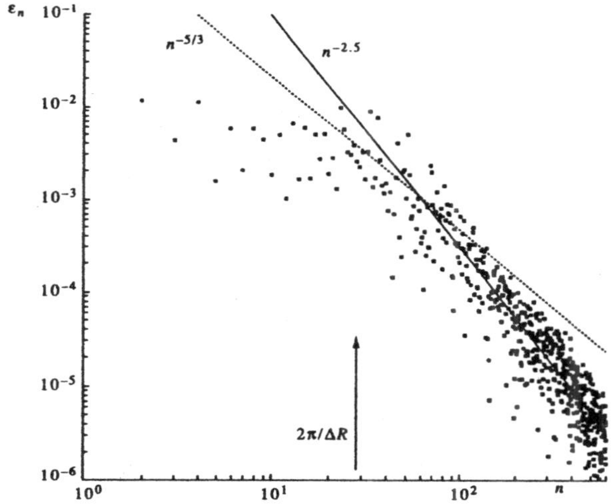

Fig.3.42. Spectrum of kinetic energy for variant 3 at time t = 840. The solid line shows the graph $n^{-5/2}$. The spectrum differs from the Kolmogorov–Obukhov spectrum $n^{-5/2}$

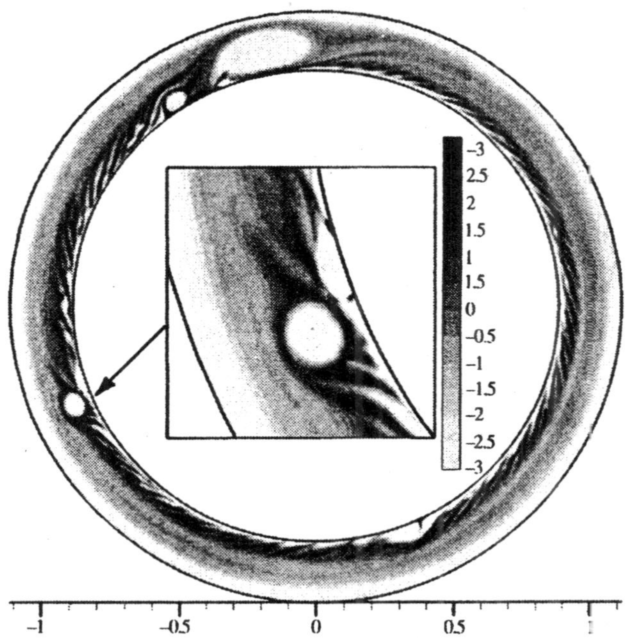

Fig.3.43. Structure of vorticity for variant 3. The centre shows the enlarged fragment of the structure in the vicinity of a circular vortex (grey colour refers to low values of absolute vorticity, lighter and darker tones to vorticity in the clockwise and anticlockwise directions, respectively), at t = 840.

tones indicate the vorticity in the clockwise and anticlockwise direction, respectively. It should be mentioned that the finer structure (vortices) concentrates in the vicinity of the large structures and in the vicinity of the inner cylinder.

The aim of our calculations was to show the evolution of the spectrum with the development of turbulence. This relates in particular to the third case in which the criterion of stability (3.19) or (3.21) was violated. However, in all these equations, the main large structures, examined in the calculations, had the scale comparable with the size of the gap in the cylinder. This is clearly indicated in Fig.3.35, 3.38 and 3.41. The high-frequency part of turbulence also appeared in our calculations but it was localised mainly in the region between the large vortices and the walls of the cylinders. In the remaining part, the flow was almost laminar with a small degree of vorticity, associated with the inner cylinder. For example, the third variant in the inner cylinder shows the structure of small-scale vortices (the high-frequency part of the turbulence spectrum).

The kinetic energy of turbulent motion is associated with large-scale vortex structures. It is assumed that the high-frequency part of the spectrum is generated in the nonlinear interaction of the large-scale structure with each other and with the walls. The role of the nonlinear interaction is not only in increasing the effective viscosity (more accurately speaking, this viscosity is not turbulent but 'dynamic', using the terminology introduced in [148]), greatly exceeding molecular viscosity. In contrast to molecular viscosity which does not depend on the scale, effective ('dynamic') viscosity is sensitive to the dimensions of the structures (vortices) of the flow. The larger structure results in a higher 'dynamic' viscosity. 'Dynamic' viscosity characterises the extent of nonlinear interaction. As indicated by the calculations, the nonlinear interactions are characterised by different degrees of development in different periods of the integration range. Therefore, we can discuss here the local nature of 'dynamic' viscosity.

As mentioned previously by Zel'dovich [58] and confirmed by our calculations, the structure of the flow between the cylinders is represented by alternating turbulence, i.e. the regions of almost laminar flow are replaced by regions of turbulence (large vortices). Possibly, the relationship between these regions is determined by the Reynolds number but isotropic turbulence does not appear at all.

The direction of our further investigations should be associated with increasing grid resolution. The second approach to the investigation of this process is associated with examination of the evolution of a large vortex on detailed grids.

3.5. NUMERICAL SIMULATION OF THE PROBLEMS OF DEVELOPMENT OF HYDRODYNAMIC INSTABILITIES AND TURBULENT MIXING

1. The Rayleigh–Taylor instability (RTI) arises in many nonstationary physical phenomena in which fluids (gases or other media), separated by the contact surface, are subjected to the effect of the volume force (for example, gravitational force), directed towards the lighter medium (or vacuum). The Richtmyer–Meshkov instability (RMI) develops in pulsed acceleration of the contact boundary – during the passage of a shock wave through the boundary. The increase of the intensity of perturbations as a result of the Kelvin–Helmholtz instability (KHI) takes place in the presence of a discontinuity in the velocity component, tangential in relation to the contact surface. In most cases, KHI accompanies the development of RIT and RMI. Jets of a heavier substance, penetrating into a lighter substance during the development of RTI or RMI, have the typical mushroom shape and this is also the consequence of the development of KHI. The mushroom shape becomes less evident with an increase of the ratio the densities and disappears in the limiting case in which there is vacuum on one side of the contact boundary.

These instabilities are reflected in a large number of varying natural phenomena and technological applications, for example, pulsed compression of targets of inertial confinement fusion, turbulence, shock wave experiments, explosive flows, compression of the magnetic field by a metallic liner, sustainment of plasma by the magnetic field, explosion of supernova and collapsing stellar cores in astrophysics, ionospheric phenomena in geophysics, formation of colonies of microorganisms in water, and other phenomena. Recently, special interest has been paid to the examination of RTI and RMI in connection with the problems of inertial confinement fusion [106]. In the actual conditions, different factors may influence greatly the development of RTI and RMI: surface tension, compressibility, viscosity, heat conductivity, diffusion, shock waves, the nonstationary nature of the acceleration regime, heterogeneities, physical properties of media, etc. [226].

It is generally recognised that in examining many complicated phenomena it is not possible to restrict examination to only experimental and analytical investigations. The rapid increase of the productivity of computers in the last couple of decades has stimulated the development of computing methods in the mechanics of fluids generally and for the examination of the problems of hydrodynamic instabilities, in particular. The application of numerical models makes it possible to eliminate secondary factors and concentrate on the main factor, which is sometimes difficult to achieve in the actual experiments [6,8,83,95,134,193,197,215,217]. In addition to this, the numerical model provides more complete information on the phenomena in comparison with the experiments. A large number of computing studies have been carried out with the two-dimensional numerical simulation of RTI (for example, [24-27,32,162,164,167,242]). In a number of cases, these results make it possible to interpret quite efficiently the natural experiments, but they do not explain many important details. The increase in the dimension of the phenomena (transition from two-dimensional to spatial flows) is accompanied by physical effects which in problems with a smaller size either do not occur or are manifested in a quantitatively different degree. Successes in the development of supercomputers have made it possible to transfer successfully to the third measurements – nonstationary spatial calculations are now more common [18,26,52,60,66,67,99,103,134,172,175,189,190,191, 215,235,240,243].

It should be mentioned that, at present, there are no universal computing methods in the computing dynamics of fluids which would be characterised by satisfactory accuracy and reliability. This method has its own region of application, its advantages and disadvantages. In the given case, the range of the investigated problems is restricted by the problems of development of hydrodynamic instabilities. These are difficult tasks – the behaviour of the contact boundary in nonlinear and turbulent stages is highly irregular and is subjected to strong deformation.

In the present work, direct numerical simulation was carried out to investigate the development of instabilities and turbulent mixing in mixed liquids (gases) at high Reynolds numbers and Prandtl numbers of the order of 1. The numerical method includes implicitly some nonlinear dissipative mechanism for the fluctuations of density, as in the case of kinetic energy. This 'rational' approach to direct numerical simulation is justified in [26].

2. The numerical simulation of RTI was carried out using the

model of a compressible non-viscous multicomponent medium. The initial system for constructing numerical schemes of calculations the total system of the Euler equations, written in the Cartesian coordinates in the divergent form. These are equations for the densities of n components of the medium

$$\frac{\partial(\rho\xi_i)}{\partial t}+\mathrm{div}(\rho\xi_i\mathbf{V})=0, \quad i=1,...,n,$$

the equations for three component of pulse density

$$\frac{\partial(\rho u)}{\partial t}+\mathrm{div}(\rho u\mathbf{V})=-\frac{\partial P}{\partial x},$$
$$\frac{\partial(\rho v)}{\partial t}+\mathrm{div}(\rho v\mathbf{V})=-\frac{\partial P}{\partial y},$$
$$\frac{\partial(\rho w)}{\partial t}+\mathrm{div}(\rho w\mathbf{V})=-\frac{\partial P}{\partial z}-\rho g,$$

and, finally, the equation for the density of total energy

$$\frac{\partial(\rho E)}{\partial t}+\mathrm{div}((\rho E+\rho)\mathbf{V})=-\rho g w.$$

Here t is the time, (x, y, z) are accordingly; $V = (u, v, w)$ is the velocity vector; ρ is density; $\xi_1,\xi_2,...,\xi_n$ are the mass concentrations of the components; $\sum_{i=0}^{n}\xi_i=1$, $E = e + \mathbf{V}^2/2$ is the specific total energy and e is the specific internal energy; $g > 0$ is the absolute value of the gravitational force. Here, we assume that the direction of the gravitational force is opposite to the direction of the axis z, directed upwards. To close the system of the equations, the equations of state for each component is required. In the present study, in all calculations, we use the equation of state of the ideal gas $P = (\gamma-1)\rho e$ for each component (any other equations of state, including tabulated ones, can be used). In all the calculations, presented below, the integration range has the form of a rectangle or right-angled parallelepiped. The boundary conditions of periodicity (or the conditions of corresponding symmetry) are specified at the lateral side of this formation, whereas the non-flow-through conditions are specified on the upper and lower boundaries of the region. No additional conditions are specified on the surface, separating the light and heavy medium. Although, no front tracking is carried out, as is the case in two- and three-dimensional problems in the numerical investigation of the development of RTI (for example, [162,164,175]). In the case of three-dimensional problems of RTI, this type of tracking is extremely difficult and relatively expensive, especially in the case of multimode problems (and cannot be used in the turbulent stage). In the present case, the contact surface is implicitly controlled by the variations of density (concentration) in each of node of the grid. Our mathematical model does not take into account the real viscosity and surface tension. Nevertheless, the very design of the scheme with the requirement for monotonic nature ensures some nonlinear dissipated mechanism which leads to the decay of shortwave harmonics. In other words, the harmonics with the wavelength smaller than some effective wavelength λ_{cr} are extinguished. This is confirmed by our calculations. Evidently, λ_{cr} is approximately equal to several steps of the computing finite-different grid. Youngs use the same approach in his study of turbulent mixing [243]. In order to extinguish shortwave harmonics and ensure the possibility of stable counting, Li was forced to introduce the finite viscosity in his three-dimensional calculations [189,190].

3. The set of the programmes, prepared for this application, is used actively in the numerical modelling of various problems, associated with the examination of hydrodynamic instabilities and turbulent mixing. We shall mention a number of investigations carried out by the authors together with N.A. Inogamov, A.Yu. Dem'yanov, S.I. Abarzhi and other colleagues in the last couple of years [20,53,54,60,66–68,102,103,178,179,208,210]. As an illustration, we shall mention only the problem of gravitational turbulent mixing. In gravitational turbulent mixing, pure heavy and light substances are separated by the turbulent mixing zone (TMZ). In [53,60,103, 210] investigations were carried out into the field of velocities, acceleration, densities and other parameters of the TMZ, and information is obtained on the statistical properties of the zone.

The urgent problem of turbulent mixing has been investigated in a number of investigations [14,64,70,98,130,135,138,164,207,217, 225,243]. Usually, special attention is given to the vertical profile $\rho(z)=\langle\rho\rangle_{\perp}$, obtained by averaging in respect of the transverse coordinate (sign $\perp$). It is used to determine the mixing coefficients

α_+ and asymmetry coefficients As = α_+/α_-.[1] In transverse averaging, the important numerical information on the dominant structures, responsible for the mixing dynamics, is lost. In the studies in which investigations were carried out into the spectra and transverse structures, the authors confined themselves firstly to the comment on the enlargement of these structures[2] and, secondly, to the examination of the short-scale asymptotics of the spectrum. In the current stage, qualitative statements are insufficient. In the present work, we proposed, firstly, the approach for quantitative description of enlargement. It is based on the construction of the spectre of the quantities $f = \{\rho, u, w, p\}$ and self-similar variables $\tilde{f}_{\hat{n}}$ in respect of the numerical data, where ρ, u, w, ρ is the density, the horizontal and vertical velocity, pressure, respectively. Secondly, investigation of the short scale asymptotics, important for explaining the problems of the fineness (dispersion) of mixing and Kolmogorov dissipation, is supplemented by the examination of the long-wave region. This is the most important problem, because the long-wave amplitudes in particular determine the rate of expansion of the mixing layer and, consequently, the coefficients $\alpha_\pm$.

The characteristic calculation of the two-dimensional problem on a detailed grid (1600 × 1600) is shown in Fig. 3.44. In this problem, the density of the heavy substance $\rho_h = 1$ (the dark colour in Fig. 3.44a–c), and the light substance $\rho_l = 0.33$ (light colour). The ratio of the planes $\mu = \rho_l/\rho_h = 1/3$. The Attwood number is At = $(\rho_h+\rho_l)/(\rho_h+\rho_l) = 0.5$. The intermediate shades correspond to intermediate density. The force of gravity g is directed downwards. The dimensions of the region are $L_x \times L_y$. For the single normalisation of units, it was assumed that $L_x = 2\pi$, $g = 1$, and in the given case $L_x = L_y$. At $t = 0$, we specified the short scale perturbation in the form of the subsurface field of velocities:

$$\mathbf{v} = -\nabla\varphi, \quad \varphi = \sum \varphi_n, \tag{3.23}$$

---

[1]The mixing layer in the incompressible medium, stationary at infinity, is restricted at the top and bottom by fronts or mixing boundaries propagating in the heavy (density ρ_h) and light (ρ_l) liquids, respectively. The total thickness of the layer h consists of the sum $h_+ + h_-$ of displacements of the upper (+) and lower (−) fronts from the plane of the non-perturbed position of the boundary of the media. In self-similar case, we write $h_\pm = \alpha_\pm At\, gt^2$, where $At = (1-\mu)/(1+\mu)$, $\mu = \rho_l/\rho_h$ is the ratio of densities, g is acceleration. This leads to the definition of the coefficients α_+ and As.

[2]This phenomenon was predicted theoretically [64,136]. In [136], the authors derived a self-similar equation $\bar{\lambda} \sim t^2$ determining the scaling of the transverse dimensions of dominant structures controlling the accerelation of the upper front.

$$\varphi_n = \text{sign}\, z\left(a_n^0 \cos nx + b_n^0 \sin nx\right)\exp(-n\,|\,z)/n,$$

where n is the number of harmonics, $n = 50 \div 100$. At $t = 2 \div 3/\sqrt{\text{At}}$, the transition stage is completed and the self-similar regime is established. The existence of the transition stage (Fig.3.44d) is determined by the non-self-similar nature of the initial data.

It is interesting to investigate the horizontal–vertical structure of the mixing layer or the 'internal mechanics' of Rayleigh–Taylor turbulence. There is one horizontal layer of the cells. The horizontal and vertical dimensions of the cell are of the same order of magnitude. It is interesting to look 'inside' the cell. The cell is rather complicated. We shall present description of the cell in the 2*D*-case, following from direct numerical modelling.

Tens of numerical calculations have been analysed. The ratio of density μ, the density of the grid N_x, (the number of nodes in the horizontal $10^2 \div 10^3$) and geometrical characteristics of the

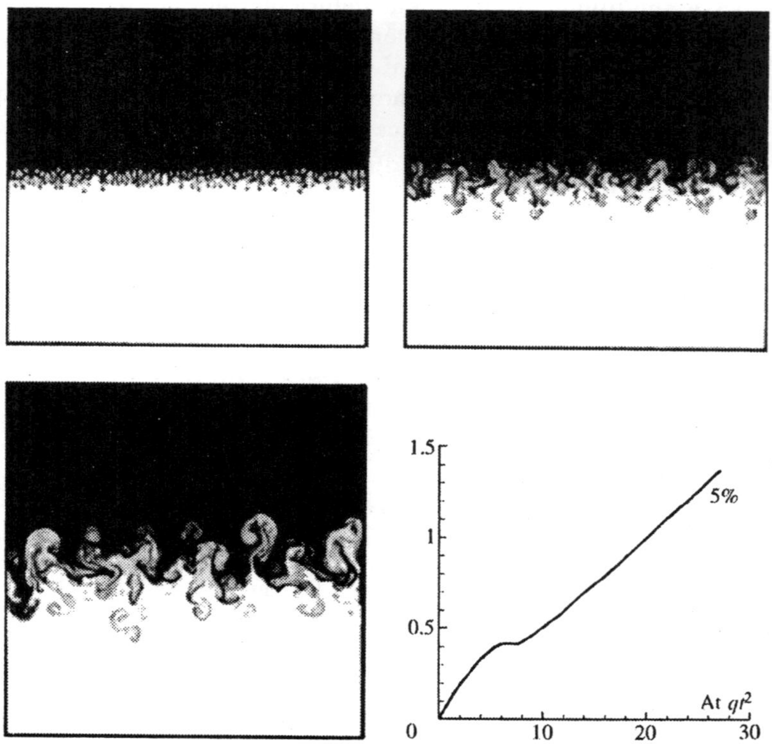

Fig.3.44. Numerical modelling of spontaneous gravitational turbulent mixing: a–c) distribution of density for moments t = 2, 4, 6; d) height of penetration (h_+) of light substance into heavy one in relation to At gt^2.

calculation configuration were varied. This characteristic is given by two parameters $(L_h+L_l)/L_x$ and L_l/L_h, where $L_x = 2\pi$ is the width of the calculation region on the horizontal coordinate x, and L_h and L_l are the initial thicknesses of the heavy and light layers. We use the values of (i) $1-\mu \sim 0.1$, belonging to Bussinesq's case of symmetric $h_+ = h_-$ expansion, (ii) mean ($\mu = 1/2 \div 1/3$) and (iii) low ($\mu = 1/10$) ratios μ. The upper (in the heavy liquid) and lower (in the light liquid) boundaries of the mixing layer are dencted by h_+ and h_-.

The beginning and end of the self-similar stage of mixing was determined in [68]. From the calculation region $L_x \times (L_l + L_h)$ we cut out a square or a rectangle $\Delta L_x \times \Delta L_z$ with a width of $(2 \div 4)$ h_+ and the height $-h_- < z < h_+$, which included the investigated self-similar 'vortex cell'. The point is that in the self modelling stage, the cells are still relatively small in comparison with the width L_x (at least 4–5 cells in L_x to exclude the effect of the side boundaries and the set of the long-wave statistics). It was required to increase the image of the cell for detailed analysis (it is irrational to consider the entire calculation region). In addition to this, the cells grow (inverse cascade). Therefore, without scaling on h_+ it is difficult to compare the self-similar cells relating to different moments of time. The centre of the rectangle $\Delta L_x \times \Delta L_z$ was selected arbitrarily on the section L_x on the level $z = 0$ (the initial position of the contact boundary). Attention was given to the fields ρ, the volume fraction of the light liquid $c = (\rho_h - \rho)/(\rho_h - \rho_l)$, the vertical velocity w, vorticity $\omega = \text{rot } \mathbf{v}$, enstrophy ω^2 and vertical acceleration of the liquid particles $Dw/Dt = w_t + (\mathbf{v}\nabla)w$.

The typical cells are shown in Fig. 3.45, $\mu = 1/3$, $N_x \times N_z = 1600 \times 1600$. In the calculation field at moments $t = 3$, 4, and 6 squares $2.5\ h_+ \times 2.5\ h_+$, $h_+ = 0.05/(1 + \mu)\ g_A t^2$, $g_A = (1-\mu)g$ were cut out. They were selected randomly with the centre on the level $z = 0$. We shall examine Fig.3.45 in greater detail. The powerful jets of the heavy liquid travel downwards. They formed after merger in the previous stages of the inverse cascade and represent conglomerates of several jets. The size of of the conglomerate is of the order of $h_+ \times h_+$.

The conglomerates of the jets are divided by large bubbles with the size of $h_+ \times h_+$ (on the right, $t = 3$; the left, $t = 4$; on the side $t = 6$). They 'loose' in the position on the vertical to the 'old' bubbles. On the other hand, they are large, travel at high velocity and (in contrast to the 'old' bubbles) are efficiently bonded with the main bulk of the light liquid (wide mouth). Consequently, there are prospects for their inflation and acceleration. The 'old' bubbles, losing the bond with the main bulk, transform to floating vortex pairs. These pairs 'hover' at the front h_+. It is important to note that these pairs or dipoles are floating (like thermics) and by this they differ from the dipoles in a homogeneous fluid (in a homogeneous case there is no baroclinic generation of vorticity).

The bubbles suck in the light liquid. This is caused by the difference in the pressures Δp outside the mouth of the bubble and inside the bubble. The value Δp may be high. Therefore, the flows of the light liquid are relatively fast. The velocity of the light liquid is so high that its dynamic pressure drags in the fragments of the

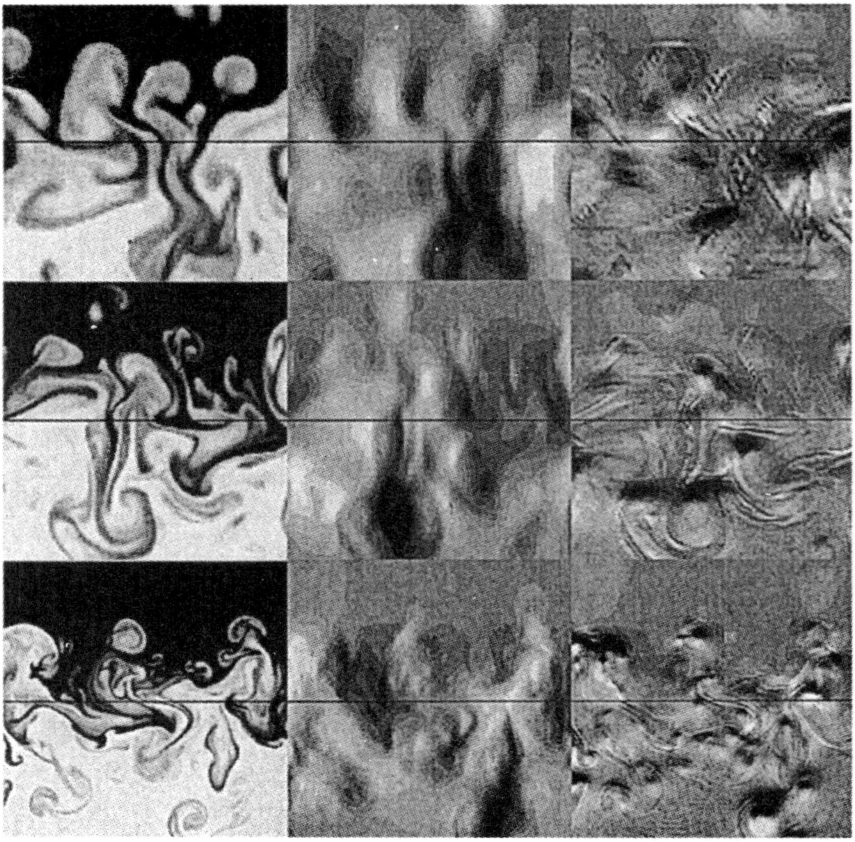

Fig.3.45. Characteristic structure of the 'mean cell' in the mixing layer. Horizontal straight line – level $z = 0$. Upper row – $t = 3$, central row $t = 4$, lower row $t = 6$. Left – density ρ. Black – $\rho_h = 1$, white – $\rho_l = 1/3$. Centre – vertical velocity w. Black – downward motion $w < -2.5\ h_+$, white – upwards $w > 2.5\ h_+$. Right – vertical acceleration of liquid particles Dw/Dt. Black – negative acceleration $Dw/Dt < -3g_\Lambda$, white – positive acceleration $Dw/Dt > -3g_\Lambda$. The z axis directed upwards.

jets of the heavy liquid into the mouth of the bubble. The fragments of the heavy jets in the mouth complicated supply of the light liquid to the bubble. The competition between the breaks in the light liquid into the bubble and the process of extending of the mouth continues for some period of time. The closure of the mouth is the reason for the formation of vortex pairs.

High absolute values of the vertical acceleration $Dw/Dt = \frac{\partial w/\partial t}{w_t+(\mathbf{v}\nabla)w}$ are obtained (i) in the 'impact' of the heavy liquid, (ii) in the floating of the bubbles upwards and in the 'impact' of the light of jets, and (iii) in the 'sucking' of the light liquid into the bubbles (high positive acceleration in the mass of the bubble). The case (i) will be investigated. The heavy liquid is accelerated in stages, descending to large depths, $\sim 1.5\ h_+$. This results in the formation of high velocities, 1.5–2.5 times higher than the speed of the front $\dot{h}_-$. If the rate of acceleration of the heavy substance is low, deceleration of the heavy substance to the velocity of the front takes place relatively rapidly ('impact'). This is associated with considerable deceleration of the heavy substance. Another interesting detail, associated with the 'mushroom' at the tip of the heavy jets, is the marked acceleration of the light liquid downwards behind movement of the heavy substance. This takes place in the region behind the cap of the mushroom in Fig.3.45.

If the separated bubbles transform to floating vortex pairs (light thermics), then the mushroom-shaped formations, which complete the heavy jets, represent vortex pressing the form of sinking thermics. In Fig 3.45, acceleration Dw/Dt is compared with the Archimedes acceleration g_A. The lifting of the light thermals is associated with the positive and negative acceleration of the order of g_A. It is important to stress that the values of g_A are an order of magnitude higher than the mean acceleration $\sim \ddot{h}_+ \approx 2\alpha + g_A/(1+\mu)$, $\alpha_+ \approx 0.05$. Consequently, it should be concluded that the pattern of the flow in the central large-scale cell of the single-layer chain is relatively nontrivial. This relates to the comparison of the actual pattern in a description of turbulent diffusion on the basis of the semi-phenomenological approach. In turbulent diffusion, the acceleration should be of the order of $\sim \ddot{h}_+$. Of course, it should be taken into account that the range of high acceleration and deceleration occupies a relatively small fraction of the mixing region.

4. We shall examine the problem of the characteristic or mean

transverse scale $\langle\lambda\rangle$ in two-dimensional geometry. Figure 3.45 shows that the scale $\langle\lambda\rangle$, affecting the rate of expansion of the turbulent layer, is relatively large ($\sim 2h_+$). Of course, in addition to the large-scale components, the pattern of the flow has a number of fine components, smaller than h_+. The transverse Fourier spectrum is shown in Fig. 3.46 (for the vertical component of the speed, calculated in the section $z = 0$). For the arbitrary function $f(x,...)$ its spectrum $f_n(...)$ is calculated from the equation:

$$f_n = \sqrt{a_n^2 + b_n^2}, \quad a_n = \pi^{-1}\int_0^{2\pi} f\cos nx dx, \quad b_n = \pi^{-1}\int_0^{2\pi} f\sin nx dx,$$

$$f(x...) = \sum (a_n \cos nx + b_n \sin nx).$$

The values $nt^2 = 100$, 200 and 300, correspond to the values of $\langle\lambda\rangle/h_+ = 2\pi/\alpha_+$ At $gnt^2 = 2.5$, 1.3 and 0.8. The harmonics $nt^2 = 100\div300$ have maximum amplitudes.

The self-similar spectrum is computed using the equation of the following type:

$$\tilde{p}_{\hat{n}}(\hat{z}) = \frac{\rho_n(z,t)}{\rho_h t}, \quad \bar{w}_{\hat{n}} = \frac{w_n(z,t)}{t^2}, \quad \tilde{p}_{\hat{n}}(\hat{z}) = \frac{p_n(z,t)}{\rho_h t^3}, \quad \tilde{H}_{\hat{n}} = \frac{H_n(t)}{t^3}. \quad (3.24)$$

Conditionally, it may be said that they are obtained from the equations of self-similar formulation in the coordinate their presentation by dividing by $\sqrt{\bar{n}}, \bar{n} \propto 1/t^2$. This is associated with the delta-correlation of the Fourier images of the modelling functions (discreteness of their computer presentation) and with the special nature of integration of such delta-correlated functions (the placement of the linear differential dk by the root differential $\sqrt{dk}$, cee [66,177]). The relationships (3.2) are important. For example, they give the nontrivial conclusion according to which the maximum spectral amplitude of the fluctuations of pressure increases in proportion with t^3.

The longwave section of the Fourier distribution decays slowly. This is associated with the scatter of the values of $\langle\lambda\rangle$. In [68], the authors obtained theoretically the exponential angle of inclination of $w_n \propto \sqrt{n}$ of the longwave section of the distribution. The calculations carried out on the maximum grids $N_x \times N_z$, were used in particular for the verification of this law. The appropriate straight line is plotted in Fig. 3.46. This straight line approximates quite satisfactorily the calculated data.

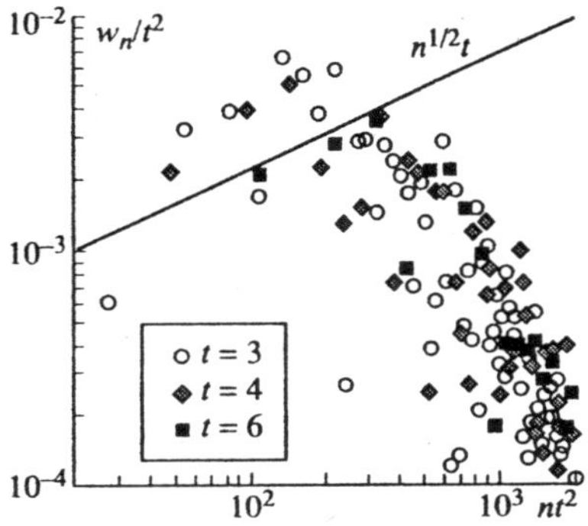

Fig.3.46. Fourier spectrum w_n of vertical velocity w (x, z = 0, t) on horizontal coordinate x for three moments of time.

5. On the basis of calculations of a large number of variants we investigated the stimulation of mixing by longwave perturbations (these calculations were carried by A.Yu. Dem'yanov using the method of large particles [32]). The initial perturbation was specified in the form of speed and represented the sum of the short scale and wide range perturbations. The wide-range perturbation, homogeneous in respect of the wave numbers, did not introduce the separated scale and, consequently, retains the quadratic self-similar nature (possibly with different coefficients α_+). We shall write the initial wide-range perturbations also in the form of expansion in respect of harmonics (3.23). Let it be that:

$$a_n^0 = (w_{sat})_n \cdot \xi \cdot \varepsilon_{wb}, \quad b_n^0 = (w_{sat})_n \cdot \xi_n' \cdot \varepsilon_{wb}, \tag{3.25}$$

where ξ_n, ξ'_n are independent random numbers, uniformly distributed in the range [–1, 1]. The initial field (3.23), (3.25) is determined by three factors. Firstly, it is the rate of saturation $(w_{sat})_n = F\sqrt{(1 \pm \mu) g/n}$, $F \approx 0.6$ for the two-dimensional case and of the order of 1 for the three-dimensional case ([66]), secondly, it is the random factor and, thirdly, dimensionless parameter ε_{wb}. This perturbation does not have a specific scale. It is defined by a single parameter, amplitude ε_{wb}, of random wide-range noise. Figure 3.47 shows the results of the calculations for different values of ε_{wb}. It may be seen that the stimulation may greatly increase the mixing coefficient α_+. These calculations indicate the existence of the threshold $(\varepsilon_{wb})_{thr}$ in approximately 10 pro mille. At amplitudes smaller than the threshold, mixing takes place in the spontaneous regime, and the effect of noise can be ignored. In the case of super-threshold amplitudes, the flow is rearranged. The spectral 'hillock' is displaced to the longwave side. This reflects the enhancement

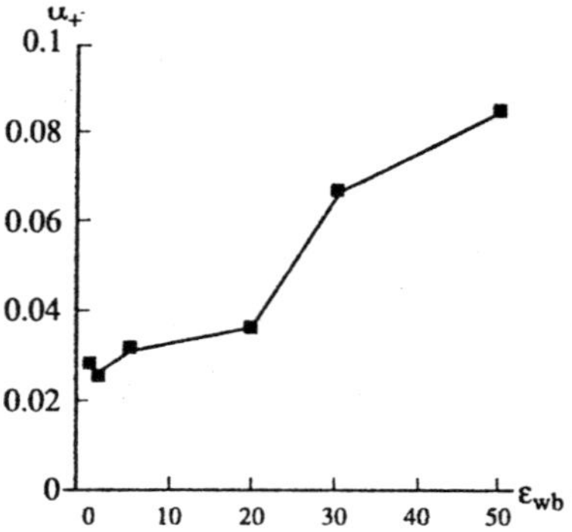

Fig.3.47. Dependence of the mixing coefficient A_+ (t = 5) on amplitude ε_{wb} in pro mille.

of the sub-harmonics. The increase of the longwave section is associated with the intensification of mixing, reflected in the increase of coefficient α_+. If in the spontaneous process the origin of sub-harmonics is associated with the central hillock in the spectrum, then in the case of stimulation, part of the sub-harmonics is taken from the wide-range noise, present in the initial stage. The threshold amplitude is high and very important. For example, it is not rational to decrease the amplitude of the noise by means of some expensive technological measures (of the type of polishing) below the threshold amplitude.

The side boundaries of the calculation region restrict the horizontal size of the large structures (clusters), complicate mixing and, consequently, decrease the value of the turbulent mixing coefficient. In the calculations, this 'constriction' effect is especially evident in the modelling of evolution of short-scale perturbations after reaching the time $t = 6 \div 7/\sqrt{\mathrm{At}}$. N.A. Inogamov analysed the problem of the asymptotics of Rayleigh–Taylor mixing at long times and extended calculation regions (pipes). He obtained the 'constricted' asymptotics $h_+ \propto t^{2/5}$, instead of the free asymptotics $h_+ \propto t^2$. We assume that in these calculations, the moment $t = 6 \div 7/\sqrt{\mathrm{At}}$ corresponds to the start of inhibition by the side boundaries and to the rearrangement of the flow with the free asymptotics to 'constricted' asymptotics.

6. It is important to mention again the following most important moments of these investigations.

1. It has been established that in the narrow jets the vertical velocity is several times higher than the speed of expansion of the mixing zone $\dot{h}_+$ in the direction of the heavy liquid.

2. Various methods (correlators, spectra) have been used to analyse the characteristic scales $\langle\lambda\rangle$ of the structures on the

horizontal. It has been established that the typical scales are relatively large $\langle\lambda\rangle \approx h_+$ and the scatter of the scales is wide.

3. This is caused by the slow decay of the longwave tail of the Fourier spectrum in respect of wave numbers n. For example, the spectrum of vertical velocity at $n \to 0$ decreases only by $n \propto \sqrt{n}$ (see the straight line in Fig. 3.46).

4. The exponents of the self-similar scaling of the Fourier amplitudes have been determined. In contrast to the scaling of the wave number $n \to nt^2$, this is a relatively non-trivial problem, because the dependences $w(x,t)$, $\rho(x,t)$, $p(x,t)$ in relation to the horizontal coordinate x are random-periodic functions (the vertical coordinate is assumed to be fixed inside the mixing layer). In particular, it is assumed that $w(n,t)$ is the amplitude of the Fourier harmonics $n = L/\lambda_n$, where L is the width in the direction of the horizontal line of the mixing region, λ_n is the wavelength of the n-th harmonic. Consequently, the self-similar Fourier amplitude is not specified by the equation $w(nt^2,t)/t$ as it appears at first sight, and in the case of the two-dimensional problem we can use the equation $w(nt^2,t)t^2$. After the establishment of the self-similar regime, the dependence on time is no longer valid and the spectrum $w(nt^2,t)t^2$ become stationary (Fig.3.46) in the self-similar variables.

5. The stimulation of mixing by longwave noise has been investigated. The results show that in the case of typical homogeneous noise with no specified scale, the self-similarity $h \propto t^2$ is retained. The threshold amplitude of random wide-range noise was determined. Below this amplitude, this noise can be ignored, Fig. 3.47. The inhibition of mixing by the side boundaries has been investigated. The effects of stimulation and inhibition strongly influence the mixing coefficient α_+ by increasing and decreasing it, respectively.

3.6. NUMERICAL MODELLING OF THE PROCESSES OF PROPAGATION OF AN IMPURITY IN THE ATMOSPHERE FROM A LARGE-SCALE SOURCE

The authors of [76] investigated numerically the structure of of convective columns formed during a fire. The maximum height of the convective columns and the height of spreading for different values of energy generation and the size of the region have been determined. The dependence of the results on the constants of the algebraic model of turbulence has been investigated. The results of calculations have been compared with the data obtained in laboratory experiments. Modelling calculations have been conducted in the three-dimensional formulation.

1. In the last decade, the effort of a number of scientific teams has been directed to the development of a mathematical model of large-scale fire, primarily a city fire. In the monograph [92], the authors investigated and classified the main reasons for the formation of such fires, determined the definition of the large-scale fire, differing from the commercial industrial (or forest) fire by the high intensity of combustion and/or the rate of spreading. The authors of the monograph [42] discuss the possible global climatic consequences of the fires as a result of the ejection into the atmosphere of aerosol particles in the form of soot and ash. Estimates of the reserves of the 'potential fuel' of such a fire and the estimates of the mass of the smoke in different combustion conditions are presented. Various aspects of the modelling of forest fires have been investigated [50]. The authors of [49,78,229] proposed calculation models, based on the systems of the Navier–Stokes equations with constant transfer coefficients. Subsequently, the authors of [47,48,97] used different variants of the algebraic models of turbulence. In [212], the authors presented the results of a fire in Hamburg and the results of numerical modelling of this fire which took into account the actual values of energy generation for the case of a city fire. In the same study, the authors presented the data on an oil fire in Long Beach (1958) which had a radius of 0.5 km and was regarded as a point source. In [97,141] in numerical and laboratory experiments it was shown that at a radius of the region of fire greater than 8 km, the equations for the point source are not valid for the calculation of the height of hovering and spreading of the convective columns. The region of the fire cannot be regarded as a point source.

This section of the book is concerned with a more detailed examination of this problem. In particular, we match more accurately in a wide range the controlling parameters of the numerical and laboratory experiments (in particular, the value of energy generation in the region of the fire). Calculations were carried out with different values of the constant in the equation for the path length of mixing $K = 0.125$ (according to the results obtained in [47,48] $K = 0.2$ [212] ad $K = 0.4$). The results of the comparison of the data obtained in laboratory experiments and the calculations show that the best agreement is recorded in the case with $K = 0.2$. The possibilities of using, in the calculations, the differential models of turbulence on the basis of the procedure, proposed and used in [74,75,77], have been discussed. As shown previously in [48], the phase transitions, determined by the presence of moisture in the atmosphere, greatly affect the parameters of lifting, hovering and emission of the aerosol into the atmosphere. In the present study (as in [97]) these effects are not taken into account. This is explained, on the one side, by the need to investigate the main relationships governing the flow of the model of the dry atmosphere and, on the other side, by the orientation of the calculations to the data of laboratory experiments (instead of the full-scale experiments), in which the moisture content of the atmosphere was not taken into account.

It should be mentioned that the investigated physical phenomena of the large-scale fire has a number of important special features which are not reproduced when using two-dimensional (flat or axisymmetric) models. In this work, we present several results of the calculations of the region of fire in the three-dimensional formulation of the problem (including the case of side wind) which have not as yet been carried out. This is associated with the high requirements imposed by the problem on the resolution capacity of the algorithm both in the immediate vicinity of the region of fire and in the region of convective rising flow and in the region of hovering and spreading of the convective columns.

2. In the present section, we investigate the system of equations of turbulent motion of compressible gas, written in relation to the equations averaged out according to Favre (with the exception of density and pressure) [159] and supplemented by the algebraic model of turbulence [97]:

$$\frac{\partial\rho}{\partial t}+\nabla\cdot(\rho\mathbf{V})=0,$$

$$\frac{\partial\rho\mathbf{V}}{\partial t}+\nabla\cdot(\rho\mathbf{V}\mathbf{V}+p\mathbf{I}-\mathbf{R})=\rho g,$$

$$\frac{\partial\rho E}{\partial t}+\nabla\cdot\left(\rho\mathbf{V}\left(E+\frac{p}{\rho}\right)-\mathbf{R}\cdot\mathbf{V}+\mathbf{q}\right)=\rho g\cdot\mathbf{V}.$$

The equations are closed on the basis of the Bussinesq approximation according to which the tensor of the Reynolds stresses is proportional to the tensor of the mean strain rates:

$$\mathbf{R}=-\frac{2}{3}\mu_t\nabla\mathbf{V}\mathbf{I}+2\mu_t\mathbf{S},$$

where **R** is the tensor of the Reynolds stresses; **S** is the tensor of the strain rate, **I** is the unit tensor, and also by the relationship for the heat flow:

$$\mathbf{q} = -\frac{\mu_t}{\sigma_t}\nabla e, \quad \sigma_t = 0.9.$$

We shall use the equation of state $p = (\gamma-1)\rho e$ with the parameter of the adiabate $\gamma = 1.4$. We shall use the cylindrical system of the coordinates r, φ, z, and the plane $z = 0$ corresponds to the surface of the earth, and the z axis is directed in the vertical direction.

The values of the coefficient of turbulent transfer are calculated in accordance with the algebraic model [97] ((this model was also used in [47,48]) and according to this model, in the axisymmetric case turbulent viscosity μ_t is determined from the equation:

$$\mu_t = \frac{\rho l^2}{2}\left(\left[\frac{\partial v}{\partial z}+\frac{\partial u}{\partial r}\right]^2 + 2\left[\left(\frac{\partial v}{\partial r}\right)^2+\left(\frac{v}{r}\right)^2\right]\right)^{1/2},$$

$$l = K\left(\left[v^2+u^2\right]^{1/2} B^{-1/2} + B^{1/2}\left[\left(\frac{\partial^2 v}{\partial r^2}+\frac{1}{r}\frac{\partial v}{\partial r}-\frac{v}{r^2}\right)^2 + \right.\right.$$

$$\left.\left. + \left(\frac{\partial^2 v}{\partial z^2}\right)^2 + \left(\frac{\partial^2 u}{\partial r^2}+\frac{1}{r}\frac{\partial u}{\partial r}\right)^2 + \left(\frac{\partial^2 u}{\partial z^2}\right)^2\right]^{-1/2}\right),$$

$$B = \left(\frac{\partial v}{\partial r}\right)^2 + \left(\frac{\partial v}{\partial z}\right)^2 + \left(\frac{v}{r}\right)^2 + \left(\frac{\partial u}{\partial r}\right)^2 + \left(\frac{\partial u}{\partial z}\right)^2,$$

In these equations, l is the path length of the mixing, K is the empirical constant. In the numerical calculations, the Prandtl number was assumed to be Pr = 1, and the value of K was varied. (It should be mentioned that the system of the equations does not contain the coefficient of dynamic viscosity μ, because in the investigated problem, the relationship $\mu_t \gg \mu$ is valid).

As in [97], the diffusion equation for the impurity concentration was not investigated. The impurity was simulated by the introduction of a specific number of particles on the external side of the region of fire by means of the fixed periods of time and by the calculation of the movement of these particles on the basis of the kinematic equations of motion (for more details see [75]). According to the experience obtained in previous investigations, this approximation of the 'passive' impurity makes it possible to describe with sufficient accuracy the propagation of the emissions of smoke and aerosol in the case of a large-scale fire.

The calculation region in the plane r, z has the form of a rectangle with the dimensions $2R_0$ (in respect of r) and $H = 24$ km (in respect of z). The calculations were carried out considering the standard atmosphere [97]. At the initial moment of time $t = 0$ the static pressure in the entire calculation region is assumed to be equal to the atmospheric pressure at the height z above the level of the fire. The source of energy generation was specified in the form of a cylinder with a radius of R_0 and the height h of 100 m. The intensity of the volume source Q^* varied with time in accordance with a linear law to the value Q^*_{max} which was obtained after 30 min as in the case of studies in [47,48,229]. The intensity of the source was assumed to be constant. It should be mentioned that the variation of the time required by the source to reach the maximum power ($0 < t < 30$ min) shows that the height of hovering and spreading of the convective column is almost independently of this time.

3. The calculations are carried out using the TVD scheme of the of the second or third (for the three-dimensional formulation) problem of the approximation order in respect of spatial variables (with the exception of the points of the extreme of the characteristic variables) and the first order in respect of time for the Euler part of the equations [77]. For the approximation of the processes of diffusion transfer we use the scheme of the first order in respect of time and of the second order in respect of space.

The given method of linearisation of the hyperbolic part of the equations [159] ensure the three-diagonal form of the block matrix of the coefficient, corresponding to each coordinate direction. In the calculation of the Jacobi matrix – the vectors of the source of the Reynolds part of the equations (Reynolds stresses, the heat flows and the dissipative function), we carried out the linearisation of only the differentiation operators providing a contribution to the main diagonal of the matrix of the coefficient, and the coefficients of viscosity were taken from the lower time layer.

In order to decrease the volume of the calculations when solving the linearised system of the equations in each step in respect of time, we use the method of approximate factorisation, consisting of the representation of the operator in the left part of the equations in the form of the products of two one-dimensional operators.

In the axisymmetric formulation, the calculations were carried out using the 81 × 121 stationary grid, at the maximum Courant number of 0.75 for the flow field. The Courant number is restricted by the requirement on accuracy, and a further increase of the number, according to the calculations, results in a rapid increase of the dissipative error of the system. The grid became denser in the direction of the surface of the earth on the z axis. In particular, the calculation region was divided into four parts – the uniform grid 40 × 6 was specified in the region, the 40 × 115 grid was specified above the region. This grid was uniform in respect of the z axis and extended in respect of r to the given value of the upper boundary; on the right of the region, we specified the 40 × 6 grid, uniform in respect of z and extended in respect of r to the value of the outer boundary; on the right and then at the top of the region which specified the 40 × 115 grid, extended in respect of r and z. In three-dimensional formulation, the difference grid contained 50 × 50 × 100 nodes. The boundary conditions in the calculations were specified in accordance with [97]. The accuracy of the calculations was inspected on the basis of the balance of mass and energy. In addition to this, the calculations were carried out on the grids with different numbers of the nodes, and the previously presented dimensions of the grid were selected on the basis of the need to ensure accurate calculations and the independence of the results of the calculations on the definition of the outer boundaries. The calculations were carried out in a Pentium Pro 150 personal computer (with a 32 MB operating memory). The computing time of 1 h of physical time was equal to approximately 30 h of processor time (for the axisymmetric formulation).

4. One of the main aims of the present investigation was the determination of the possibility of obtaining an agreement between the calculated results and the data recorded in laboratory experiments [77,159]. This comparison would make it possible to determine the permissible boundaries of the application of the algebraic models of turbulence and the need for transition to differential models. Consequently, calculations were carried out for different values of the specified empirical constant in the equation for the mixing length, included in the given algebraic model. Previously, it was shown that the results of calculations in the problem of the rising of a single thermic in a stratified atmosphere

[74] strongly dependent on the value of this constant. Table 3.4 shows the values of the controlling parameters of the problem which were varied in the course of the calculations. The given values of the energy generation corresponded to the data obtained in laboratory experiments [77,159] and to the data obtained by other authors. For example, in [49] Q = 0.23 MW/m^2 at a lower value of t^*= 300 s, and in [48,229] Q = 0.05 MW/m^2 at t^*= 1800 s, and the authors of [229] also investigated the cases with Q = 0.1 MW/m^2 and Q = 0.025 MW/m^2. The latter value t^*= 1800 s is in good agreement with the estimates of the time to the establishment of the maximum energy generation in a large city fire. At the same time, the value Q = 0.05 MW/m^2, corresponding to complete combustion for 1 hour of wood materials with the area density of loading of 10 kg/m^2 (at the heat forming capacity of the inflammable material $Q = 19.6 \times 10^6$ J/kg) is obviously too low for a large city fire. For example, according to the data presented in [212] on a fire in Hamburg in the July of 1943, the power is estimated as 1.7×10^6 MW for an area of approximately 12 km^2. This gives the power of the source of Q = 0.14 MW/m^2 for the radius of the zone of $R \simeq 2$ km. It should also be mentioned that according to the same data, the wind at the moment of formation of the fire was only slight and the moisture content of the atmosphere was low. Other parameters were also similar to the parameters used in modelling: the tropopause on the level of 11.5 km, the temperature in the stratosphere 216 K, the temperature at the surface of the earth 288.1 K. The following results were obtained: the height of the smoke column 12 km, the region with the highest content of the smoke several hours after the start of the fire 8–9 km. These data are in good agreement with the results of the calculations [212]. In addition to this, the calculations gave the maximum vertical velocity of the convective flow of 68 m/s.

Table 3.4

Amount of energy generated in the region Q = 0.073 MW/m^2						
R_0 = 5 km		R_0 = 12 km				R_0 = 22 km
K = 0.125	K = 0.2	K = 0	K = 0.125	K = 0.2	K = 0.4	K = 0.1

Amount of energy generated in the region Q = 0.24 MW/m^2				
R_0 = 1 km	R_0 = 2 km	R_0 = 5 km	R_0 = 12 km	R_0 = 22 km

The main stages of the modelled gas-dynamic flow will be investigated. Attention should be given to the fact that the initial stage of formation of the flow is physically similar to the problem of surface thermics [74]. The periphery of the region is characterised by the start of development of the wall flow, directed into the region. With time, this flow is converted into a vortex. In contrast to the case of the surface thermal, where this vortex affected the entire region of the flow, in the examined problem the situation is far more complicated. As a result of the large dimensions of the region (at least 5 km), the peripheral vortex did not manage to move to the near-axial region with the formation of its own vortex, initially less developed than the peripheral vortex, but subsequently rapidly propagating, and in subsequent stages the near-wall vortex transforms to a convective column. A similar flow pattern was also recorded in [97]. In [97] it was established that at a radius of the region of fire greater than 10 km (i.e. the characteristic height of the tropopause $H^* \simeq 11$ km for mean widths), the initial stage of the flow becomes even more complicated and is characterised by the presence of more than two vortices. This multi-vortex cellular structure of the fire with a radius of more than 10 km results in the situation in which the complicated vortex flows absorb a large part of the energy, generated during the fire. This leads to a decrease in the rate of formation of the convective column and in the later stage of the development of the flow to a lower height of hovering and spreading of the convective column.

The results of calculations for $K = 0.2$ and the data of the laboratory experiments, processed in the form convenient for comparison, are presented in Table 3.5. Here H_{max} is the height of hovering, H^* is the height of spreading of the convective column, h_{max} and h^* are the appropriate height of hovering and the height of spreading, converted from the results obtained in the laboratory experiments using the similarity theory [141]. Table 3.5 presents the mean values of the height of hovering of the upper edge for the time of 30–60 min, because the upper edge carries out oscillations around the equilibrium position.

As indicated by Table 3.5, the calculated data of the laboratory experiments are in satisfactory agreement. Nevertheless, although the calculations made it possible to select the constant in the algebraic model $K = 0.2$, resulting in good agreement with the laboratory experiments, on the whole this approach is disputable. (In a definition in the calculations of the values of $K = 0.125$ we

Table 3.5

R, km	Q, MW/m^2	H_{max}, km	h_{max}, km	H^*, km	h^*, km
5	0.073	8.3	8.3	5.1	5.3
12	0.073	11.2	9.7	6.0	5.2
22	0.073	12.0	11.8	6.0	5.6
5	0.24	12.1	13.4	8.0	9.3
12	0.24	17.2	16.5	11.0	9.5
22	0.24	18.8	18.7	10.5	9.7

obtain the values 30–35% higher than the experimental values, and at K = 0.2 the values 40–50% lower than the experimental data were obtained). The main problem, associated with the application of algebraic models, is that they cannot be used to describe complicated effects associated with the considerable curvature of the flow lines and with intensive vorticity. However, many foreign and Russian investigators used in calculations of this type varieties of algebraic models without analysing their effect on the resultant solution. Therefore, the present work was concerned with the detailed analysis of the application of one of the most widely used algebraic models. One of the parameters of the model was varied and the results were compared with the laboratory experiments. The main conclusion was that by testing the model of the experimental data it is possible to select a satisfactory parameter for such a model. The results of the calculations of the propagation of an impurity for different technological processes have an acceptable error [151]. For the fundamental investigation of the given problem it is necessary to increase greatly the size of the grids and/or use the differential models of turbulence. Until recently, these models have not been used in the calculations of large-scale fires. The problems associated with the application of models of this type are the subject of further investigations. On the level of the technological estimate of the propagation of the impurity we can restrict ourselves to simpler approaches. Below, all the results of the calculations are given for the variants with K = 0.2 or a different value of K is presented.

Figure 3.48 shows the isolines of temperature and the field of distribution of the impurity for a variant with the radius of the region of R_0 = 5 km and Q = 0.24 MW/m^2 at the moment of time

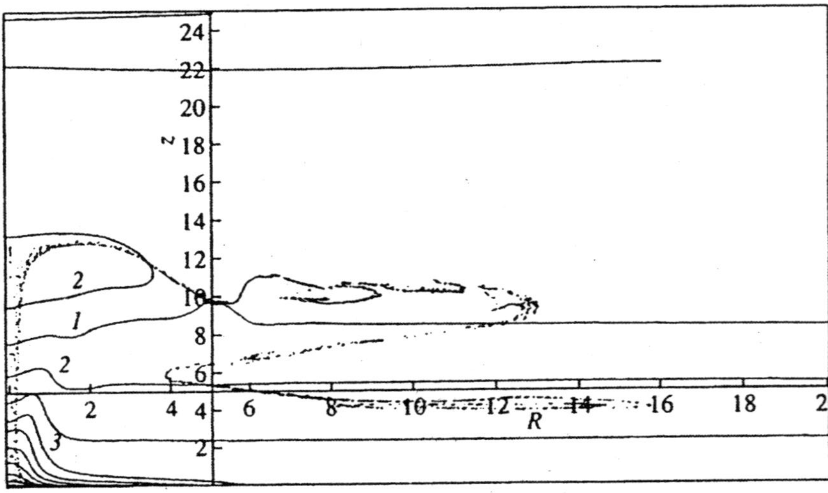

Fig.3.48. Isolines of temperature (solid lines) in the field of distribution of the impurity for R_0 = 5 km and Q = 0.24 MW/m^2 and moments of time t = 60 min. The isoline 1 – T = 230 K, 2 – T = 250 K, 3 – T = 270 K, and so on, with ΔT = 20 K (T_{max} = 430 K).

t = 60 min. The given pattern is characteristic of the stage of hovering of the convective column and the spreading of the tongue of the impurity cloud.

Figure 3.49 shows the isolines of temperature and the field of distribution of the impurity for the variant R_0 = 12 km and Q = 0.24 MW/m^2 for the moment of time t = 40 min. It may be seen that the convective column has already reached the height of hovering, although the temperature in the top part of the column is still higher than the temperature of the surrounding medium. However, in the region of the tongue of the impurities, there are clearly visible complicated vortex structures. Figure 3.50 shows the moment of time t = 60 min for the variant R_0 = 12 km and Q = 0.24 MW/m^2, characterised by the quasi-stationary distribution of the impurities. Figure 3.51 shows for comparison (Fig. 3.43) the form of the convective column for R_0 = 12 km, Q = 0.073 MW/m^2 and t = 40 min for the case with K = 0 (assuming the absence of the effect of turbulence).

Investigations were carried out into the effect of lateral wind on the parameters of the convective column, in particular the height of emission of the passive impurity from the region of fire into the atmosphere. The calculations were carried out using a simple model

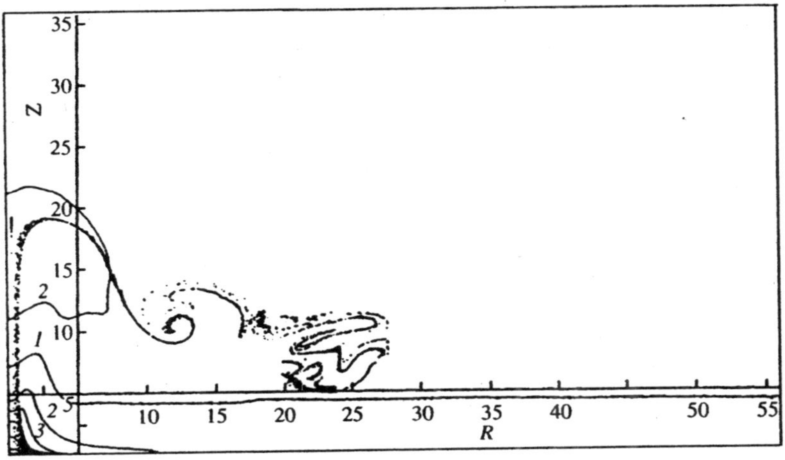

Fig.3.49. Isolines of temperature (solid lines) in the field of distribution of the impurity for R_0 = 12 km and Q = 0.24 MW/m^2 and moments of time t = 40 min. The isoline 1 – T = 255 K, 2 – T = 305 K, 3 – T = 355 K, and so on, with ΔT = 50 K (T_{max} = 720 K).

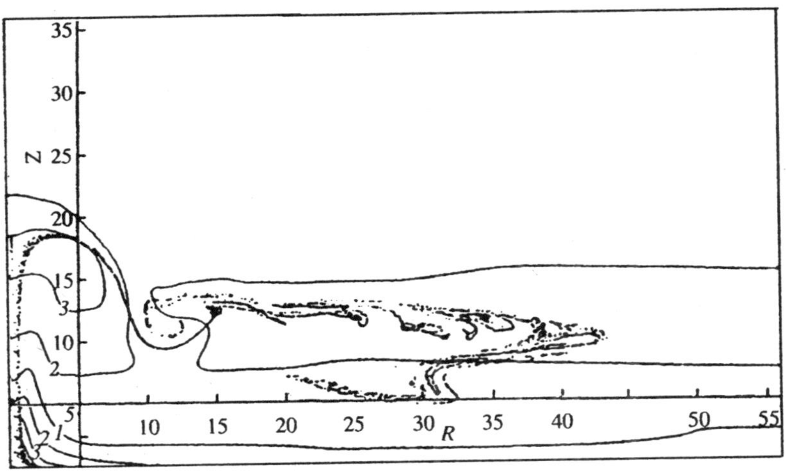

Fig.3.50. Data identical with those in Fig.3.49, for time t = 60 min, R_0 = 12 km, and Q = 0.24 MW/m^2. The isoline 1 – T = 260 K, 2 – T = 285 K, 3 – T = 310 K, and so on, with ΔT = 25 K (T_{max} = 650 K).

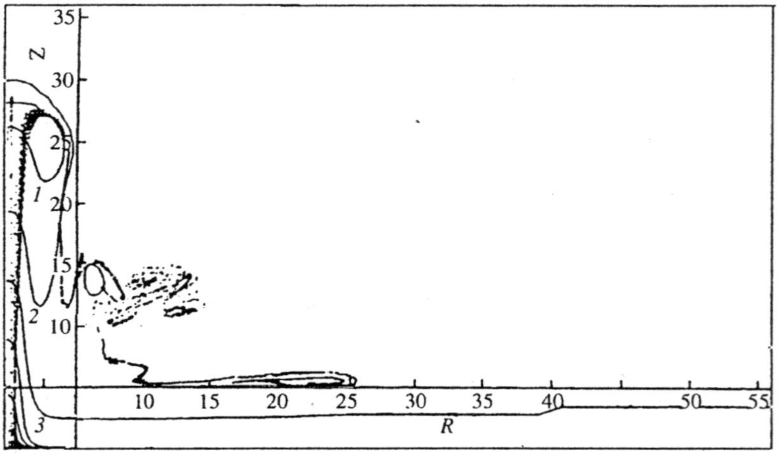

Fig.3.51. Data identical with those in Fig.3.49, for the variant R_0 = 12 km and Q = 0.073 MW/m^2, corresponding to time t = 40 min assuming no effect of turbulence (non-viscous calculations). The isoline 1 – T = 200 K, 2 – T = 235 K, 3 – T = 270 K, and so on, with ΔT = 35 K (T_{max} = 550 K).

of the region of fire in the form of a spatial cylindrical source (energy and impurity) with the energy generation constant in volume and with respect of time.

The initial parameters of the problem were as follows: the radius of the region R_0 = 10 km, the power of the source Q = 0.05 MW/m^2, the speed of wind 10 m/s (stratification of the wind in respect of height was not specified). The integration of the equations of motion of the non-viscous gas was carried out using the MUSCL TVD scheme [174], with the third order of approximation in respect of spatial variables and the second order in respect of time.

The initial stage is characterised by the formation of a toroidal vortex at the periphery of the region of fire. After 15 min, the region above the source is characterised by the intensive lifting of the heated gas and the propagation of the cloud in the radial direction at a height of the order of 4.5 km. The flow was visualised by introducing a source of a passive impurity into the region of fire. Figure 3.52 shows the projection of the cloud (top view) above the region of fire at t = 20 min (in the left half) and with the wind (in the right half). The direction of the wind is indicated by the arrow. It is interesting to note that although the source of the fire is a figure of rotation, in the absence of the wind,

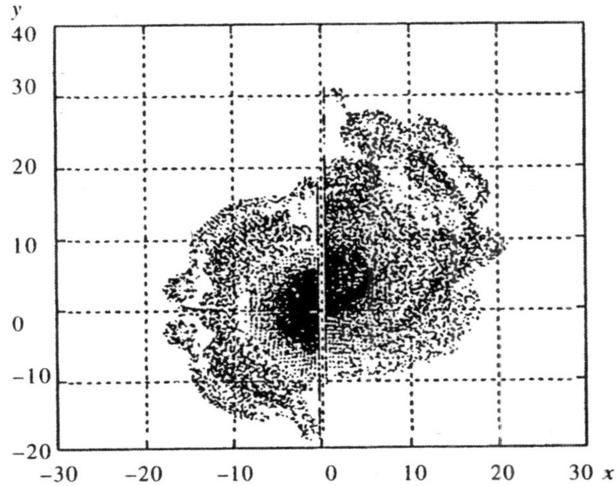

Fig.3.52. Projection of a cloud of an impurity (top view), t = 20 min, without wind (left) and with wind (right), R_0 = 10 km, Q = 0.05 MW/m². The velocity of side wind 10 m/s, direction indicated by the arrow. The figures on the axes are in km. Point (0,0) indicate the epicentrum of the fire.

the distributed impurity is not such a figure. This indicates the instability of the axisymmetric solution of the given problem. Figure 3.53 shows the side view for the same moment of time in the presence of wind (directed from left to right). The scale of the axes in both graphs is given in kilometres. It should be mentioned that these calculations were regarded as a first step for transition to the examination of the given problem in the three-dimensional formulation. Therefore, at present we can only talk about the qualitative structure of the flow in the three-dimensional case, and not about specific quantitative results. The complete model will take into account the turbulence characteristics of movement of the gas, the diffusion equation for transfer of the impurity and the stratification of the wind in respect of height.

In conclusion, Fig. 3.54 shows a possible variant of the visualisation of evolution and propagation of a cloud of smoke under the effect of lateral wind. In this case, the initial formulation of the problem includes the equation for the transfer of the impurity (there may be several of them). The isosurfaces of the concentration of the impurity are also shown in the graph.

the stability parameter depends on the averaging scale gas of the

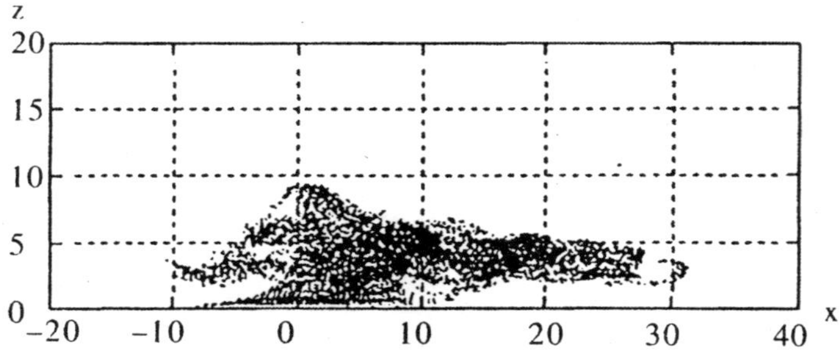

Fig.3.53. For the same variant of calculations and time as in Fig.3.52, side view of the region of fire (in plane xz in the presence of wind (its direction coincides with the direction of the z axis).

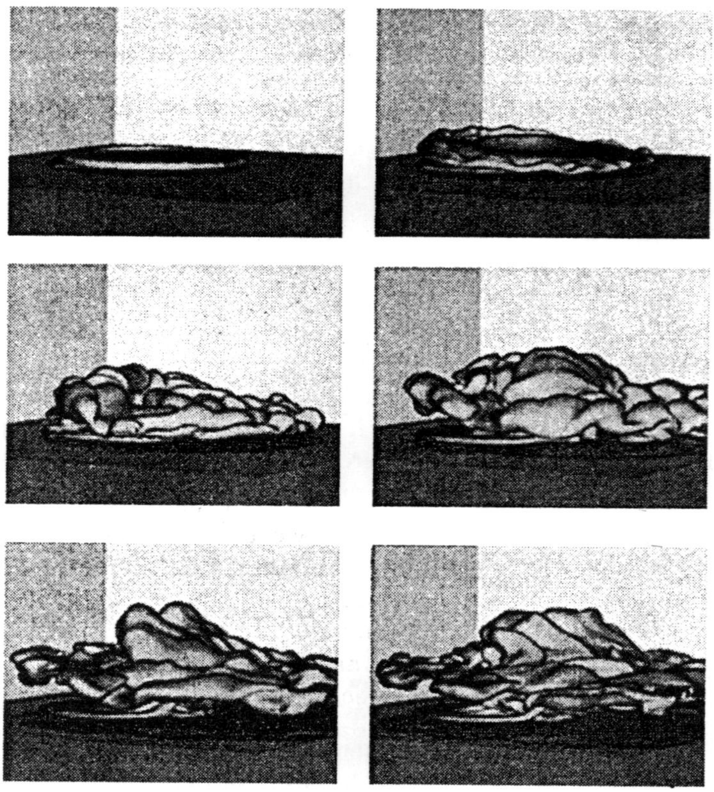

Fig.3.54. Dynamics of development and propagation of a cloud of impurity from a large-scale source of buoyancy under the effect of wind. R_0 = 10 km and Q = 0.05 MW/m^2. The velocity of side wind 10 m/s. Moments of time 10, 20, ..., 60 min.

Chapter 4

Astrophysical turbulence, convection and instabilities

4.1. TURBULENCE

Traditionally, astrophysical objects are very large. For example, the dimensions of a solar spot of the order of 10^5–10^7 cm, the radius of the sun is ~10^{11} cm, the radius of a red giant ~10^{15} cm, the size of an accretion disc may exceed these dimensions, and the radius of the galactic system reaches 10^{21}–10^{22} cm. It may be seen that if we estimate the Re number Re = $RL=/\nu$ at these dimensions of the flow, it will be considerably higher than the critical number for the shear flow equal to approximately 2000. Consequently, the problem of the turbulence of the flow is very urgent in the astrophysical conditions. After all it is clear that the characteristic size cannot be represented by the size of the investigated object and it is necessary to use the characteristic size of the heterogeneity which itself should be determined in the calculations. Viscosity is significant only for fine-scale pulsations ($Re_{10} \approx 1$). For the scales with the characteristic wave numbers $\lambda \gg \lambda_0$, the structures cannot depend on viscosity. In turn, the dissipation of turbulent motion is determined by high frequencies. However, for the Kolmogorov–Obukhov spectrum it is possible to determine the rate of dissipation through the large-scale L: $\varepsilon = (\Delta u)_3/L$, where L is the characteristic size of the region containing the turbulent flow, Δu is the variation of the mean speed in the dimension L. According to the theory of statistical turbulence, the ratio of turbulent viscosity and molecular viscosity is $\nu_{tur}/\nu_{boil} \approx$ Re, and the pressure gradient (momentun transfer) is $\Delta P \approx -\rho(\Delta u)^2$.

Qualitatively, the turbulent flow differs by the fact that its thermodynamic and the hydrodynamic characteristics (vector of the speed, pressure, temperature, concentration of impurities, density, speed of sound, the refractive index, etc) are subject to random fluctuations, generated by the presence of a large number of vortices of different scale and, consequently, they change very chaotically and irregularly with time and in space. In addition to this, it is important to examine the problem of the intermitting structure of turbulence. All these factors make it necessary to carry out numerical modelling of turbulence in the astrophysical conditions for every object!

An important special feature of the astrophysical conditions is the presence of gravitational forces. It is possible to derive a criterion characterising the local stratification of the flow

$$N^2(z) = g\partial\rho_* / \rho_*\partial z,$$

where ρ_* is the potential density. If $N^2 > 0$, the Archimedes force is backmoving, and turbulence should use energy for work against the effect of the Archimedes force and, consequently, it develops only slightly and often concentrates only in the individual thin layer. In this case, the Richardson number is a criterion:

$$\mathrm{Ri} = -\frac{g\left(\frac{dp}{dz}\right)}{\rho\left(\frac{du}{dz}\right)}.$$

In subsequent stages, the concept of the α-disc has been introduced with special reference to accretion discs. The main assumption of the model of the α-disc is the assumption according to which the sink of the kinetic energy of turbulence and the thermal energy, generated in the transition of kinetic energy to thermal energy as a result of viscosity, are equal. This is one aspect.

On the other hand, the increase of the angular moment with the radius stabilises turbulence. For example, this problem was investigated in the work of Ya.B. Zel'dovich [158]. He introduced the Taylor criterion in the local examination of the criterion of development of instability:

$$\mathrm{Ty} = \frac{\dfrac{d\left(\omega r^2\right)^2}{dr}}{r^5\left(\dfrac{d\omega}{dr}\right)^2},$$

Here $\omega = \dfrac{d\varphi}{dt}$ is angular velocity, r is radius (the cylindrical coordinates r, φ, z are used). This number differs from the criterion introduced by Bradshaw [144] but is more suitable for astrophysical situations because it should provides an universal dependence for different geometrical situations. The Taylor criterion was derived from the local examination of flow stability. If the rotation laws are known, it is possible to derive the Taylor criterion for the entire flow. It is assumed that high values of Ty correspond to more rigid conditions of turbulence of the flow. For example, for the exponential law of the dependence of angular speed on radius ω = const r^n, the Taylor criterion has the following form:

$$\mathrm{Ty} = \frac{8(n+2)}{n^2}.$$

The stabilisation of the flow in relation to turbulisation of curves occurs at $n > -2$ which corresponds to the value Ty ≥ 0. For a Keppler accretion disc $n = -3/2$, which gives Ty = 16/9. If $n = 0$, we are concerned with solid-state rotation and, in this case, there is no shear flow and, consequently, no energy is available for the development of the turbulent flow and this is reflected in the value of the Taylor criterion Ty = ∞.

If we remember Taylor's studies [233] where attention was given to the flow between rotating cylinders with the gap ΔR, then at $\Delta R/R = 0.222$ in the Taylor's studies $\mathrm{Re}_c \approx 2\times10^5$, which corresponds to the value of the Taylor criterion for the Keppler disc. This value is hundred times higher than the critical value for the Kuetta flow, but is still considerably smaller than the Reynolds numbers characteristic of the astrophysical conditions.

The problem of turbulence is very important for accretion discs. The observed x-ray irradiation in accretion of matter on a relativistic object in double stellar systems is determined by the rate of accretion:

$$L \approx \eta M c^2 \text{ erg/s},$$

Here η is the efficiency of conversion of gravitational energy, M is the rate of accretion of matter on the relativistic object, c is the speed of light. For a neutron star $\eta \approx 0.1 \div 0.2$, for a black hole $\eta \approx 0.06 \div 0.4$. The rate of accretion is determined by the mechanism of transport of the angular moment on the disc to the outside. There are several possibilities for the redistribution of the angular moment. Evidently, molecular viscosity is not important in the astrophysical conditions. However, as discussed previously, turbulent viscosity $\tau_m \approx$ Re may operate [84]. However, according to Taylor's studies for the laws of rotation with increase of the angular moment to the outside, the instability to turbulisation of the flow may be suppressed. However, nonlinear interactions may generate turbulence in the flow [58]. In particular, large-scale structures may form [26,148]. In these conditions, the transfer of the moment is associated with large-scale vortices and does not result in heating of the accretion disc away from the gravitating solid.

The traditional model of the α-disc links the viscosity with the temperature of the matter of the disc. This is characteristic of the Kolmogorov–Obukhov spectrum where the small scales of turbulence annihilate into heat. In this model, the efficiency of the transport of the angular moment is characterised by the parameter α:

$$\alpha = \frac{V_t}{V_s} + \frac{H^2}{4\pi\rho V_s^2}$$

where $\frac{\rho V_s^2}{2} = \frac{3}{2}\rho\frac{kT}{m_p} + \varepsilon_r$ is the thermal energy of the matter, ε_r is the density of radiation energy, V_s is the velocity of sound and V_t is turbulent velocity. In this assumption, the tensor of the tangential stresses $w_{r\varphi}$ is written in the following form:

$$w_{r\varphi} = \nu_t R \frac{d\omega}{dR} \approx \alpha\rho V_s^2.$$

Usually, it is assumed that parameter α is in the range

$10^{-15}\left(\frac{M}{M_{cr}}\right)^2 < \alpha < 1$. In fact, this parameter should be determined on the basis of modelling the turbulent flow in the investigated conditions. The definition of parameter α also includes the strength of the magnetic field H.

The magnetic field may influence the transport of the angular moment. However, the following problem must be taken into account. If the density of the energy of the magnetic field is higher than the kinetic energy, the structure of the flow is determined by the magnetic field and there is no turbulence. On the other hand, if the density of the energy of the magnetic field is smaller than the kinetic energy, the magnetic field has no effect on the flow. Magnetic viscosity becomes considerable if the values of the kinetic energy and the density of the energy of the magnetic field are of the same order of magnitude. This situation exists in the case of magnetic dynamo.

However, we know from examination that the transport of the angular moment outside the disc does take place. Thus, we must face the problem of finding the mechanism of transport of the angular moment in the accretion disc.

4.2. THE STRUCTURE OF THE ACCRETION DISC

The structure of the accretion disc has been studied in a large number of investigations, starting from different analytical and simple modelling constructions up to the calculations of the structure of the accretion disc in the three-dimensional approach. In this book, we shall describe the structure of the accretion disc taking into account the study [4].

The accretion discs or gas formations in the vicinity of gravitating bodies represent an interesting object for examination in astrophysics and theoretical physics. These formations are typical of the nuclei of galactics and double stellar systems where the processes in the disc control the evolution of the binary system as a whole. Special attention in these objects is paid to the examination of the mechanisms of the loss of the angular moment by the matter of the disc and subsequent accretion of the matter on the compact gravitating object. The rate of transfer of matter on the central body from the accretion disc determines the nature of radiation and this is especially important for investigations.

The methods of mathematical modelling, based on different

physical models and computation algorithms, are used widely at present for examining the processes in accretion discs, together with analytical methods. For example, the authors of [196,202,224] carried out numerical calculations and the results of these calculations shows the formation, in an accretion disc, of a double stellar system of spiral shock waves. It has been reported that these formations represent an important mechanism of the loss of matter by the disc. In addition to this, the redistribution of the angular moment is caused by viscosity, the magnetic field, and in our opinion, the formation of large-scale vortices.

The authors of [35,38,40] modelled the formation of a disc and the formation of its structure taking into account the intercalation of matter between the components of the binary system. In these studies, the calculation region contains both components of the binary system and, consequently, the vicinity of the star-accretor where the disc forms, contains a relatively small number of calculation cells of the difference grid. Consequently, it is difficult to resolve in detail the special features of the spatial structure of the flow in the disc. At the same time, it is not possible to produce large-scale vortices because of the high numerical viscosity which depends on the size of the calculation cells (or, in a more general case, one should talk about dissipation processes).

Therefore, for more detailed examination the processes in the discs it is desirable to separate of the disc as an individual element. There is a principal physical difference in this formulation of the problem: there is no accretion of matter on the gravitating centre and on the outer boundary of the disc. In this case, it is possible to carry out a more detailed examination of the spatial formations in the disc, formed under the effect of gravitational forces of both components and rotation of the binary stellar system as a whole [3]. The structure of the flow in the disc is modelled in the two- and three-dimensional spatial approximations.

We examine a system consisting of two stars with the masses M and m, rotating around the centre of the mass of the system with constant angular velocity ω. The star with mass M will be referred to as the primary component, and this, with mass m as the secondary component (the star-donor) (Fig. 4.1).

According to the third Kepler law, distance A between the components is linked with the angular speed of rotation of the system by the following relationship

$$\omega^2 A^3 = G(M + m). \tag{4.1}$$

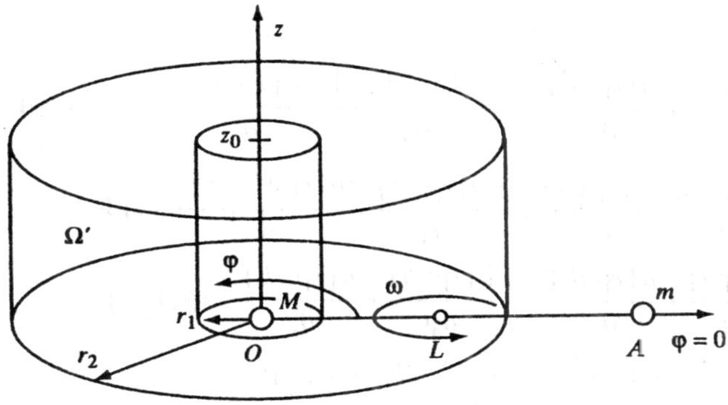

Fig.4.1. A system consisting of two stars with masses M and m, rotating around the centre of masses of the system with constant angular velocity ω. The star with mass M will be referred to as the primary component, the star with mass m as secondary (star-donor)

Here G is the gravitational constant. Distance L between the primary component and the centre of the mass of the system has the following form

$$L = Am/(M + m). \tag{4.2}$$

We introduce a cylindrical coordinate system (r,φ,z), rotating with angular velocity ω, with no inertia. The origin of the coordinate of the system coincides with the centre of the primary component, and the axis $\varphi = 0$ passes through the secondary component (Fig. 4.1). It is assumed that in the region $\Omega = (r_1 \le r \le r_2)\times(0 \le \varphi \le 2\pi\times(-z_0 \le z \le z_0)$, representing a cylindrical ring whose axis of rotation coincides with the axis z, we specify the stationary (in the absence of the secondary component) gas configuration of the accretion disc in which the gas rotates around the primary component. Self-gravitation of the gas is not taken into account because it is assumed that the mass of the gas in the disc is considerably smaller than the mass of the first component. The gas is compressible, ideal and its behaviour is described by a system of three-dimensional equations of gas dynamics in Euler variables [84]:

$$\begin{gathered}\frac{\partial(d\rho)}{\partial t}+\frac{\partial(r\rho u)}{\partial r}+\frac{1}{r}\frac{\partial(r\rho v)}{\partial\varphi}+\frac{\partial(r\rho w)}{\partial z}=0,\\ \frac{\partial(r\rho u)}{\partial t}+\frac{\partial\left(r\rho u^2+rp\right)}{\partial r}+\frac{1}{r}\frac{\partial(r\rho uv)}{\partial\varphi}+\frac{\partial(r\rho uw)}{\partial z}=p+\rho v^2+r\rho\mathbf{F}_r,\end{gathered} \tag{4.3}$$

$$\frac{\partial(r\rho v)}{\partial t}+\frac{\partial(r\rho vu)}{\partial r}+\frac{1}{r}\frac{\partial\left(r\rho v^2+rp\right)}{\partial\varphi}+\frac{\partial(r\rho vw)}{\partial z}=-puv+r\rho\mathbf{F}_\varphi,$$

$$\frac{\partial(r\rho w)}{\partial t}+\frac{\partial(r\rho vu)}{\partial r}+\frac{1}{r}\frac{\partial(r\rho wv)}{\partial\varphi}+\frac{\partial\left(r\rho w^2+rp\right)}{\partial z}=r\rho\mathbf{F}_z,$$

$$\frac{\partial(r\rho e)}{\partial t}+\frac{\partial(r\rho uh)}{\partial r}+\frac{1}{r}\frac{\partial(r\rho vh)}{\partial\varphi}+\frac{\partial(r\rho wh)}{\partial z}=r\rho(\mathbf{F},\mathbf{v}),$$

$$e=\varepsilon+u^2/2+v^2/2+w^2/2,\quad h=e+p/\rho;$$

with the equation of state

$$p=(\gamma-1)\rho\varepsilon. \tag{4.4}$$

Here r is radius, φ is the polar angle, t is time, ρ is the density of the gas, p is pressure, ε is the specific internal energy, e is the total specific energy, γ is the adiabate parameter, h is the total specific consulting, $\mathbf{v}=(u,v,w)^T$ is the speed of the gas, u is the radial component of the speed, v is the azimuth component, w the component of the speed on the axis z, F is the total specific external force with the component $(\mathbf{F}_r,\mathbf{F}_\varphi,\mathbf{F}_z)$ acting on the gas particle.

In the dimensionless variables, where the scale multipliers are represented by the values

$M=1\times10^{33}$ g, $R=10^{11}$ cm, $G=6.67\times10^{-8}\ \mathrm{cm^3\cdot g^{-1}\cdot s^{-2}}$,

The equation for the component of the external force, which takes into account the effect of the forces of gravitation of the first primary and secondary component, the centrifugal force and the Coriolis force, have the form

$$\mathbf{F}_r=-\frac{r}{\left(r^2+z^2\right)^{3/2}}+\frac{md}{\left(d^2+z^2\right)^{3/2}}\cos(\alpha+\varphi)-\omega^2 s\cos(\eta+\varphi)+2\omega v,$$

$$\mathbf{F}_\varphi=-\frac{md}{\left(d^2+z^2\right)^{3/2}}\sin(\alpha+\varphi)+\omega^2 s\sin(\eta+\varphi)-2\omega u,$$

$$\mathbf{F}_z=-\frac{z}{\left(r^2+z^2\right)^{3/2}}-\frac{mz}{\left(d^2+z^2\right)^{3/2}},$$

where

$$d=\sqrt{r^2+A^2-2Ar\cos\varphi},\quad \sin\alpha=r\sin\varphi/d,\quad \cos\alpha=(A-r\cos\varphi)/d,$$
$$s=\sqrt{r^2+L^2-2rL\cos\varphi},\quad \sin\eta=r\sin\varphi/s,\quad \cos\eta=(L-r\cos\varphi)/s. \tag{4.5}$$

It should be noted that the equations for the forces in the selected coordinate system do not depend on the solution and, consequently, on time, with the exception of the last member (containing speed), associated with the Coriolis force.

The initial data in the calculations, described below, were represented by the stationary gas configuration. It was thus possible to exclude the effect of gas-dynamic processes, caused by the nonequilibrium nature of the initial state of the disc. In Ref. 2, detailed investigations were carried out into possible equilibrium of cylindrically symmetric configurations of the gas, rotating in the vicinity of the gravitating centre; its pressure and density are linked by the polytrope equation:

$$p=k\rho^{\gamma},\quad k,\gamma=\text{const}.$$

In particular, special attention has been given to selecting equilibrium configurations of the gas with a boundary of the type:

$$Z(r)=\pm\alpha r\exp\left(-\beta(r-r_0)^2\right);\quad \alpha,\beta,r_0>0 \tag{4.6}$$

and equations of the equilibrium distribution of the functions of velocity $v_{\varphi}(r, z)$ and gas density $\rho(r, z)$ are presented. By means of analytical numerical investigations it is shown that this configuration of the gas may be regarded as equilibrium and that it can be used as initial data in modelling the processes in the accretion disc, in both the three-dimensional and two-dimensional approximation.

The distance between the components in the binary stellar system was assumed to be A = 4, the mass of the secondary component in the main variant of the calculation was m = 1. In this case, the total rotation of the binary system takes place during the period $T_s \approx 35$. The gas was assumed to be ideal with the equation of state (4.4) and the parameter of the adiabate γ = 5/3. The parameters in equation (4.6), which determine the initial equilibrium configuration of the disc, were assumed to be equal to α = 0.2,

$\beta = 9$, $r_0 = 0.8$. At these values of the parameters, the main mass of the gas is concentrated in a cylindrical ring $0.5 < r < 1.1$, $-0.2 \leq z \leq 0.2$, and the total mass of the disc is $M_{gas} \approx 0.1$, the complete rotation of the gas particle with equilibrium configuration takes place in the period $T_1 \approx 0.6$ and $T_2 \approx 10$ for the internal and external edge of the disc, respectively.

Taking into account the symmetry of the problem in relation to the plane $z = 0$, the calculations were carried out in the region $\Omega' = (r_1 \leq r_2) \times (0 \leq \varphi \leq 2\pi) \times (0 \leq z \leq z_0)$ with the appropriate selection of the boundary conditions at the 'lower' boundary of the region. In the remaining boundaries of the region we specified the 'free' boundary conditions of the type

$$\frac{\partial f}{\partial r} = 0\left(\frac{\partial f}{\partial z} = 0\right), \text{ where } f = r, u, v, w, e, \tag{4.7}$$

or the non-percolation condition

$$u = 0\left(w = 0\right). \tag{4.8}$$

Depending on the variance of the calculations, the parameters of the region were selected equal to $r_1 = 0.2$, $r_2 = 1.4$, and $r_1 = 0.2$, $r_2 = 2.1$, and the values of z_0 varied from 0.2 to 1.0. The size of the difference metal was from $60 \times 40 \times 3$ to $120 \times 80 \times 60$ cells. To solve the resultant equations, two different methods, described in [8,9], were used. The results obtained by these methods coincided, indicating the physical consistency of the results.

The results of numerical calculations in the two-dimensional approximation

The calculations were carried out with different parameters of the system, the boundary and initial conditions. Here, it is important to mentioned briefly only the most important results. In the process of the calculations, the general characteristic stages of the development of processes in the accretion disc were determined. In the initial stage, the solution of the system of equations (4.3), (4.4) was carried out in the absence of the secondary component ($m = 0$). In this case, the initial equilibrium condition of the gas of the type (4.6) did not change which confirms not only the stationary nature of the state but also its stability. It should be mentioned that this formulation of the problem differs from the traditional models using the Kepler law of rotation in which the disc is assumed to be very thin and cold.

In order to avoid high-intensity relaxation processes, the distance between the component was varied from $10^3 A$ to A in a relatively short period of time $T_A = 4.5$, in accordance with a linear law. The parameters of the system and the specific external force change in this case in accordance with the relationships (4.1), (4.2) and (4.5), respectively. In this stage, the initial state of the gas in the disc also changed only slightly.

In the following stage, under the effect on the field of external forces **F** and the force of the gas-kinetic pressure, the gas gradually acquires the positive radial component of the speed leading to the formation of a shock wave propagating to the periphery of the calculation region. The front of the shock wave becomes elliptical and this is a consequence of the heterogeneity, in respect of the polar angle, of the components of the 'perturbing' total force (Fig. 4.2). The lines of the level of density and the vector of the gas velocity are indicated on the graphs.

In transition through the front of the shockwave, the gas almost completely loses the radial component of the speed. Consequently,

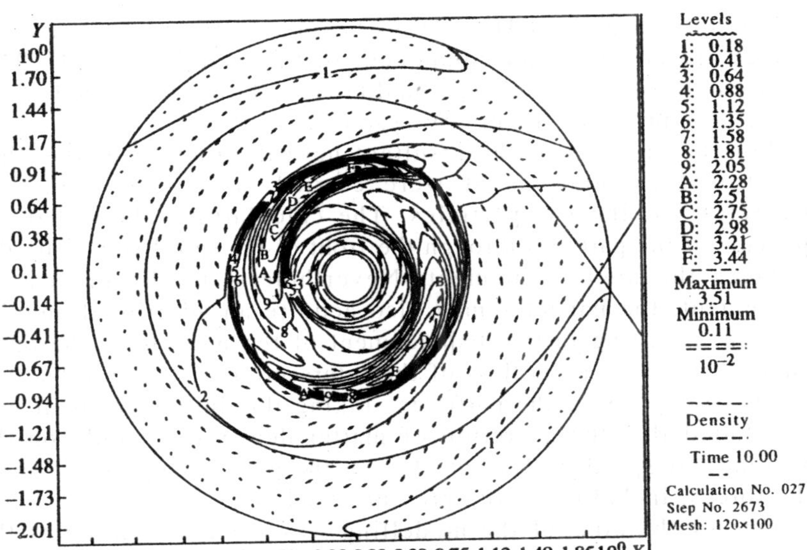

Fig.4.2. Lines of the level of density and vector of gas velocity at the moment $t = 10$. The solid line indicates the Roche surface.

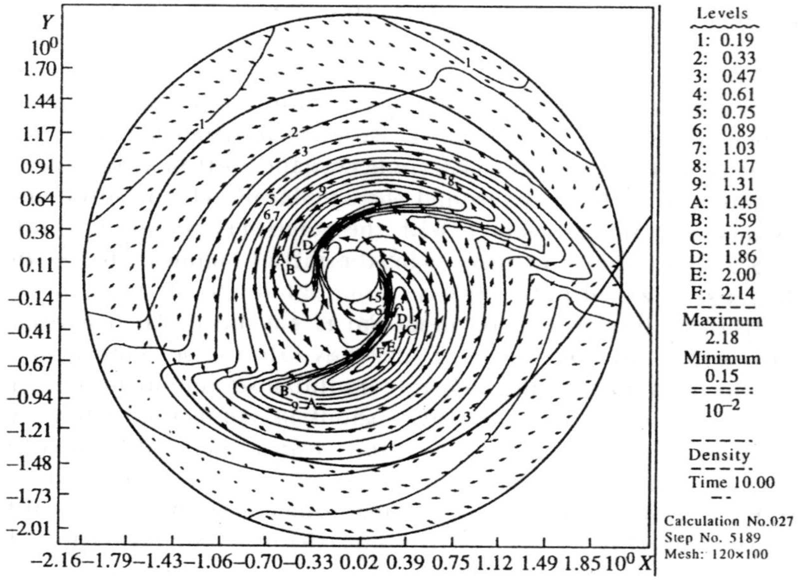

Fig.4.3. Lines of the level of density and vector of gas velocity at the moment t = 19. The solid line indicates the Roche surface.

a new quasi-stationary state forms behind the front, and the spiral waves are an important component of this state (Fig. 4.3). This configuration does not change qualitatively during the period of the order of rotation of the binary system $T_s \approx 35$.

For the analysis of the resultant helical structure of the disc, additional calculations of were carried out to determine the trajectory of the gas particles. The calculations show that the passage of the particles through the front of the spiral waves usually reduces the angular moment. Nevertheless, in the process of movement, the particles may also acquire an additional angular moment resulting in the displacement of the particle to the outer edge of the disc, regardless of the intersection of the spiral waves by the trajectory of the particles. Taking this into account, it was concluded that the accretion of matter depends not only on the presence of spiral waves but also on the general superposition of the forces, including on the pressure gradient.

One of the aims of the modelling was the examination of the possibility of formation of the stationary state of the accretion disc in the binary star system. For this reason, a series of calculations were carried out in the region containing the Roche cavity of the

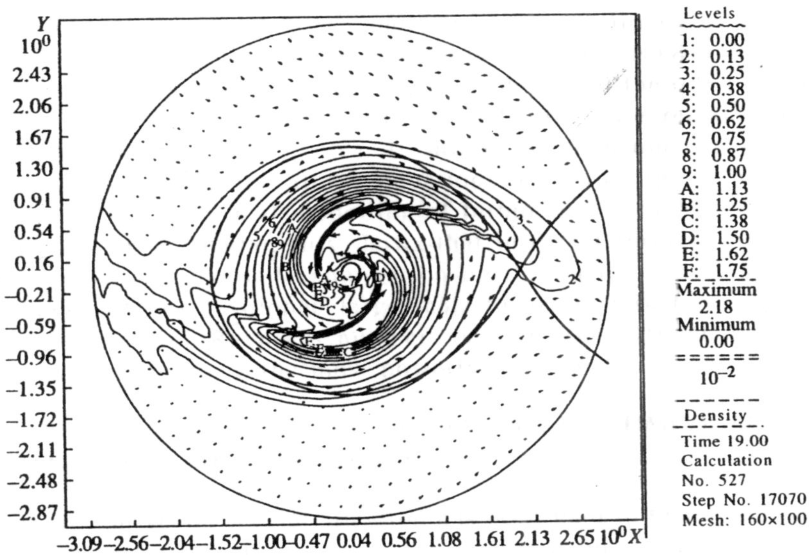

Fig.4.4. Lines of the level of density and vector of gas velocity at the moment $t = 19$ (the solid line indicates the Roche surface; the region of calculations contains the Roche cavity of the primary component and Lagrange point L_1).

primary component and the Lagrange point L_1. It has been shown that in the case of the 'free' boundary conditions (4.7) the disc is characterised by the establishment of a two-arm helical structure of the disc (Fig. 4.4) which does not change qualitatively during a long period of time, of the order of $12T_s$.

However, as previously, this structure is not stationary because there is discharge of the matter of the disc both in the direction of the internal boundary of the region (to the primary component) and to the periphery, mainly in the vicinity of the internal Lagrange point. In this case, the intensity of the spiral waves changes with time, like the shape of the fronts of the waves (Fig. 4.5).

The stationary configuration can be produced formally by avoiding the outflow of the matter of the disc by selecting the non-percolation condition (2.8) on the boundaries of the region. In this case, the structure of the disc, containing spiral waves, is also established (Fig. 4.6), and the presence of these waves in the given conditions indicates that the resultant waves are shock waves. It should be mentioned that the problem of the nature of formation of spiral waves has been examined in greater detail in Ref. 1.

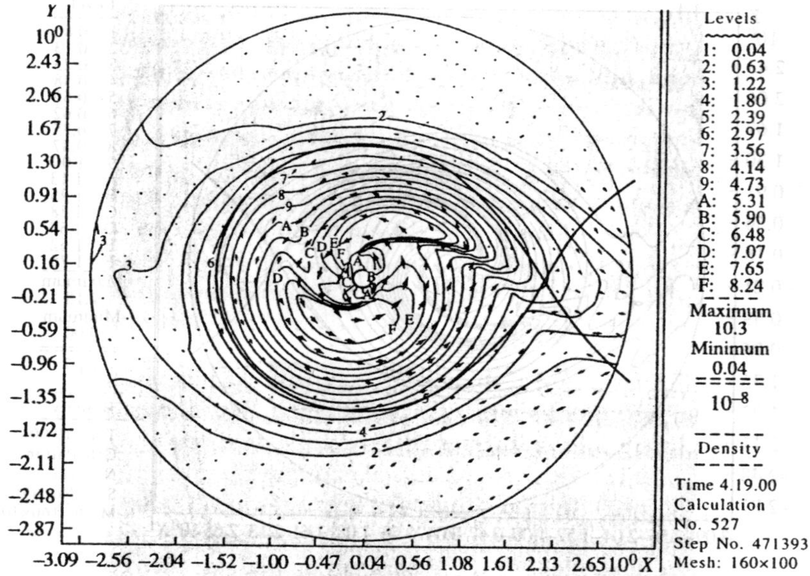

Fig.4.5. Lines of the level of density and vector of gas velocity at the moment t = 419 (the solid line indicates the Roche surface; the region of calculations contains the Roche cavity of the primary component and Lagrange point L_1).

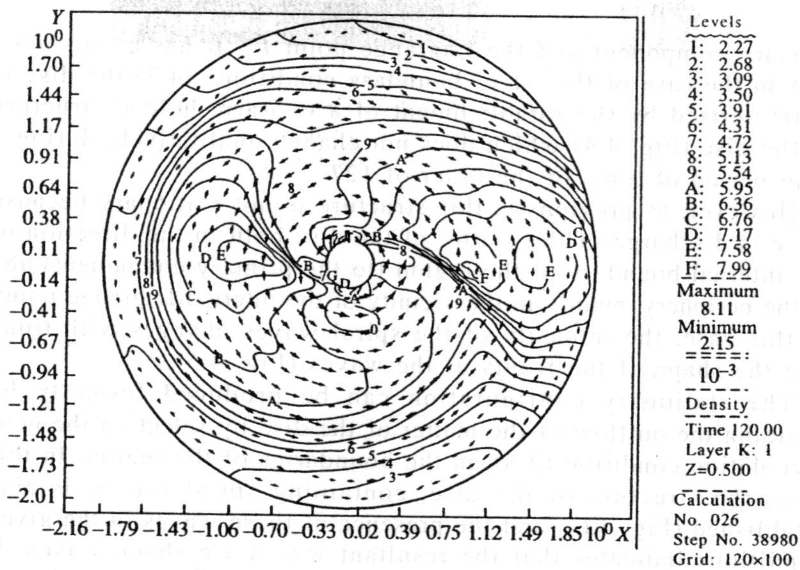

Fig.4.6. Lines of the level of density and vector of gas velocity at the moment t = 120. The stationary configuration of the disc was produced by selecting the non-percolation condition at the boundaries of the integration region.

The results of numerical calculations in the two-dimensional approximation

The studies of many authors (for example, [41]) indicated the possibility of large differences in the results of mathematical modelling of astrophysical problems in the two- and three-dimensional approximation. Taking this circumstance into account, it was necessary to carry out three-dimensional modelling of the processes in the accretion disc which in other aspects are completely identical with the calculations carried out in the previous paragraph. As shown by the calculations, three-dimensional modelling is characterised by the same stages of development of the helical structure of the disc, as mentioned previously. These stages will be demonstrated on an example of the variant of calculations with the mass of the secondary component equal to the mass of the primary component (m = 1), and with the parameters of the calculation region r_1 = 0.2, r_2 = 2.2, z = 0.2, containing the Roche cavity of the primary component and the Lagrange point L_1. The dimensions of the difference grid in the given calculation variant. were 120 × 80 × 60 cells of the grid in respect of the coordinates, respectively.

The initial state of the gas was, as previously, the equilibrium configuration (in the absence of the secondary component) of the type of (2.6) with the parameters $\alpha = 0.2$, $\beta = 9$, $r_0 = 0.8$. At these values of the parameters, the density of the gas differs from zero only in the part of the calculation region determined by the equation of the boundary (4.6). In the remaining part of the region, we specified the 'background' values of density and pressure, equal to $\rho_b \approx 1 \times 10^{-6}$, $\rho_b \approx 1 \times 10^{-7}$, whereas the density and pressure in the equilibrium configuration were characterised by the values of $\rho_0 \approx 1\times10^{-1}$, $p_0 \approx 1\times10^{-3}$, respectively. On the 'side' boundaries of the calculation region we specified the 'free' boundary conditions (4.7), and the 'upper' boundaries the non-percolation condition (4.8), and on the 'lower' boundary, as already mentioned in the paragraph 2, we specified the conditions of symmetry of the flow in relation to the plane z = 0.

As in the two-dimensional approximation, the external effect on the gas particle, determined by the relationships (4.1), (4.2), (4.5), gradually increased to the calculated value by decreasing the distance between the components from 10^3 A to A in accordance with the linear law in the period of time T_A = 4.5. Figures 4.7 and 4.8 show the lines of the level of gas density in the section of the calculation region for the fixed angle φ = 0 and the projection of

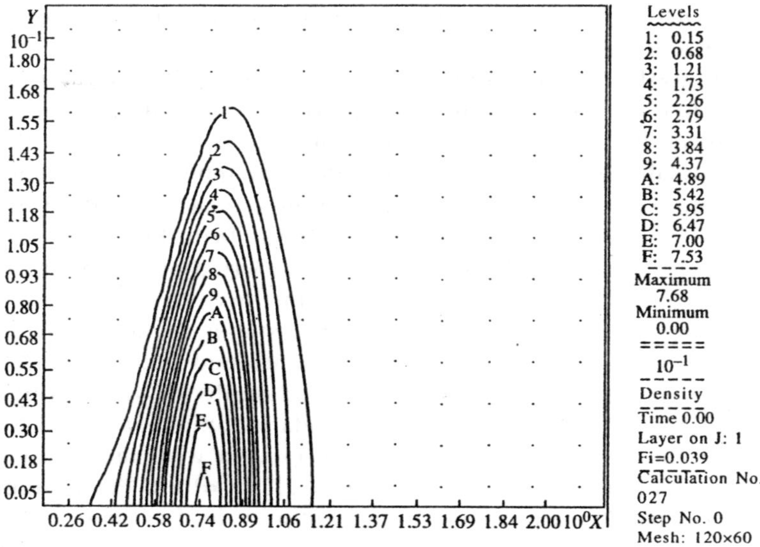

Fig.4.7. Lines of the level of gas density in the cross section of the calculation region at fixed angle $\varphi = 0$ and the projection of velocity vectors on the plane of the section at time $t = 0$.

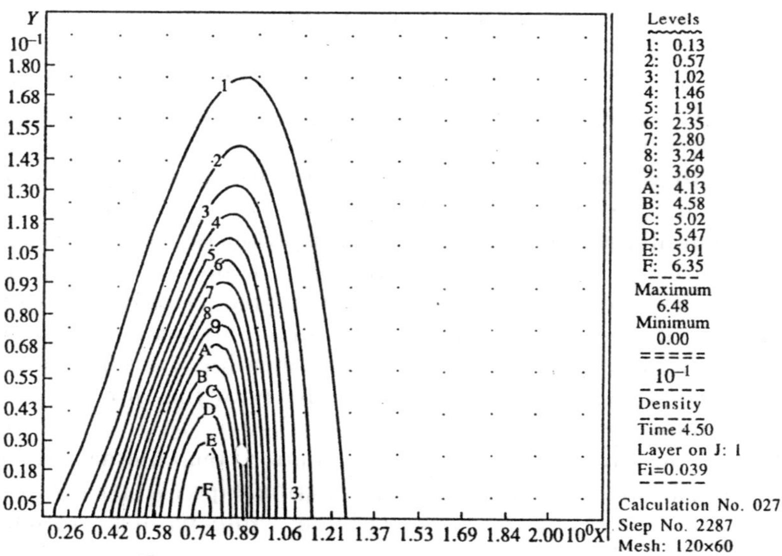

Fig.4.8. Lines of the level of gas density in the cross section of the calculation region at fixed angle $\varphi = 0$ and the projection of velocity vectors on the plane of the section at time $t_A = 4.5$.

the vectors of the speed and the plane of the section for the moments of time $T = 0$ and T_A, respectively.

As indicated by Fig. 4.7 and 4.8, the initial configuration changes only slightly in this period confirming the equilibrium nature of the configuration.

The next stage is characterised by the formation of a radial shock wave, propagating in the direction to the periphery of the calculation region which as a result of the heterogeneity of the field of external forces (4.5) in respect of the polar angle gradually acquires the elliptical form. The new core of the disc, containing two distinctive helical structures, forms behind the front of the radial wave. It should be mentioned that the radial component of the velocity of the gas and also the component of velocity on the z axis, in the forming core are small in comparison with the azimuthal (rotational) component. In this sense, we can talk about the quasi-stationary nature of the flow of the gas in the centre of the disc. To illustrate these processes, one should examine Figs. 4.9 and 4.10, which show the lines of the level of density and projection of the field of the speed of the gas on the planes of the sections of the calculation region at fixed values of $r = 0$ and $\varphi = 0$, respectively.

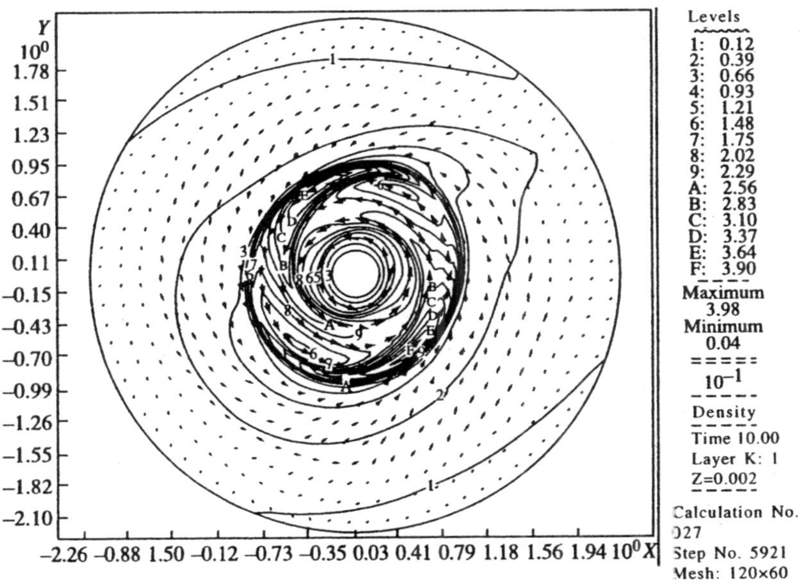

Fig.4.9. Lines of the level of density and projection of the field of gas velocities on planes of sections of the calculation region at $r = 0$.

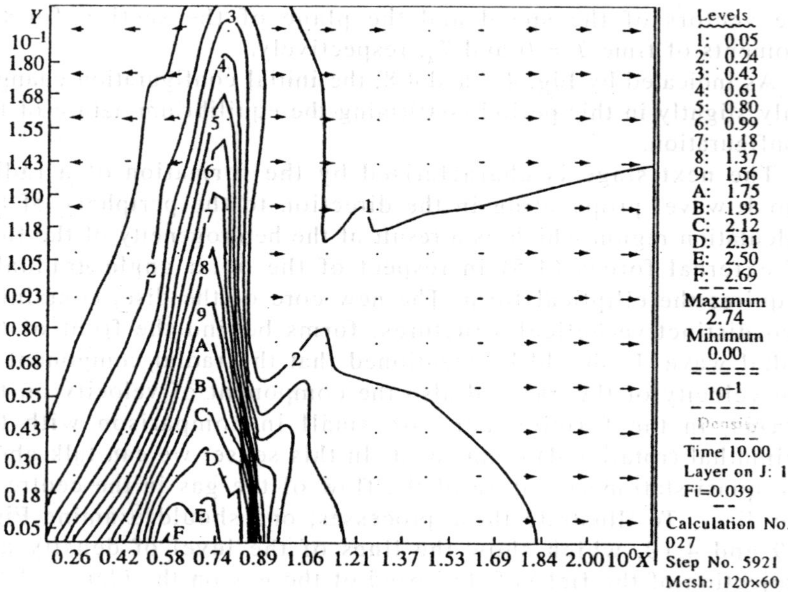

Fig.4.10. Lines of the level of density and projection of the field of gas velocities on planes of sections of the calculation region at $\varphi = 0$.

When the radial wave leaves the boundaries of the calculation region, the spiral structure of the disc is established. In the sections at fixed values of r the structure is identical with that observed in two-dimensional modelling (see the section at $r = 0$, shown in Fig.4.11). This structure of the disc does not differ qualitatively during the period of time from T_s to $2T_s$ (depending on the calculation variant). As previously, in this case it can be concluded that the observed flow is the quasi-equilibrium configuration of the disc in the field of external forces, determined by (4.5).

Thus, one can note the qualitative agreement of the results of two- and three-dimensional modelling in the stage of formation of the quasi-equilibrium helical structure of the accretion disc. Quantitative comparison is also interesting. Using this comparison, comparison was made of the patterns of the flow for the same moment of time in the variant of three- and two-dimensional calculations corresponding to each other. For this purpose, the three-dimensional configuration was integrated in the direction of height (axis z) and was subsequently presented in the form of the lines of the level in the two-dimensional ring-shaped region, like the two-dimensional configuration. As shown by Figs. 4.12-2D, 4.13-3D,

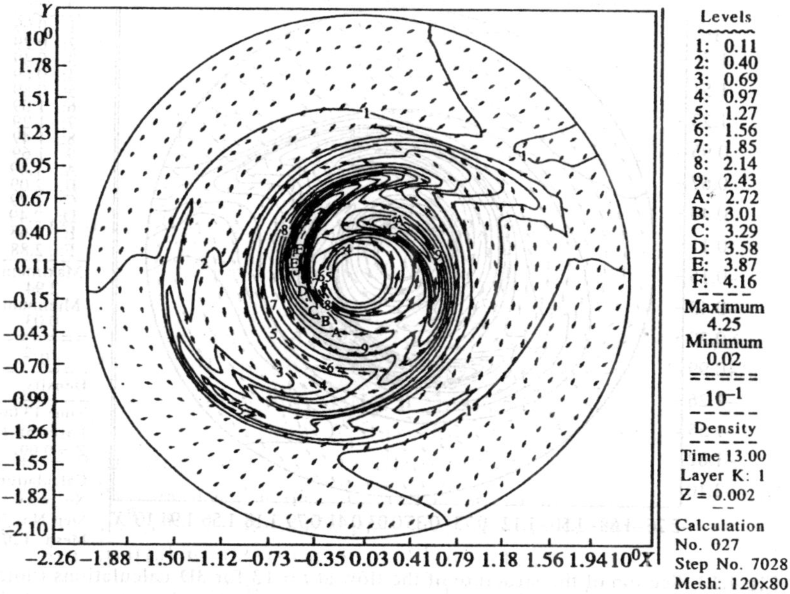

Fig.4.11. Section of the structure of the flow at t = 13 (notations as in previous figures).

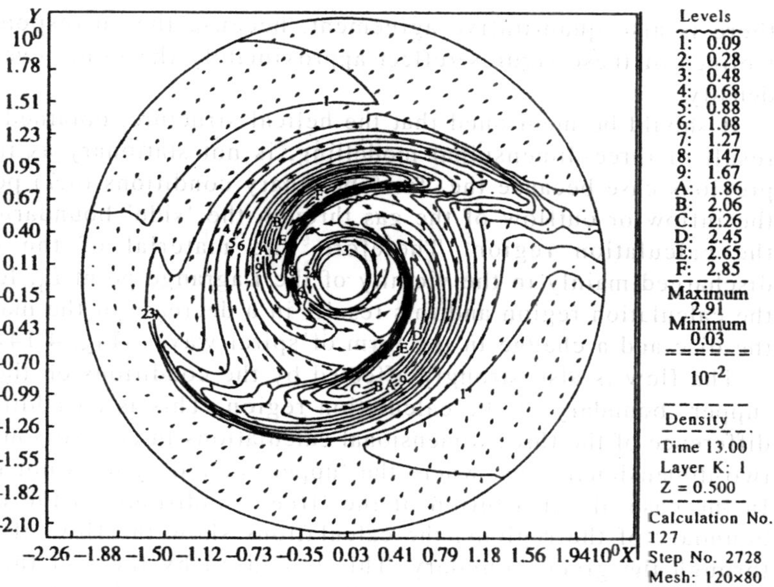

Fig.4.12. Section of the structure of the flow at t = 13 for 2D-calculations (notations as in previous figures).

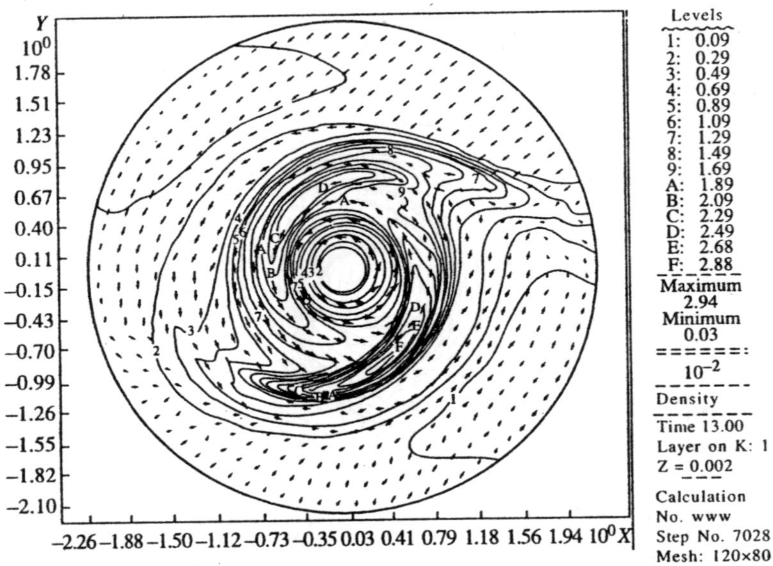

Fig.4.13. Section of the structure of the flow at t = 13 for 3D-calculations (notations as in previous figures). In this case, the 3D-configuration was integrated 'in the direction of height' (z axis) and then depicted in the form of lines of the level in the two-dimensional ring-shaped region, like 2D-configuration.

there is also quantitative agreement, because the corresponding isolines of these figures reflect approximately the same levels of density.

It should be mentioned that the helical structure, obtained as a result of three-dimensional modelling, is not stationary as in the previous case because the 'free' boundary conditions (4.7) permit the inflow or outflow of the gas through the 'side' boundaries of the calculation region. Therefore, the material of the disc, discharged mainly in the vicinity of the Lagrange point L_1, leaves the calculation region and this results in a decrease of the mass of the disc and a change in the form of spiral waves (Fig. 4.14).

The flow is also strongly affected by the conditions on the left 'upper' boundary of the calculation region. This is an additional difference of the three-dimensional calculations in comparison with two-dimensional ones, where the 'upper' boundary does not form. In the case of substitution of the 'free' conditions on the upper boundary of the regions, the calculations show the flow of mass through the given boundary. This is caused by the fact that the component of the external force $\mathbf{F}_z$, determined by the relationship (4.5), is of fixed sign and specifies the direction of the vector of

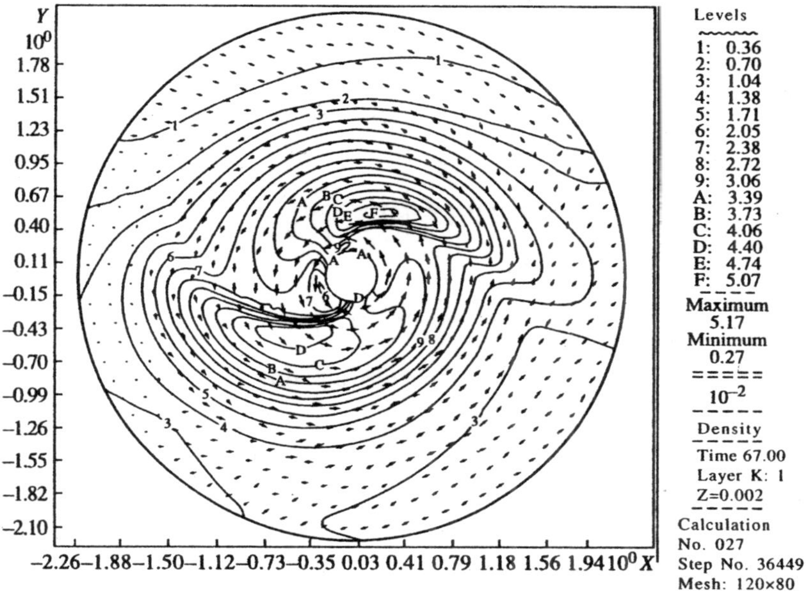

Fig.4.14. Section of the structure of the flow at $r = 0$ at $t = 67$ (notations as in previous figures).

the force in relation to the equatorial plane $z = 0$. This effect also influences the part of the calculation region in which the 'background' values of the pressure and density ρ_b, p_b are initially given, and the pressure gradient, compensating the effect of the external force in the equilibrium concentration, does not form.

Thus, in three-dimensional modelling it is also not possible to obtain the stationary configuration of the accretion disc within the framework of the examined formulation of the problem. The calculations show that an important factor is the mechanism of exchange of matter between the accretion disc and the shell of the binary system which greatly controls the processes taking place in the disc. In the simplest case, this mechanism is represented by the complete absence of exchange of matter between the shell and the disc, by formulating the conditions of non-percolation (4.8) in all boundaries of the region. In this case (as in two-dimensional modelling) one can obtain a stationary flow in the disc containing spiral waves (Fig. 4.15) but this is nevertheless slightly artificial.

The agreement the between the results of the two- and three-dimensional modelling, obtained in the calculations, is very interesting for subsequent modelling of the processes of development of

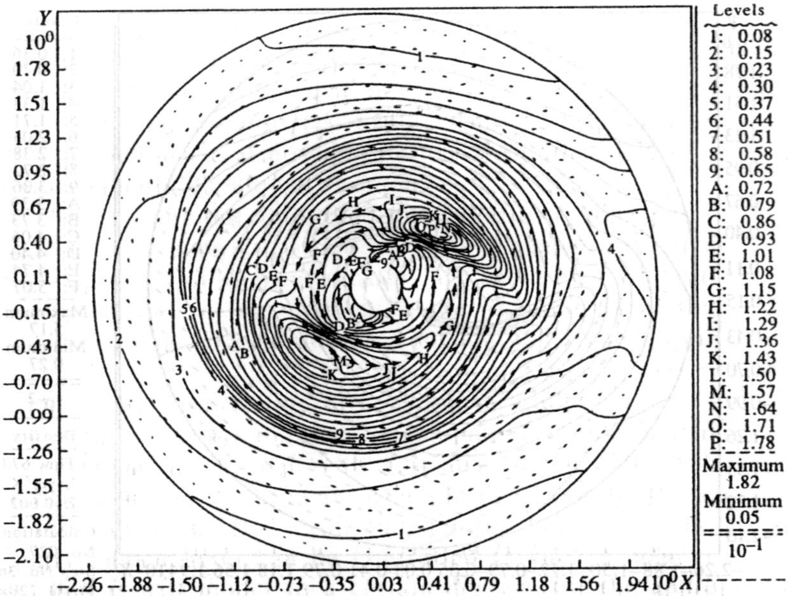

Fig.4.15. Lines of the level of density and vector of gas velocity in stationary solution for 3D-calculations. The stationary configuration of the disc was produced by selecting the non-percolation conditiona at the boundaries of the region.

turbulence in the accretion disc. The large-scale structure of the vortices is of the two-dimensional type. Consequently, it is possible to carry out modelling of the development of turbulence in the two-dimensional approach so that we can use numerical grids with sufficient detail.

It is interesting to note that in studies with the exchange of mass in the binary system and, consequently, with the accretion process, the three-dimensional and two-dimensional calculations give differences in the structures of the flow and agreement in the result is recorded only for the values of γ close to 1. This is caused by the importance of the centrifugal barrier in the accretion regime. Matter does not accrete without the loss of the angular moment and the newly falling matter with a smaller angular moment is forced to flow around the resultant torus thus increasing the thickness of the disc and changing the structure of the flow. In the unidimensional case, the centrifugal barrier is not permeable [41]. Another interesting result is associated with the gradual decrease of the intensity of helices with subsequent loss of one of them (upper one). The authors of [37] also detected only one of the helices because the internal helix was disrupted by the flow of accreting substance. In the present work, the single-sleeve structure is associated with a low density of matter.

Analysis of the results of these calculations show that the hydrodynamic processes in the disc are determined by the complicated force field resulting in the formation of a structure containing spiral formations with increased values of density and pressure. However, without explicit consideration in the mathematical model of the mechanisms of exchange of matter between the accretion disc and the shell of the binary system it is not possible to produce the stationary structure of the disc. To achieve the stationary state of the disc, it is necessary to ensure the flow of mass from the donor star, for example, within the framework of the complete problem [38,40,41].

4.3. THE EFFECT OF VISCOSITY ON THE MORPHOLOGY OF THE FLOW OF MATTER IN SEMI-DIVIDED BINARY SYSTEMS. RESULTS OF THREE-DIMENSIONAL NUMERICAL MODELLING

Investigations of the effect of numerical viscosity on the structure of the flow were conducted in [41]. In this section, we shall present mainly the results of this investigation. Previously in [36,38–41, 146], the authors investigated the morphology of the flow in semi-divided binary systems. Within the framework of the three-dimensional numerical model, it was established that the presence of a diluted inter-component gas greatly changes the structure of the gas flows in the system. In particular, it was established that the self-consistent solution of the problem does not contain the impact interaction of the jet of matter, discharged from the internal Lagrange point, with the resultant accretion disc ('hot spot'). However, these solutions were obtained for the relatively high numerical viscosity in the disc and this was determined by the need for the application of a coarse calculation grid because of the limited possibilities of the power of the computing techniques used. In terms of α-viscosity, the numerical viscosity corresponded to [39]. At the same time, the problem of the structure of the flow at low viscosity values is of considerable interest, as confirmed by appropriate experiments. This problem is especially important in connection with the large-scale structure, possibly existing in the accretion disc.

In the present work, as in [36,38–41,146], the process of the flow of the gas is described by a system of Euler equations disregarding physical viscosity. The presence in the model of only numerical viscosity restricts the possibilities of detailed investigation for different values of viscosity, beçause the minimum value is restricted by the limit of numerical viscosity. Nevertheless, changes in the structure of the flow with the variation of viscosity in a wide range may be obtained on the qualitative level. Since the numerical viscosity dependence (for the selected difference scheme) on the time and spatial resolution, in order to obtain the dependence of the solution on the viscosity, a sequence of calculations with a decreasing step of the grid was used.

Investigations were carried out using a semi-detached binary system, consisting of a donor with mass M, filling the Roche cavity, and an accretor with mass M_{ac}. The masses of the components of the system were assumed to be equal. In order to describe the flow of the gas, the numerical solution of the system of hydrodynamic Euler equations in the three-dimensional space was used.

The adiabatic approximation was used in the solution. However, the qualitative consideration of the radiation losses of the energy may be taken into account by introducing the parameter of the adiabate close to 1. This parameter corresponds to the case close to the isothermal one. Consequently, the given physical–mathematical model of the disc can be regarded as a system characterised by energy losses [35,223].

As mentioned previously, the numerical viscosity in the investigation was reduced by changing the step of the calculation grid. Unfortunately, the limited power of the computer used in the study did not make it possible to model the flow in a large range on a dense grid. Consequently, like in [139,183,195,202], the flow of the gas was modelled in a limited region consisting of a parallelepiped (because of the symmetry of the problem in relation to the equatorial plane, the calculations were carried out only in the upper half space). A sphere with a radius of 0.01 was cut from the calculation region. The sphere represented the accretor. To provide more information, the parameters of the calculation region in the equatorial plane are shown in Fig. 4.16 (the solid line) and, in addition to this, the dotted line shows the Roche potential, the dot-and-dash line shows the calculated region used in [26,38–41,146].

The initial conditions were represented by a rarefied gas with a density 10^{-5} times lower in comparison with the density of the gas in the jet. The boundary conditions were determined on the internal Lagrange point where the conditions of injection of the jet of the

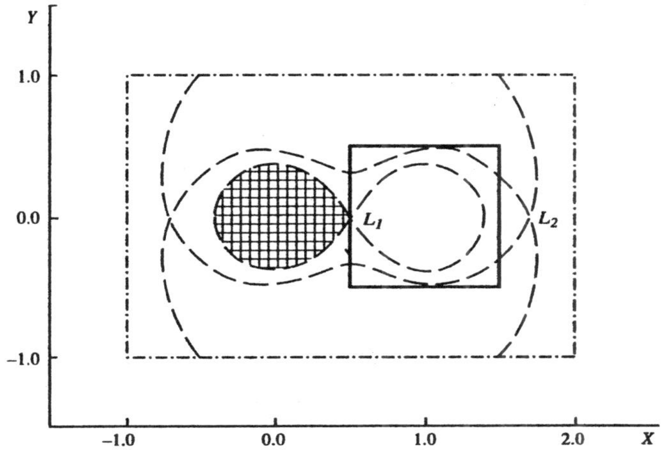

Fig.4.16. Calculation region for the 'complete' (dot-and-dash) and 'restricted' (thick line) problems. The cross-hatched region shows the donor. The broken lines show Roche equipotentials.

substance in the vicinity of the accretor were specified. The velocity of sound in the boundary cell was $c(L_1) = 10^{-1}$. The initial ('background') values of the parameters were maintained throughout the calculations at the fictitious nodes situated inside the accretor, and also at the fictitious nodes in the remaining part of the external boundary. Finally, the boundary conditions were determined as a result of solving the problem of the breakdown of a discontinuity (the Rieman problem) between the parameters of the gas at the fictitious node and the parameters of the gas at the calculation node closest to the given point.

The results obtained in previous investigations [see: 36,38–41, 146] showed a significant role of the inter-component shell in the formation of the flow structure. The application, in the calculation model, of the restricted calculation region does not make it possible to take into account efficiently the inter-component shell and, correspondingly, the 'restricted' solution may greatly differ from the 'complete solution'. We shall examine the effect of the restriction of the calculation region on the solution. For this purpose, which we compare the structure of the flow from the 'complete' model [41] with the calculations, obtained in the same grid in the restricted region. Figures 4.17 and 4.18 show the isolines of the density and the vector of the speed in the equatorial plane of the system for the 'complete' and 'restricted' calculations, respectively. Figures

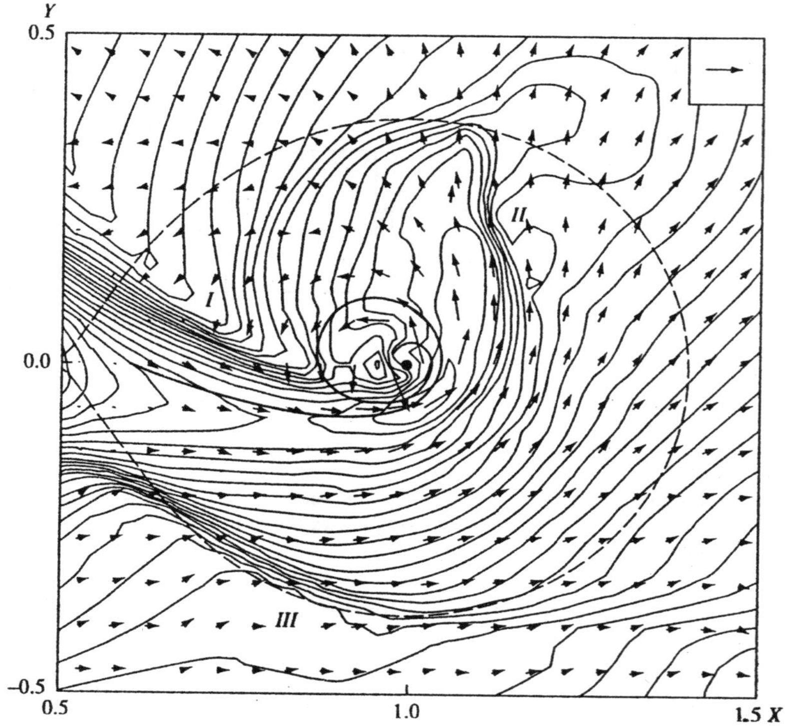

Fig.4.17. Isolines of density and vector of velocity in the equatorial plane for calculations in the 'complete' region. The solid dot in the centre indicates the accretor, broken lines the Roche equipotentials. The vector in the upper right corner correspond to dimensionless velocity $u = 4$. The thick line shows the boundary ('last') line of flow from those through which the flowing matter falls directly into the disc.

4.17 and 4.18 also show the boundary ('the last') lines of the flow through which the matter penetrates directly into the disc. Comparison of the presented results shows that the morphology of the flow in the region around the disc in the 'restricted' problem reflects the main special features of the flow of the 'complete' problem. In particular, the system is characterised by the formation of an accretion disc with approximately the same linear dimensions. Shock waves I and II form. These waves are caused by the interaction of the gas of the intercomponent shell with the jet of the matter discharged from L_1. At the same time, the effect of the intercomponent shell in the 'restricted' problem is not taken into account completely correctly. In particular, there are no flows of

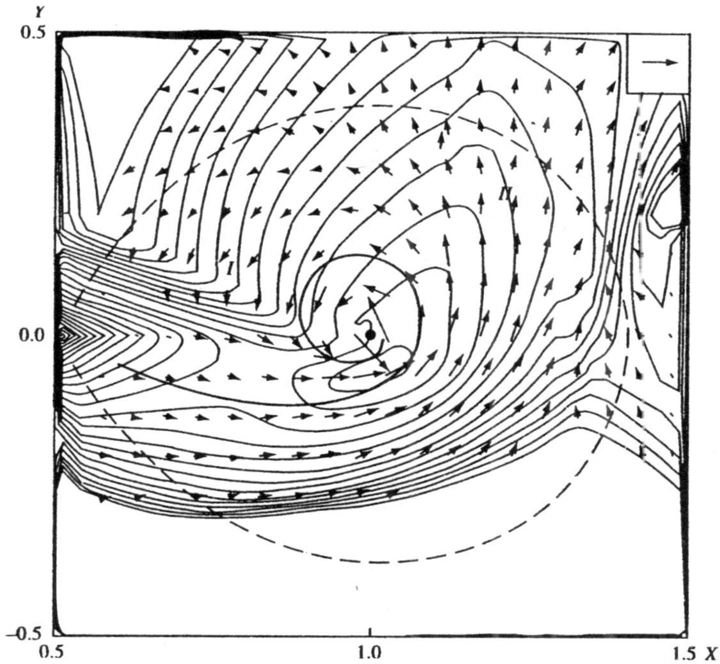

Fig.4.18. Isolines of density and velocity vector in the equatorial plane for calculations in the 'restricted' region (symbols are the same as in Fig.4.17).

the substance along the Roche plane of the donot star which, as shown in [40], results in the collapse of the atmosphere of the mass-losing star, causing changes in the structure of the flow in the vicinity of point L_1 and a significant increase of the temperature of transfer of matter in the system. This circumstance greatly changes the parameters of the structure of the 'restricted' problem in comparison with the 'complete' problem. This may be illustrated by Fig. 4.19, showing the distribution of density in the equatorial plane of the system along the line, passing through L_1, parallel to the Y axis. The results presented in Fig. 4.19 show that in the 'restricted' problem, the jet of the substance, discharged from L_1, expands to the characteristic size $\varepsilon = c(L_1)/(A\Omega)$ [199]. The density in the cross-section of the jet decreases exponentially [199], whereas in the 'complete' problem, the density of the matter of the jet is greatly determined by the effect of the collapse of the atmosphere of the donor star. In addition to this, the 'restricted' problem does not contain the flow of the intercomponent gas, returning to the system

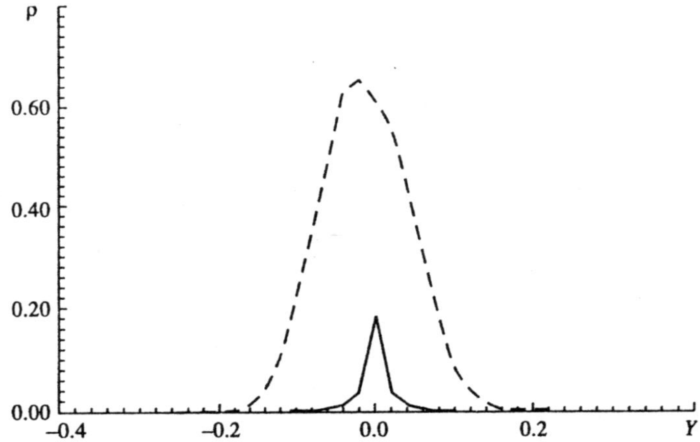

Fig.4.19. Profiles of density in the equatorial plane along the line passing through L_1 ($x = 0.5$, $y = 0$, $z = 0$) parallel to the Y axis for the 'complete' (broken line) and 'restricted' (solid line) problems.

under the effect of the Coriolis force (from the side of the jet, opposite to the direction of orbital motion). This circumstance results in the disappearance, in the 'the restricted' problem, of the shockwave III which forms in the 'complete' formulation. The following conclusions can be made:

1. In the 'restricted' problem, examination of the effect of the general shell on the solution is not completely correct and this results in both qualitative changes in the structure of the flow – the system does not contain flows of the intercomponent gas (lines of the flow *a* and *b* in Fig. 4.20) and, consequently, shockwave III disappears, – and also in quantitative changes – because the collapse of the atmosphere of the donor star is not considered, the parameters of the jet change and, in particular, the thickness of the jet greatly decreases.

2. The morphology of the flow of matter in the vicinity of the accretor in the 'restricted' and also 'complete' problems on the qualitative level is similar. By partially taking into account the intercomponent shell – i.e. the substance rotating around the accretor and interacting with the jet (flow lines *c* in Fig. 4.20), in the 'restricted' problem we can obtain a solution corresponding to the 'complete' solution in the region in the vicinity of the accretor. In particular, as in the 'complete' problem [36,38–41,146], in the 'restricted' solution, the jet, deflected under the effect of the gas

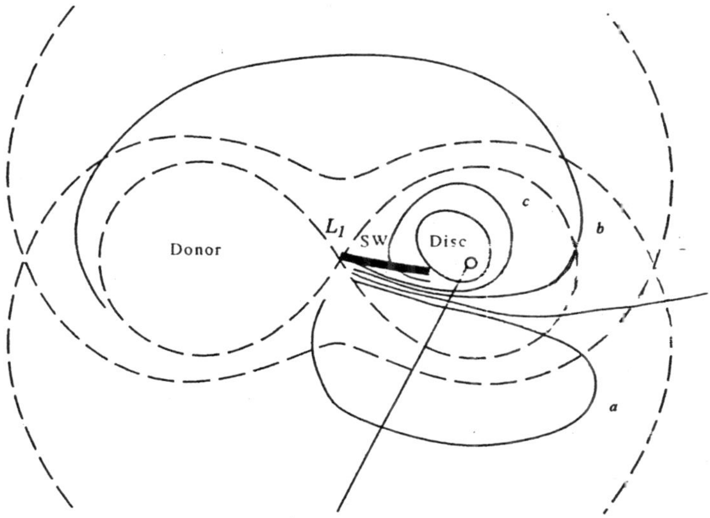

Fig.4.20. Schematic representation of special features of gas-dynamic flow of matter in semi-divided close binary systems. The figure shows Roche cavities (broken lines), position of the accretor and Lagrange points and also the accretion quasi-elliptical disc. The shock wave, formed as a result of interaction of the gas of the general shell with the jet, is denoted SW and indicated by the thick line. The flows *a,b,c* form the general shell of the system. Flow line *d* corresponds to the flow of matter leaving the system.

of the intercomponent shell, approaches the disc along the tangent and does not cause any impact effect on the edge of the disc (the effect of the 'hot spot'). The region of generation of energy is situated along the edge of the jet facing the orbital motion where the interaction of the general shell with the jet results in the formation of an elongated impact wave I (Figs. 4.17, 4.18, 4.20).

The result of comparison of the 'restricted' and 'complete' problems show that the interaction of the jet with the accretion disc can also be examined, with certain reservations, in the 'restricted' formulation. Because of this circumstance, it is possible to carry out calculations on denser grids and, consequently, at lower viscosity. It also makes it possible to investigate the effect on viscosity on the structure of the flow of matter in the vicinity of the accretor and examine the problem of the existence of the 'hot spot' at low viscosity.

Three series of calculations were carried out with different spatial resolutions of the grids: 31×31×17.61, 61×61×17 and 91×91

×25 (they will be referred to as the calculations A, B, and C, respectively). In all variants of the calculations, the grid was uniform. In terms of the α-disc, the numerical viscosity of the variants of the calculations A, B and C corresponded approximately to $\alpha \approx 0.08 \div 0.1$, $0.04 \div 0.06$ and $0.01 \div 0.02$.

Comparison of the results of the calculations A, B and C together and also with the results presented in [195] (where a 200×200×50 grid was used in the same formulation) makes it possible to examine the effect of viscosity on the solution. Figures 4.21, 4.22 and 4.23 show the fields of density and the vector of velocity in the equatorial plane of the system for the variants A, B and C and also the images of the flow lines *a* restricting the accretion disc. Numerical analysis of the solutions shows that the interaction between the jet and the disc is free of impacts in all calculation variants. The morphology of the jet and of the accretion disc is of the same type and, consequently, 'hot spots' do not form in the system. This fact may be illustrated by Fig. 4.24 showing,

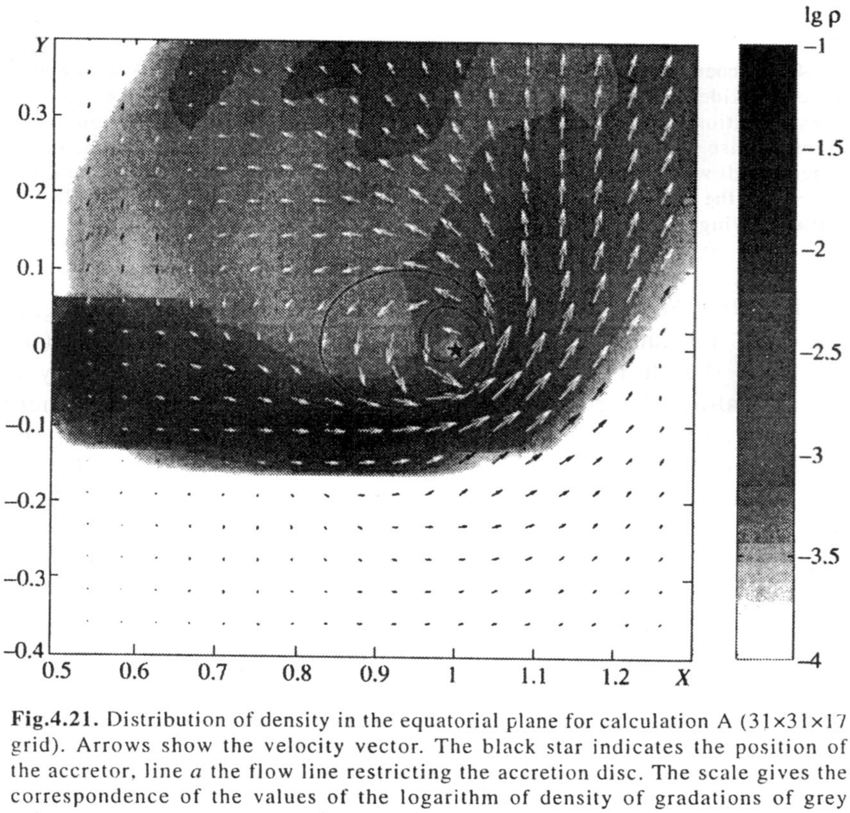

Fig.4.21. Distribution of density in the equatorial plane for calculation A (31×31×17 grid). Arrows show the velocity vector. The black star indicates the position of the accretor, line *a* the flow line restricting the accretion disc. The scale gives the correspondence of the values of the logarithm of density of gradations of grey colour.

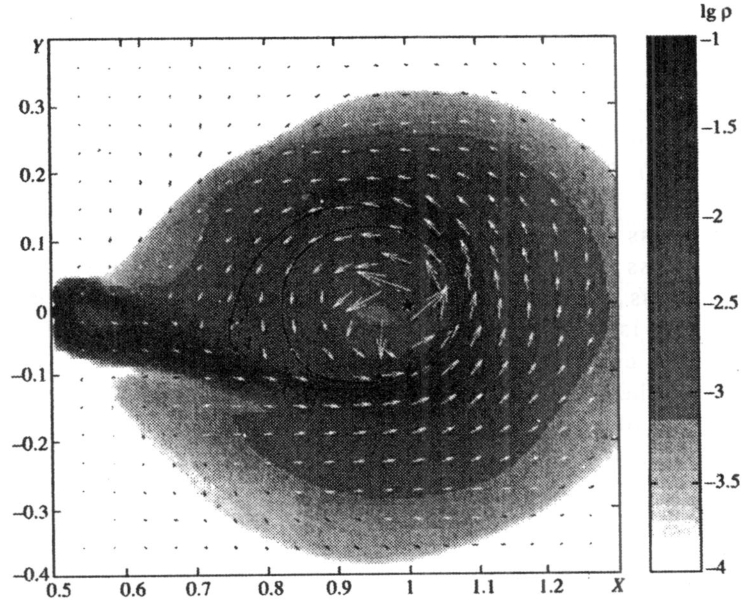

Fig. 4.22. The same as Fig.4.21, for calculation B (61×61×17 grid).

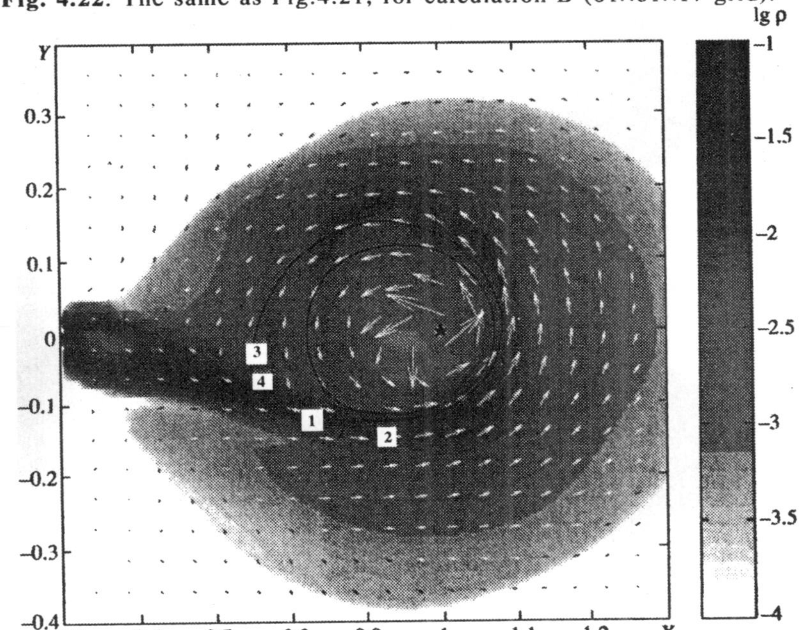

Fig.4.23. The same as Fig.4.21, for calculation C (91×91×25 grid). In addition to flow line *a*, restricting the accretion disc, the graph shows the flow line *b* passing through the shock wave. Section 1–2 on the line *b* corresponds to the area of contact of the jet of matter from L_1 and the accretion disc, section 3–4 corresponds to the are of interaction of the inter-component gas with the jet (shock wave).

for the calculation variant C (with the minimum value of viscosity), the distribution of the dimensionless temperature in the XY plane for four values of the coordinate Z: $z = 0$ is the equatorial plane, and $z = h$, $2h$, $3h$, where h (the size of the calculation grid in the direction of height (corresponds to the scale $0.01A$. The dimensionless temperature at point L_1 is 10^{-2}. For transition to dimensionless temperature, it is necessary to multiply the dimensionless values by $GM/(AR)$, where G is the gravitational constant, M is the total mass of the system, A is the distance between the components of the system, R is the gas constant. Analysis of the temperature field, presented in Fig. 4.24, indicates the absence of the region of energy generation in the area of contact between the jet and the disc, i.e. the absence of the 'hot

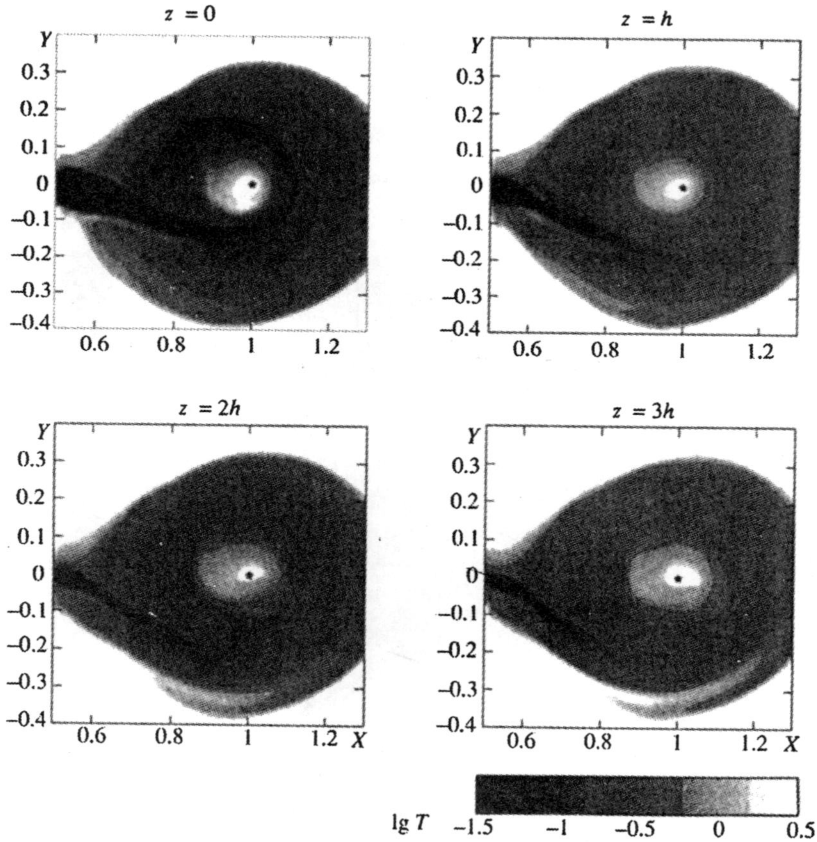

Fig.4.24. Distribution of dimensionless temperature for calculations B (91×91×25 grid) in the equatorial plane ($z = 0$) and also in planes parallel to this plane $z = h$, $z = 2h$, $z = 3h$, where $h = 0.01$ is the size of the calculation grid in the direction of height. The black star indicates the position of the accretor.

spots' for all values of z. Examination of the variation of the gas parameters along the flow lines confirms this conclusion. In particular, Fig. 4.25 shows the variation of dimensionless temperature along the flow line b (Fig. 4.23). Analysis of the variation of temperature makes it possible to draw conclusions on both the absence of 'hot spots' in the area of contact between the jet and the disc (region 1–2) and on the position of the region of energy generation in the area of formation of shockwave I (region 3–4 in Fig. 4.23 and 4.25).

Comparison of the resultant structures of the flow shows (see, for example, Fig. 4.21, 4.22, 4.23) that the improvement of the spatial resolution of the grid and, consequently, the decrease in the numerical viscosity reduces the extent of diffusion spreading of the jet discharge from the internal Lagrange point. In the variants of the calculations with a finer jet (variants of B, C and also calculations in [195]) a significant role in the system is played by the gas of the intercomponent shell flowing around the jet at the top and bottom. This fact is illustrated in Fig. 4.26 where, for calculating C, the field of the density and vector of velocity in the cross-section YZ, produced through the point (0.9; 0.0; 0.0), is shown. The variation of the density (increase in the thickness of the flow) in the region of flow of the gas of the shell around the jet, is also shown in Fig. 4.27 where the frontal plane (XZ) shows the field of density for calculations C. The presence in the system of the gas flowing around the jet may lead to the impression of an

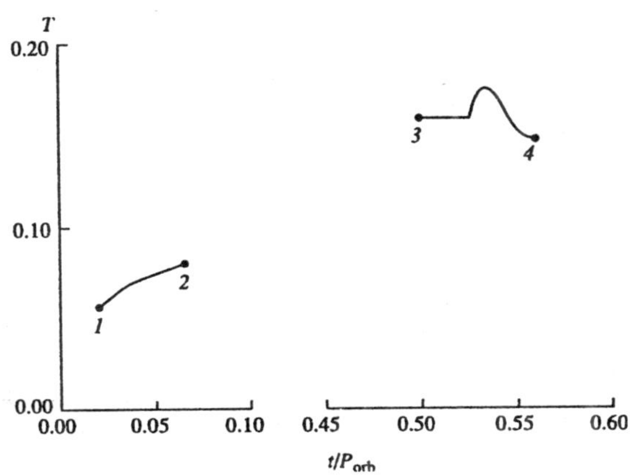

Fig.4.25. Dimensionless temperature along flow line b in Fig. 4.23. Section 1–2 corresponds to the area of contact of the jet and the accretion disc, 3–4 corresponds to the shock wave.

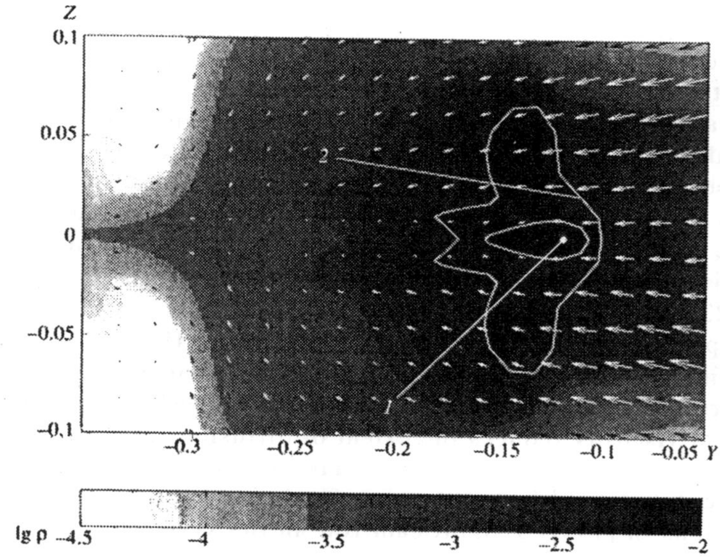

Fig.4.26. Distribution of density and velocity vector in plane *YZ*, drawn through the point (0.9; 0; 0), for calculations B (91×91×25 grid). Marker 1 shows the position of the central part of the jet restricted by the isoline lg $\rho = -2.3$, marker 2 shows the front of the shock wave.

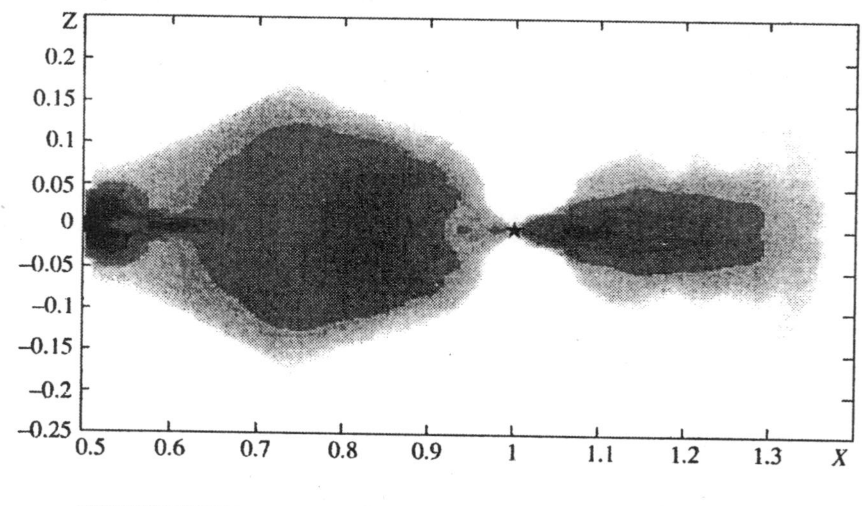

Fig.4.27. Distribution of density and velocity vector in the frontal (*XZ*) plane for calculations B (91×91×25 grid). The black star shows the position of the accretor.

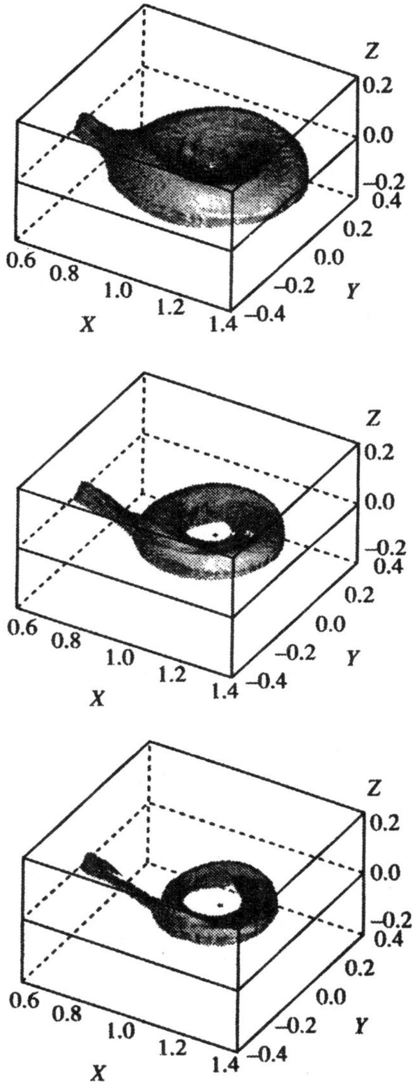

Fig.4.28. Spatial image of density isosurfaces.

impact interaction between the jet and the disc (see, for example, the three-dimensional image of the isosurface of density in Fig. 4.28 and also the results in [195]), although analysis of the solution indicate unambiguously the smooth nature of the flow.

The structure of the image of the flow was characterised in [36, 38–41,146] in the following manner:

- the disc – matter of the jet which is immediately trapped by the gravitational attraction of the accretor and in subsequent stages accretes on the central body;
- the intercomponent shell – all remaining matter, and part of this matter rotates around the accretor, interactions with the jet and can be accreted in subsequent stages.

Thus, the matter of the jet was divided on the basis of the physical feature: if the gas ieaves the system or interacts in subsequent stages with the initial jet, this matter does not belong to the disc. In solutions with low viscosity another part of the shell appears, i.e. the matter flowing around the jet at the top and bottom. Analysis of the flow lines shows that this part of the shell does not belong to the disc. A large part of this matter leaves the system. The second part of matter gradually approaches the equatorial plane, rotating around the accretor, and in subsequent stages interacts with the jet. In order to supplement the definitions in [36,38–41,146], it is efficient to introduce the additional term –the hallo around the disc, describing the substance which

1 Rotates around the accretor (gravitationally bonded);

2 Does not belong to the disc (in physical formulation);
3 Interacts with the jet (collides with the jet or flows around it);
4 In subsequent stages, after the interaction, either takes part in accretion or leaves the system.

The variation of viscosity in the system leads unavoidably to changes in the rate of accretion of the matter. Analysis of variants A, B and C shows that, as expected, in the calculation with low viscosity the rate of accretion decreases. Unfortunately, detailed examination of this problem requires extremely long computing times greatly exceeding the available resources. The point is that in the calculations with high viscosity [36,38–41,146], the viscosity time of the disc

$$\tau_V = \frac{R^2}{\nu} = \frac{R^2}{\alpha c_s H},$$

where R is the radius of the disc, c_s is the velocity of sound, H is the characteristic thickness of the disc, did not greatly exceed the hydrodynamic scale of time for the disc $\tau_{ga} = \frac{R}{c_s}$ so that it was possible to obtain a stationary solution at times of the order of several orbital revolutions of the system. With a decrease of viscosity, finding the stationary solution is associated with problems because the limited power of computes does not make it possible to carry out calculations at times longer than τ_V. Analysis of variant C, i.e. examination of the dependence of the mass of the disc and the hallo on time, shows that even at times exceeding 15 orbital periods the solution is not yet stationary. Because of this circumstance, it was not possible to obtain reliable numerical estimates of the rate of accretion of matter in systems with low viscosity.

Analysis shows that in the systems with different values of viscosity, the jet, deflected under the effect of the gas of the intercomponent shell, approaches the disc along the tangent and does not cause any impact disruption of the edge of the disc ('a hot spot'), and the region of excessive heat generation is situated in the shockwave along the edge of the jet, facing against the direction of orbital movement. This circumstance made it possible to formulate a conclusion on the qualitatively similar pattern of the flow at different values of viscosity in the system. At the same time, in variants with low viscosity (characterised by low values of diffusion and, consequently, the small thickness of the jet) the flow pattern showed an important role of the gas of the intercomponent shell flowing around the jet at the top and bottom. An additional term ' the hallo around the disc' was introduced to describe this detail of the flow pattern.

4.4. LARGE-SCALE STRUCTURE OF TURBULENCE IN ACCRETION DISCS

Interesting investigations of the development of turbulence in accretion discs were carried out in [188] where it was shown that classic hydrodynamic turbulence cannot in principle affect the transport of the angular moment in a Kepler disc. In a formal examination, the following system of equations for the transfer of the angular moment of forces F_R and F_Φ can be obtained for the two-dimensional case from the equations of hydrodynamics in averaging in respect of the turbulence spectrum:

$$\frac{\partial}{\partial t}\left\langle \frac{1}{2}\rho U^2 \right\rangle + \langle \nabla F_R \rangle = 2\omega\langle \rho U_R U_\phi \rangle + \left\langle \rho \frac{\partial U_R}{\partial R} \right\rangle + \nu\langle \rho |\nabla U_R| \rangle,$$

$$\frac{\partial}{\partial t}\left\langle \frac{1}{2}\rho U_\phi^2 \right\rangle + \langle \nabla F_\phi \rangle = -\frac{k^2}{2\omega}\langle \rho U_R U_\phi \rangle + \left\langle \rho \frac{\partial U_\phi}{\partial \phi} \right\rangle + \nu\langle \rho \|\nabla U_\phi| \rangle$$

These equations show that, in contrast to the flat flow, sources of turbulence may include Coriolis forces – the first member in the first equation in the right-hand part. The member containing epicyclic frequency $k^2 = \dfrac{2\omega}{R}\dfrac{dR^2\omega}{dR}$, the first member on the right-hand part of the second equation, is a stabilising factor. Consequently, it is necessary to solve the problem of the possibility of development of turbulence in the nonlinear regime. These investigations can be carried out only using methods of mathematical modelling.

As already mentioned, the methods of mathematical modelling are restricted by the problem of numerical viscosity or dissipation (for schemes of a high order). In addition to this, it is usually argued that this instability cannot be detected because it requires high Reynolds numbers. For example, for an accretion disc $Re_c \approx 8 \times$

10^5, which is considerably higher than the Reynolds number in the flat flow. At the same time, computing experiments carried out in [188] showed that where there is instability to the development of turbulence this instability is detected, and in other cases the flow remains laminar. In this study, calculations were carried out of the evolution of the kinetic energy of turbulence for different laws of rotation. The results show that for the exponent in the law of rotation $n > -2$, the kinetic energy of turbulence in the disc decreases and the flow transfers to laminar regime. In a study by Zel'dovich [58], similar results were obtained for the Taylor criterion for liquids in the gap between rotating cylinders.

As mentioned previously, it is necessary to solve the problem of numerical viscosity in multidimensional calculations, modelling the development of the nonlinear phase of turbulence. In traditional experimental hydrodynamics, the presence of viscosity might prevent the development of turbulence. In other words, the flow in a disc with low Reynolds numbers remains laminar, whereas the development of turbulence requires high Reynolds numbers so that the problem of numerical modelling of turbulence is very difficult to solve because of the high values of numerical viscosity.

However, a solution was found in the following direction. The numerical viscosity in solving Euler equations is not identical with small Reynolds numbers. The point is that Euler equations have their dynamic instabilities from which turbulence structures may form in calculations, whereas the application of the Navier–Stokes equations could not result in the formation of such structures because of high numerical viscosity and restricted wavelength (in comparison with the size of the difference grid), existing turbulent instability. In 1990, Porter [214] carried out test calculations for comparison of the code of Navier–Stokes equations and the cold of Euler equations. It was shown that the code for the Euler equation resolves far more efficiently the fine structure and short waves in comparison with the Navier–Stokes equations at Reynolds of numbers. It was concluded that the Euler equations may be used for modelling with high Reynolds numbers with much better resolution. Identical conclusions were obtained by Belotserkovskii in 1985 [16].

Recently, the problem of the coherent transport of the angular moment has been discussed in a number of studies. As indicated by the previous discussion, the problems of the structure of the accretion disc and the mechanism of the transport of the angular moment has a controlling effect on the structure of the flow in the disc. The application of high values of the turbulent viscosity or high numerical viscosity leads, as in the model of α-viscosity, to rapid heating of the matter of the accretion disc. The problem of the high-temperature of the disc has been reflected in a large number of investigations into the advection of energy in accretion discs (see the review in [147]). The redistribution of the angular moment through large-scale structures does not increase temperature. In this case, the structure of the disc will be calculated using efficient difference grids enabling the development of turbulence of large-scale structures in the calculations.

The authors of [152] investigated the mechanism of coherent transport of the angular moment as a result of the development of large-scale vortices. The initial instability of the shear flow is explained by the development of Rossby vortices [188]. The authors of [152] investigated the development of Rossby vortices as a result of the tangential injection of compressed air resulting in a radial pressure gradient. In particular, this gradient leads to the formation of a heterogeneity in the distribution of vorticity and increases the intensity of Rossby instability. Thus, it may be assumed that the non-monotonic form of the distribution of density, entropy and pressure gradient lead to the formation of Rossby vortices. They in turn lead to the development of turbulence in accretion star discs [188].

4.5. CONVECTIVE INSTABILITY IN ASTROPHYSICS

In astrophysical conditions, the problem of mixing plays the controlling role in many objects and processes. For example, the problem of mixing, as a result of the development of convection, is important in explosions of supernovas, in the evolution of stars, in thick accretion discs and other objects and processes.

The development of convection is associated with Archimedes forces and energy generation processes. Convection occurs where the diffusion processes cannot ensure efficient heat conductivity. This results in the formation of a classic situation with a heavy fluid positioned above a light fluid in the field of gravitational forces, resulting in the development of Rayleigh–Taylor instability. A similar situation forms in the deceleration of shockwaves in mass-losing hells of supernovas or in the scraped up interstellar gas.

Instability also develops during the passage of shockwaves through the layer structure of pre-supernovas. The Richtmyer–Meshkov instability may develop in this case. It is necessary to solve the problem of the geometrical instability of the fronts of thermonuclear combustion in supernovas. Instability may be associated with both detonation and deflagration fronts of thermonuclear combustion.

At present, the possibility of the development of Richtmyer–Meshkov instability is indicated by the detection of the early yield of γ-radiation from the breakdown of radioactive Co^{56} [156,232]. Examination of the mixing processes resulting from the development of convective instability is important for determining the occurrence of heavy elements and further processing of the mechanisms of explosion of stars essential for investigating the physical properties of matter under extreme conditions.

Initially, we shall pay attention to the development of convection during evolution of stars. The widely used Schwartzschild criterion of convective instability says that the temperature gradient at every point must not exceed the adiabatic gradient. Consequently, the stability equation has the following form:

$$-\left(1-\frac{1}{\gamma}\right)\frac{T}{P}\frac{dP}{dr}>-\frac{dT}{dr}, \quad -\left(1-\frac{1}{\gamma}\right)\frac{T}{P}\frac{dP}{dr}>-\frac{dT}{dr},$$

where T is temperature, P is pressure, r is the actual radius in the star, γ is the adiabatic indicator equal to the ratio of the specific heat capacities $\frac{c_p}{c_V}$. The fulfilment of the criterion is usually verified using the following procedure. Attention is given to a small isolated element of the volume of matter and whether it is subjected to small perturbations and whether the element interacts with the surrounding matter. The justification of the criterion is usually regarded as convincing but, in reality, it is not suitable for the accurate evaluation of the stability of the system in terms of the normal modes and the initial conditions because the system reacts to the loss of stability and to a perturbation as an **integral unit**. From this viewpoint, the qualitative arguments, used for the justification of the Schwartzschild criterion, are no more than an assumption that for the spherical configuration, the instability can be described by the modes of a relatively high order (i.e. the modes belonging to the spherical harmonics of high order l and radial functions with many zeroes) which appear if the criterion is violated in any *small* vicinity of the isolated region. It may be asserted that the local approach does not satisfy the conditions of development of convection having the limiting values of the scale for characteristic unstable modes. The main assumption, representing the basis of the given approach, is the ignorance of the variation of the gravitational potential resulting from the perturbation. Although this assumption can be efficient for examination of high order modes, it is hardly suitable for the modes belonging to $l = 1$ resulting from the perturbation [150]. These modes are most important for the recognition of convective instability in many important cases.

The first accurate examination of non-radial adiabatic pulsations [211] shows that the instability, formed in a homogeneous compressed sphere, has the modes of all moments, including $l = 1$. The main reason for the instability is that the homogeneous density is super-adiabatic everywhere for any finite range of the the adiabate exponent γ. Consequently, it is fully justified to expect the occurrence of large-scale convection with the modes $l = 1$, for example, for the case of a sphere with a viscous liquid, heated from the inside [149]. The considerations that are valid for the homogeneous sphere must also be valid for the sphere with the equation of the state of matter described by a polytrope in the sense that in the present case, convective instabilities with the modes $l = 1$ also develop. This conclusion becomes evident if it is noted that the polytropic gas, for which the pressure and density are linked by the equation $P = \text{const}\,\rho^{1+\frac{1}{n}}$, is super-adiabatic everywhere where $\gamma < 1 + 1/n$. Consequently, if $1 + 1/n$ exceeds γ, the Schwartzschild criterion will be violated simultaneously in the entire matter and the instability will be manifested in the form of the circulation of the largest permissible scale.

However, even in cases (for example, in the Cowlini model [153]) in which a star is examined as an object with a point source), where the effective index of the polytrope decreases from a relatively high value in the external shell of the star to a sub-critical value in its central part and the conditions, finally, are different everywhere, it is clear that the most important modes are associated with convective instabilities, again those which belong to the harmonics of a lower order, including $l = 1$, and are described by radial functions having one or several zeroes. If these assumptions are accurate, the instabilities with the modes of actually high orders do not play such a significant role in the theory of convective instability, as assumed previously. In this sense, the approximation of the length of mixing for describing convection may be insufficient [150].

Large-scale instability during the explosion of a supernova
In the last 30 years, several groups of investigators have examined the reasons why the collapse of the iron core of a star with a mass $M > 10M_{\odot}$ results in the formation of a supernova of type II and the formation of a neutron star. To explain the supernova phenomena, two main mechanisms have been proposed: the first is associated with the rapid collapse of the shell of the star during the passage of a shock wave through the shell formed as a result of the arrest of the collapse on reaching the density higher than the nuclear density [237], the second is associated with the heating of matter behind the front of the shockwave due to the neutrino, leaving the neutrino-sphere, surrounding the proton–neutron star. This results in further moment of the shock wave and the collapse of the shell [239].

However, numerical calculations show that these mechanisms are unsatisfactory because of several reasons. In the first of these mechanisms, the so-called rapid mechanism, the shock wave loses a large part of its energy in the process of splitting of the iron core into free nuclons [204]. When the shock wave reaches the neutrino-sphere, the electron neutrino carries the thermal energy and the lepton number from the front of the shockwave thus decreasing the energy of the wave and pressure in it. This results in the weakening of the shockwave and, in subsequent stages, in its arrest [239]. The second mechanism is determined by the conditions in the region between the neutrino-sphere and the shock wave and depends in a critical manner on the value of neutrino luminosity and the mean spectral energy of the neutrino. In this case, the level of luminosity, required for sustaining the divergent shock wave, can be obtained only taking into account convection both below the neutrino sphere and above it. In turn, convection may develop relatively rapidly only on the condition that heating (and the construction of the unstable stratification of matter in respect of entropy) is more rapid than the movement of the matter from the shock wave to the surface of the proton–neutron star [180].

In the period after the arrest of matter in the core and the propagation of the resultant shock wave on the shell, the core of the star may also be characterised by the development of hydrodynamic instabilities associated with convection. Examination of the rapid increase of emission from SN1987 x-ray and γ-

radiation, mixing of the external layers with the products of thermonuclear explosion, containing Ni^{56}, and the observed structure in the form of two jets in the centre show that convective motions may take place in a certain stage of explosion. The development of several convective modes is possible in the period of restoration of the shock wave.

One of the types of convection forms in the passage of the shock wave through the external part of the core and is associated with the resultant negative gradient of entropy. This convection is short-lived because the negative gradient of entropy is not maintained by the heating of the central areas of the star and is rapidly reduced during convective mixing. In addition to this, at the moment of the maximum of development of convection, the radius of the photosphere for ν_e is very close to the radius of the shock wave because the region of heating behind the shockwave is very narrow. A large part of the sources ν_e and $\bar{\nu}_e$, distributed on the appropriate neutrino-spheres, whose thermodynamic condition is established by the transfer of heat and the number of leptons by the matter, accreted on the proton–neutron star. Therefore, this type of convection may enhance the divergent shockwave [145].

The numerical modelling of convection in the vicinity and below the neutrino-sphere [200], taking neutrino transfer into account, shows that the convective speed is too low in comparison with the volume speed of inflow of matter in order to cause any significant transfer of entropy and leptons. Consideration of the transfer of energy and leptons due to the neutrino from the rising element of the liquid to the background reduces the rate of growth of entropy convection 3–50 times, lepton convection 250–1000 times for the region between the neutrino sphere and the matter with a density of $\rho = 10^{12}$ g/cm^3. However, the calculations of the evolution of the initial isothermal or adiabatic perturbation from the equilibrium position [145] shows that the time to establishment of the equilibrium state with respect to the entropy and leptons of the liquid element with the background is slightly longer than in the hydrodynamic calculations. Nevertheless, owing to the fact that convection at a negative gradient of entropy rapidly dissipates, and in the case of lepton convection, which exists for a considerably longer period of time, the time to establishment of the equilibrium state is considerably shorter than the time of growth of convective motion, the conclusion of the suppression of convection with neutrino transfer taken into account does not change in principle. As a result, the contribution to neutrino luminosity from the region in the vicinity and below the neutrino sphere is insignificantly small.

The second type of convection, the so-called neutron figers [239], forms in the region in which the negative gradient of the lepton number is stabilised by the positive gradient of entropy, on the condition that the element of the liquid comes into thermal equilibrium with the surrounding matter faster than to the equilibrium in respect of leptons. The first requirement is explained by the fact that as a result of rapid convection, the gradient of entropy becomes positive, whereas the negative gradient of the lepton number, which initially forms in the vicinity of the neutrino sphere, expands into the nucleus as a result of the diffusion of leptons outwards. Since all six types of neutrino transfer energy and only two leptons $(\nu_e, \bar{\nu}_e)$ transfer the lepton number as a result of the large difference in the path to interaction with the matter, the second condition is also fulfilled. Numeral calculations of the development of this instability show, however, that in does not develop because of the following reasons [145]:

a) the condition of instability changes, because the typical values of the lepton number in the large part of the nucleus are half the critical number;

b) the transfer of the large part of energy from the liquid element to the background takes place by means of ν_e and $\bar{\nu}_e$-neutrino and, in this case, the contribution to this process from the neutrino is small;

c) in the typical conditions, the number of acts of neutrino transfer of energy, required in the nucleus, is larger than for the neutrino transfer of the lepton number;

d) the fluxes of the ν_e and $\bar{\nu}_e$-neutrino during the large part of the evolution period of the initial perturbation are directed to opposite sides and, in this case, they are subtracted in the transfer of energy and added up during the transfer of the leptons.

Until recently, the role played by the convection in the mechanism of explosion of the supernova star has been to a large extent unclear and contradicting. To a large degree this indeterminacy is associated with the development of methods of calculating neutrino transfer, selection of the equations of the state of matter, the dimensionality of the problem being solved, numerical methods, shortcomings of adequate models of the nuclear equation of the state of matter and, possibly, the absence of accurate understanding of the interaction between the neutrino and matter at nuclear densities.

In [173], using the method of SPH modelling for two-dimensional hydrodynamic calculations it was shown that the area in the vicinity of the neutrino sphere is characterised by the development of lepton convection and that entropy convection occurs above the neutrino sphere. In turn, in [239] it was reported that the convection inside a proton–neutron star is weak and, evidently, not important. The authors of [182] examined the evolution of a proton–neutron star during the time of the order of 1 s within the framework of the two dimensional hydrodynamics in the radial diffusion of neutrino.

The results show that lepton convection continues for a long time and affects the entire nucleus after 1 s. This is accompanied by the doubling of neutrino luminosity and the mean energy of the neutrino increases by 10–20%. On the other hand, it was shown in [145] that when using the multigroup diffusion approximation for calculating neutrino transfer in the approximation of the length of the mixing path for convection, the latter is characterised by low efficiency inside and around the neutrino sphere for 30 ms, regardless of the factors that modelling lasted 0.5 s from the moment of recoil of the nucleus.

In the process of collapse of the iron core of the star, approximately 99% of the entire gravitational energy is taken away in the form of neutrino. To cause collapse of the external shell of the star it is sufficient that only part of this energy is transformed to the outer layers of the star by means of an effective and fast mechanism. Convection – both inside and from the outside of the neutrino sphere – may intensify the process of transfer of energy to the front of the shock wave but, nevertheless, requires certain conditions. For example, the characteristic time of formation of convection should be shorter than the characteristic time of accretion of the matter of the outer layers on the core. In addition to this, convection requires constant supply, as in the case of entropy convection inside a proton–neutron star. Evidently, a more realistic mechanism may be associated with the effect of hydrodynamic large-scale instabilities acting at short characteristic times of the order of 10^{-3}–10^{-1} s and supporting the powerful flux of high-energy neutrino radiation.

The authors of [122] investigated the interesting possibilities of powerful emission of neutron radiation associated with hydrodynamic motion in a proton–neutron star. In fact, this study represents a development of the concepts proposed in [59]. The authors of [59] examined the development of convective instability in a gravitating homogeneous gas sphere (star). The main idea was based on the similar time behaviour of the integral over the cross-section for the entropy and the same integral for the strength of the magnetic field. In the experiments, these instabilities were detected in a Tokamak. The authors of [43] investigated the development of helical MHD instability in a plasma cylinder. It was shown that low unstable modes, associated with large heterogeneities, as in the case of plasma instability, grow almost rapidly. Analytical estimates in the calculation of the two-dimensional problem show that the internal, hot layers are displaced of the surface of the star during the characteristic time $\tau \approx R/V_{sound}$.

The authors of [122] carried out three-dimensional calculations of the development of hydrodynamic instability in a proton–neutron star with an excess of entropy in the centre. The numerical solution of Euler equations was used to examine the time development of the scale of the heterogeneity of entropy distribution. In the formulation of the problem, the entropy shows the evolution of the rising of hotter, lighter matter. The calculations showed the development, during 4 ms, of large-scale heterogeneities of entropy with the shape resembling the 'mushroom' of a nuclear explosion in the atmosphere of the earth with a distance from the centre of the star of 20 km. The characteristic time of movement of the single resultant bubble to the surface of the proton–neutron star was 1 ms, which corresponds to a mean speed of $c/150$ (c is the speed of light). A total of 6 bubbles with a mass of $10^{-2}\ M_{\odot}$ rose. The intensity of neutron radiation was 5×10^{52} erg/s for the given process. The energy, absorbed by the matter per 1 g in the shock wave from neutron radiation was 2.3×10^{24} erg/g s which was comparable with the neutrino losses from the front of the shock wave. Consequently, it was concluded that the mechanism of explosive convection could sustain the divergent shock wave and, therefore, result in the burst of the supernova shell.

To develop further the proposed model of burst convection, investigations were carried out into the development of a hydrodynamic instability, formed in a rotating proton–neutron star [117]. It is evident that if the proton–neutron star rotates (this assumption is fully rational) the spherical symmetry in the initial distribution of the hot gas is disrupted. According to the results of analytical and numerical calculations (see below), rotation changes the pattern of rising of the bubbles of hot matter, introducing some asymmetry to the general evolution of the perturbation.

Numerical modelling of convection

In the calculation of the distribution of density and temperature inside a proton–neutron star the central density was assumed to be $\rho_c = 2\times10^{14}$ g/cm^3 and the central temperature $T = 10^{11}$K. The equation of state, used in the calculations, was used in the tabulated form in the form of the dependence of pressure on density and entropy, in accordance with [185]. The relative concentration of the electrons was assumed to be constant and equal to 0.35. At the initial moment of time in the vicinity of the centre of the star ($r_0 = 0$) the excess of entropy was defined in accordance with the Gaussian law $S = S_0 + (S_m - S_0)\exp\{-(\mathbf{r}-\mathbf{r}_0)^{2/}b^2\}$, determined by the nonequilibrium process of neutronisation of matter after arrest of the collapse. The value of initial entropy S_0 was determined from the central initial temperature and density. In these calculations, S_0 was equal to 1.6327 k_B/nuclon. The maximum of entropy of the nuclons was recorded in the centre, $S_m = 2.8$. Parameter $b = 0.02$ was selected on the basis of the condition that the size of the region of increased entropy equals 0.2 of the size of the calculation region.

The Euler equation for calculating the development of large-scale convection, used for modelling the processes taking place inside a proton–neutron star in the three-dimensional case, have the following form

$$\rho\frac{dV}{dt} = -\operatorname{grad} P - \frac{\rho GM}{r^3}\mathbf{r},$$

$$\frac{d\rho}{dt} + \rho \operatorname{div}\mathbf{V} = 0,$$

$$\frac{dE}{dt} = T\frac{dS}{dt} + \frac{P}{\rho^3}\frac{d\rho}{dt},$$

$$\frac{dS}{dt} = 0.$$

Here ρ is the density of matter, **V** is the velocity of matter, P is pressure, E is energy, S is entropy. All these quantities are functions of three coordinates and time.

Calculations were carried out using the explicit conservative TVD difference scheme of the Godunov type.

The three-dimensional space was divided into cubic cells with a constant step of the grid. The values of all variables of the vector of densities (ρ, $\rho\mathbf{U}$, ρS) were related to the centres of the cells, and the values of the flows of these variable ($\rho\mathbf{U}$, $\rho U^2 + P$, $\rho\mathbf{V}$, ρW,

ρUS) were calculated at the boundaries between the cells. The variables were calculated at the moment of time t' from the solution of a one-dimensional Rieman problem for each spatial direction. To maintain the second order of approximation, the cycling rearrangement of these directions was carried out. Subsequently, a function was added to the equations, i.e. a source determined by the effect of the gravitational field of the star which was assumed to be constant during the calculations.

At the initial moment of time an equilibrium configuration was obtained taking into account rotation, by means of the iteration method proposed in [166] and modified taking the arbitrary equation of state into account.

Calculations were carried out for two cases. The first case was without rotation, in order to test the resultant equilibrium of the star. The second case – slow solid-state rotation where the ratio of the kinetic energy of rotation T to gravitational energy $|W|$ was selected as $T/|W| = 0.01$. The calculations of the kinetic and potential energies were carried out in accordance with integral representations, taking the density profile of obtaining the calculations into account. The period of rotation of the star was 14 ms, corresponding to the ratio of the kinetic energy to the potential of 0.01. The coordinate system was selected in such a manner that the plane of rotation of the star coincides with the plane OXY. This means that the vector of the angular velocity of rotation of the star has only one component $\Omega_z = \Omega = \text{const}$.

The distribution of entropy in time is shown in Figs. 4.29 and 4.30. The entropy is represented in two sections of the star. In the first section, it is presented along the axis of rotation (Fig. 4.29), i.e. the vector of the angular velocity of rotation is in the image plane. In other words, the axis OX is plotted on the horizontal axis, and the OZ axis of our system of coordinate is on the vertical axis. Figure 4.30 shows the view in the plane of the section in the equatorial plane. The initial configuration was selected for moment t = 0.0 46 ms, and the final configuration is shown for the moment of time $t \approx 5.87$ ms. Complete calculation corresponds to 20 ms. In contrast to the model examined in [166], in this case two bubbles form initially after 3 ms and are extended along the axis of rotation in opposite directions. Four additional bubbles form slightly later, after 5 ms, and are distributed in the equatorial plane of rotation of the proton–neutron star. The bubbles, distributed along the axis of rotation, separate from the hot nucleus and rise to the surface. This takes place as a result of the fact that the gradient of density

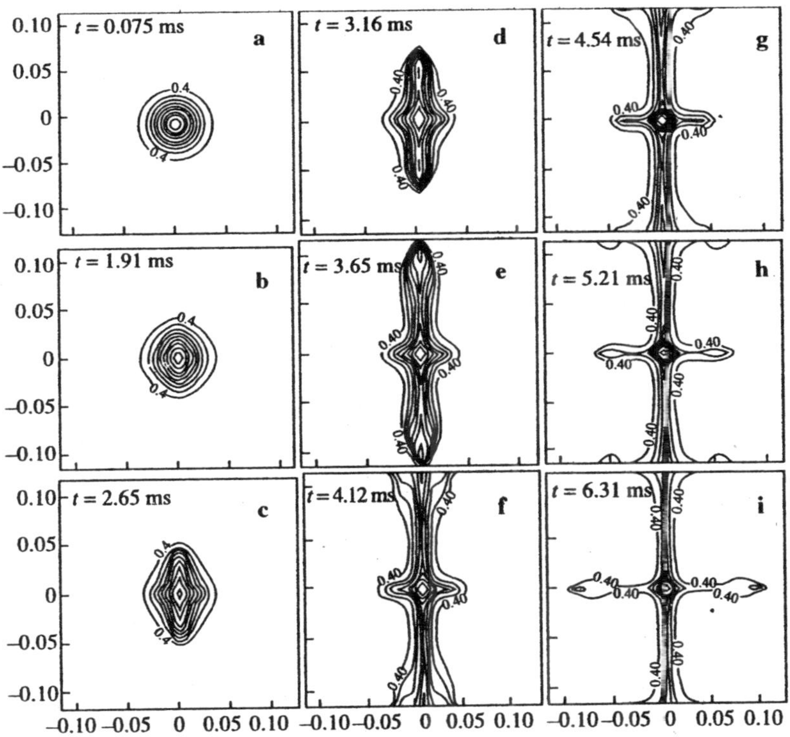

Fig.4.29. Lines of constant entropy in the *XZ* plane for nine moments of time. The marking unit of the axes correspond to 200 km. Entropy is expressed in dimensionless units. Background initial entropy corresponds to 1.6327 k_B/nucl (dimensionless value 0.37).

gradient is steeper along the axis of rotation. The cold matter descents to the centre in the gap between these bubbles.

The bubbles, distributed in the plane of rotation, separate from the hot nucleus later and also rise to the surface. Our calculations show that the formation of the first bubbles is followed by the formation of additional bubbles whose volume is considerably smaller. They also start to rise to the surface.

The first stage of development of perturbations is the increase of asymmetry along the axis of rotation of the star which continues for 3 ms (Fig. 4.29 a,b). In the plane of rotation (Fig. 4.30 a,b) the asymmetry is still very weak, and evolution is smooth. During this stage, the asymmetry of the distribution of matter is negligible and bubbles of hot matter have not as yet formed. For the stages shown in Fig. 4.29 d–f, the bubbles, moving along the axis of rotation, have

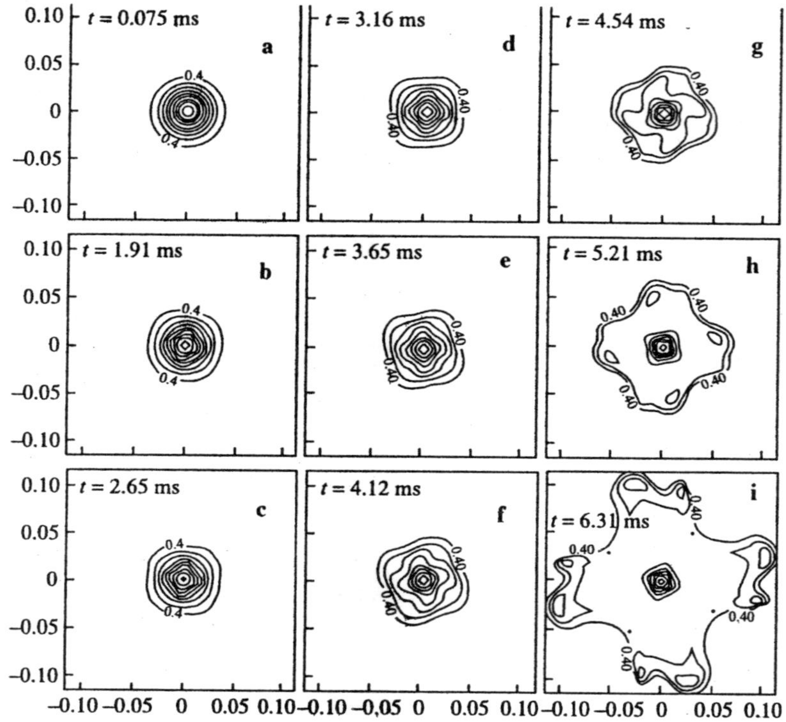

Fig.4.30. The same as in Fig.4.29 but in the equatorial plane.

already form. This stage is characterised by the start of the formation of bubbles created in the plane of rotation (Fig. 4.30 d–f). The stages, shown in Fig. 4.30 g–i, correspond to the final formation of the bubbles in the plane of the equator and to the start of rising with the bubbles to the surface. The bubbles along the axis of rotation (see Fig. 4.30 g–i) have already left the calculation region, i.e. they have travelled the distance from the centre greater 40 km. Figure 4.31 shows the three-dimensional profile of the distribution of entropy for the initial moment of time (t = 0.046 ms) and for the moment (t = 3.21 ms) when the bubbles have already formed. Figure 4.32 shows the distribution of the field of velocities. The convective cells are clearly visible, they show both the lifting of hot matter to the surface and the descent of cold matter to the centre of the proton–neutron star.

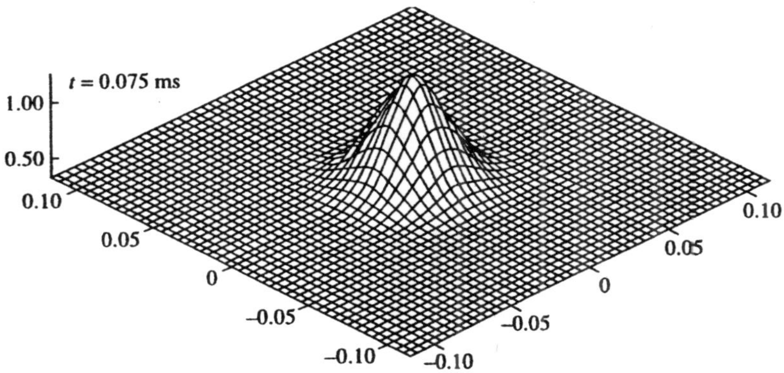

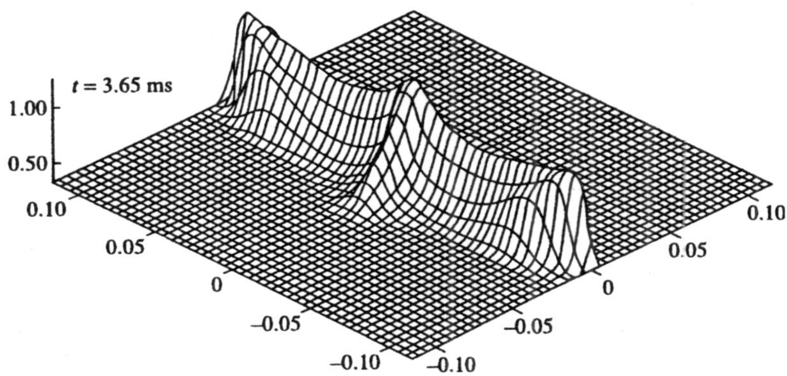

Fig.4.31. Spatial profile of entropy for two characteristic times. The scale of the axes is the same as in Fig. 4.29

Linear analysis of the problem

To evaluate the initial growth of perturbations, we shall carry out a linear analysis of the convective g-mode. We shall write all physical quantities in the form $f = f_0 + f_1$, where f_0 are the quantities corresponding to the equilibrium state, and $f_1 \approx f_1(\tilde{\omega}, z)\exp i(\sigma t + m\phi)$, where $\tilde{\omega}$ is the distance from the axis in the plane of the equator. Equilibrium equations for the functions f_0 are written in the form:

$$\frac{d\rho_0}{dz} = \rho_0 g_z,$$

$$\frac{dP_0}{d\tilde{\omega}} = \rho_0 g_{\tilde{\omega}} + \rho_0 \Omega^2 \tilde{\omega},$$

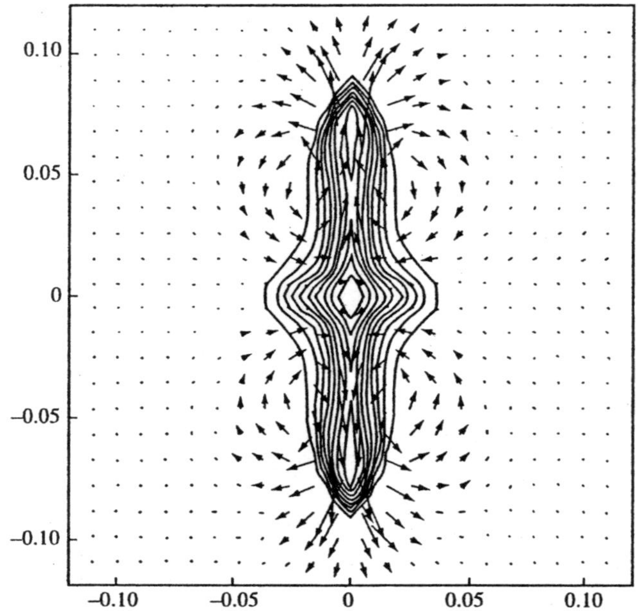

Fig.4.32. Distribution of entropy and the field of velocities in the XZ plane during the formation and start of movement of bubbles. The scales of the axes are the same as in Fig. 4.29.

where P_0 is pressure, ρ_0 is density, g is gravitational acceleration. The vector of the angular velocity $\mathbf{\Omega}$ has component $(0, \Omega\tilde{\omega}, 0)$.

The quantities f will be substituted into the initial equation and the members of the first order of smallness are retained. Consequently, we obtain a system for the amplitudes of perturbations of velocity:

$$\sigma_m V_{\tilde{\omega},1} + 2i\Omega V_{\phi,1} - \tilde{g}_{\tilde{\omega}}\sigma_m V_{z,1} / g_z = 0,$$

$$\sigma_m V_{\phi,1} - 2i\Omega V_{\tilde{\omega},1} = 0,$$

$$\sigma_m^2 V_{z,1} - g_z \mathbf{V}_{z,1}\mathbf{a} = 0,$$

where

$$\sigma_m \equiv \sigma + im\Omega, \quad \tilde{g}_{\tilde{\omega}} \equiv g_{\tilde{\omega}} + \Omega^2\tilde{\omega}, \quad \mathbf{a} \equiv (\mathrm{grad}\rho)/\rho - (\mathrm{grad}P)/\gamma P.$$

When deriving this equation, we used the approximation in which we neglected the amplitudes of perturbations of pressure and gravitational potential in order to separate the g-mode in the explicit form. The system has a non-zero solution, if the following relationship is fulfilled:

$$\sigma_m^2\left(\sigma_m^2 - g_z a_z - \tilde{g}_{\tilde{\omega}} a_{\tilde{\omega}}\right) - 4\Omega^2\left(\sigma_m^2 - g_z a_z\right) = 0.$$

In the case without rotation ($\Omega = 0$) we obtain the equation $\sigma_m^2 = g_r a_r$. The convective mode increases if the condition $\sigma_m^2 < 0$ is fulfilled.

In the presence of rotation, the condition of establishment of instability of the g-mode changes. We shall examine two limiting cases. In the direction of the axis of rotation $\tilde{\omega} = 0$, $g_{\tilde{\omega}} = 0$ we have the equation $\sigma_m^2 = g_z a_z$, coinciding with the condition of convective instability in the case without rotation. In the equator plane $z = 0$, $g_z = 0$ we obtain the equation:

$$\sigma_m^2 = \tilde{g}_{\tilde{\omega}} a_{\tilde{\omega}} + 4\Omega^2.$$

This equation shows that even small rotation increases the rate of growth of the convective g-mode. This analysis confirms the conclusion according to which the convective mode develops most rapidly along the axis of rotation and a large part of the initial perturbation rises in the form of two large-scales along the OZ axis in the opposite directions.

At the initial moment of time, the mass of matter with the excess of entropy ($S_{max} = 2.4$) is 0.07 $M_\odot$. After 4 ms 0.02 $M_\odot$ of the mass of this matter travels to the boundary of the neutrino-sphere $\rho = 10^{11}$ g/cm^3 and becomes transparent for the neutrinos situated in it. The concentration of these neutrinos with a mean energy of 30–50 MeV is sufficient for ensuring that the intensity of neutrino radiation is higher or comparable with the neutrino losses from the front of the shock wave, i.e. the mechanism of large-scale convection could sustain the divergent shock wave and cause the burst of the supernova shell. It is interesting to note that the observations of the central region of SN 1987A indicates the presence of two large-scale ejections [205].

CONCLUSIONS

Attempts to solve the problem of development of turbulence on reaching critical numbers Re_{cr} started for more than 100 years ago. The main direction of solving this problem has been focused on the stochastic nature of turbulence. It was assumed that in the presence of advanced isotropic turbulence the velocity pulsates around the mean value at every point of the flow. This assumption has enabled the stochastic description of turbulence to be used. However, the turbulent chaos is far from the total chaos of thermodynamic equilibrium and this indicates a certain degree of ordering of the structure of the flows. In particular, the often used Kolmogorov spectrum energy (and other quantities) from the wave number k is far from the uniform distribution, characteristic of thermodynamic equilibrium. The ordering at the fixed moment of time indicates the presence of coherent spatial structures.

Taking into account the ideology of constructing of the computing process, described in the book, and the results of calculations of coherent structures in turbulence, investigations were carried out in co-operation with E.G. Shifrin [150] to develop some 'axiomatic' model of the limiting developed turbulence in an incompressible liquid (to a certain degree this is theoretical justification of the approach discussed previously). Without discussing the details of this problem, we shall mention only certain assumptions of the developed theory. We propose a concept according to which the motion of the liquid at $Re \rightarrow Re_{cr}$ is characterised by the existence of some regions with a *continuous* but *not differentiated field* of velocity. In connection with this we referr to [152]: "...Doubts may arise as to whether it is generally possible to represent the velocity in turbulent motion as some continuous function of coordinates and time?... It may be necessary to conventionalise the trajectory in turbulent motion. For this purpose we must use, as the law of motion, the continuous function not in a single point and not having a derivative in respect of time, a kind of the Weierstrass function..."

The concept is based on the assumption that the controlling physical feature of developed turbulence is the property of mixing of liquid particles manifested in the disruption of the ordering of any of the systems during finite time. It expresses the fact of *non-existence of Lagrange coordinates in the sub-region of developed* turbulence in the four-dimensional physical 'space–time'.

The formation of these subregions – spots of developed

turbulence – in the region of laminar flow (with a differentiated field of velocity) is determined by the Gaussian principle of the 'smallest effect' in the situation in which the laminar flow loses asymptotic stability and transition to a wider class of flows becomes preferred from the energy viewpoint.

To describe the flows in this chaos, an expanded formulation of the problem based on the following assumptions (see a study by Shifrin in [150]) has been proposed:

1 Since the differential Navier–Stokes equations are not valid for describing the non-differentiated fields, we use *the integral laws of conservation of mass, the amount of motion* and the total energy in the most general form. They have equivalent interpretation as differential equations (macroequations) for average values in respect of the normal moving finite sub-region 'cell'. The concept of the pulsation tensor forms naturally in this case, without additional application of the so-called closure hypothesis.

2 The Navier–Stokes and Fourier laws are *expanded* for the tensor of viscous stresses and the heat flow (changing to the classic laws in the differentiated fields).

3 The *macroequations* together with the laws of equilibrium thermodynamics, expanded Navier–Stokes and Fourier laws and the averaging operator in the cell form a close system for describing the average and instantaneous (non-averaged) quantities. The initial-boundary problem at the boundaries of solids with flow-around is formulated for them.

4 *The ordered structure of the turbulent flow* is determined as the attractor of the *asymptotically stable solution* for the averaged quantities; in this case, the scale of the averaging cell is a stability parameter.

5 *Incorrectness* (non-singularity and instability of the solution) of the problem of restoration of the fields of non-averaged quantities in respect of the fields of the averaging out for the finite scale of the cell (determined by the condition of asymptotic stability of the mean quantities) is interpreted as the stochasticity of their physical realisations.

The proposed concept is regarded as an *ideological base* and justification of the validity of the direct numerical modelling of the ordered structures of developed turbulence using the algorithms approximating (or simulating) the integral laws of conservation.

The most important partial case is the 'maximumally developed' turbulence at Re$\to\infty$; this turbulence can be modelled (as the main term of the asymptotic expansion of the boundary layer type) out using the equations describing the laws of conservation (in the *Euler form*) with the boundary *condition of non-percolation*.

As indicated by the results, the approach to examination of turbulence on the basis of the solution of the Euler equations makes it possible to determine the structure of large-scale vortices. The results of calculation of the development of the large-scale structure of the flow between two rotating coaxial cylinders were presented in the third chapter.

As indicated by the results, at the end of calculations this structure is represented by one or two large vortices, surrounded by smaller ones. An almost laminar flow exists in the remaining part. The following physical model of turbulence can be proposed on the basis of the investigations.

When the critical values of the Reynolds number are reached in the shear flow, the pair of the forces exceeds the tensor of viscous stresses as a result of the pressure gradient. Large-scale vortices form in these conditions. However, the vortices are not divided any further! The interaction of the large vortices with the flow in their vicinity generates a high-frequency component of turbulence. In addition to this, because of the loss of energy for the initiation of fine-scale vorices, in some of the vortices evolution takes place in the direction of the loss of the energy of rotational motion. In turn, some vortices (both large-scale and mainly fine-scale ones) merge together, under the effect of attraction because of the Zhukovskii force. These processes are characterised by the filling of the inertia range of the turbulence spectrum. The entire pattern of developed turbulence has the form of the mixed structure.

In subsequent stages, carrying out complete physical calculations of the proposed model of turbulence, it is possible to average out the results in respect of the spectrum and obtain average flow characteristics. For example, this was carried out in chapter 3, paragraph 5. This averaging differs from the currently accepted procedure of calculation of the stochastic turbulence using expansion in respect of the moments of velocities or equations derived in expansion in respect of the Reynolds number (Hopf type).

It is important to mention that this turbulent flow does not have a single value of viscosity characterising the entire flow. Viscosity becomes a local value and, as stressed in the study by Canuto, it becomes dynamic viscosity. The same comment can be made regarding the Reynolds number. This leads to the problem of the justification of the expansion in respect of the Reynolds number and the physical substantiation of using the resultant equations.

The same concepts are used in solving the problem of development of convection. In this case, the controlling role is also played by the large scales. Attention to these phenomena was stressed as early as in 1962–1963 by Chandrasekar. It is interesting to note that the presence of large convective structures is seen in the central regions of the explosion of supernova 1987A (2 jets).

In conclusion, we shall mention the main assumptions of the given concept.

– Construction of direct calculation models makes it possible to investigate a wide range of the nonlinear problems of modern aerodynamics – the phenomena of the maximally developed shear free turbulence.

– On the basis of the experimental data on the existence of *ordered* formations with stochastic structure in shear developed turbulence the processes are divided into random processes because of their nature of movement of the vortices and non-random (organised) movements of the large-scale vortices.

– The large-scale and the organised nature of such a motion makes it possible to describe it by numerical schemes based on the *non-stationary equations of hydrodynamics* (and not statistical approaches).

– Because of the individual nature of the nonstationary ordered movement and large-scale macrostructures of turbulence, their examination in low frequency and inertia range for very high Reynolds numbers is carried out by director *numerical* modelling (*without using the semiempirical models of turbulence*) based on the examination of total smoothed-out dynamic equations for the ideal medium, namely the nonstationary Euler equations in the form of integral laws of conservation with the approximate dissipation mechanism (generated by the averaging of the parameters in the volume of the cell and ensuring the stability of calculations); the application of finite-difference 'oriented' schemes makes it possible to separate, with the required accuracy, the large-scale formations using the methods of 'rational' approximation for 'subgrid' pulsations.

– The accuracy of the following formulation of the problem has been shown: there is the principal possibility of obtaining accurate statistical characteristics of the flow which depends on large-scale turbulence, using smoothed-out equations of motion where the contribution of the small-scale vortices is represented approximately (without the requirement for the accurate calculation of true fields of pulsating quantities).

– Examination of the local defects (stochastic component) of turbulence – the nature of distribution of Reynolds stresses, the density of turbulent energy, the rate of dissipation of this energy, etc. – is carried out in the zones with steep gradients by the statistical method using kinematic models of turbulence.

– Examination of the laminar regimes of flow and phenomena of laminar–turbulent transition is carried out on the basis of complete Navier-Stokes equations.

– Numerical experiments were carried out to examine the 'scenario' of transition to chaos: the process of stochastisation of the dynamic system (the Navier–Stokes model) 'takes place' only in the presence of external mass forces (the field of gravitational forces, the effects of rotation of the earth, wind loading, etc.), generating the inertia mechanism of the systems required for these purposes.

– The interaction of perturbations is used as an example for examining the mechanism of transition and the main characteristics of the turbulent stage in the development of two-dimensional and spatial processes of Rayleigh–Taylor (in the multimode variant) and Richtmyer–Meshkov instabilities.

It should be mentioned that the problem of examining turbulence – the problem with the 'inaccurately specified information' – and the introduced 'expanded' formulation of the problem (initial system with the averaging operator) is identical to 'parametric expansion' according to A.N. Tikhonov [153]. The model itself 'should be examined not as a mathematical model of the object but as a mathematical model of examination of the object' (turbulence in the present case). The concept of the ordered structure, introduced here, realised in a stable mother as some attractor only at specific values of parameter 6' (the step of the calculation grid) is close to the concept of 'e-stability' in the solution according to Tikhonov. In other words, the ordered structure of the turbulent flow is its regularised description. (The ordered structure of the turbulent flow is determined as the attractor of the asymptotically stable solution for the average-out values; in this case, the stability parameter depends on the averaging scale – the size of the cell h).

Examination of the ordered structure in turbulence is part of the problem of self-organisation widely discussed at the present time, i.e. the process of formation of order in complex nonlinear systems and media (see, for example [154–156] and others). Without discussing the substance of these problems, we shall only mention that the ordering process is evidently associated with the *collective* (*cooperative*) behaviour of the subsystems forming the system. The models, specimens and mechanisms of the theory of self-organisation in complex nonlinear systems and media have been studied only on a small scale and this is why the application of heuristic approaches of the computation experiment in this case is especially justified. However, it is important to mention that *nonequilibrium*, *nonlinearity of the medium* and its *dissipative nature* are decisive for the formation of structures of this type. Self-organisation is the result of development of spatially heterogeneous instabilities with their subsequent stabilisation as a result of the balance between the dissipative consumption and the supply of energy from sources of nonequilibrium [156].

In the numerical approach to the examination of turbulence it is important to model the process of the initiation and evolution of the structure in respect of both time and scale. Coherent structures may exist away from equilibria only as a result of relatively efficient supply of energy and matter. In this case, the evolution of the process is interpreted as the sequence of transitions in the hierarchy of the structures of increasing complexity, and the stable solution is selected in the conditions of increasing dissipation. This in principle is the general Prigogine principle of self-organisation in nonequilibrium systems [154].

The main ideology of the developed multi-plan approach to examination of the turbulence is the development of '*rational*' numerical models adequate to the investigated phenomenon. Thus, the following answer can be given to the question: 'which models of the ideal medium, Navier–Stokes equations or which kinetic level should be used for examination of turbulent movements?'. Taking into account the *structural* representation of the turbulent flows, it is convenient to use the models which correspond to a large degree to the investigated mechanism of interaction. The appropriate *imitation* models are constructed in this way: the large-scale transfer in turbulence or in the development of the processes of instability is investigated on the basis of dynamic models for the ideal medium, laminar–turbulent flows are examined taking into account the viscosity mechanism of interaction (the Navier–Stokes equations) and the stochastic process is investigated on the kinetic level.

Thus, the methods of '*rational*' numerical modelling (BRAINWARE) of the *appropriate* mechanisms of interaction make it possible to investigate numerically a wide range of the nonlinear problems of computing mechanics – structural turbulence, transition phenomena, the processes of development of Rayleigh–Taylor, Richtmyer–Meshkov instabilities, etc. In this case it is possible (and this is very important) *to reduce greatly the level of requirements on computer resources!*

REFERENCES

1 Abakumov M.V., Mukhin S.I. and Popov Yu.P., Mat. Modelirovanie, **12**, No.3, 110-120 (2000).

2 Abakumov M.V., Mukhin S.I., Popov Yu.P. and Chechetkin V.M., Astron. Zhurn., **73**, No.3, 407-418 (1996).

3 Abakumov M.V., Mukhin S.I., Popov Yu.P. and Chechetkin V.M., *ibid*, **78**, No.5, 505-513 (2001).

4 Abakumov M.V., Mukhin S.I., Popov Yu.P. and Chechetkin V.M., *ibid*, (2002).

5 Abramovich G.N., Theory of turbulent jets, Publ Nauka, Moscow (1984).

6 Aleshin A.N., Demchenko V.V., Zaitsev S.G. and Lazareva E.V., Izv. AN SSSR, Ser. MZhG, No.5, 163-174 (1992).

7 Andrushchenko V.A., Gorbunov A.A., Paskonov V.M. and Chudov L.A., Mat. Modelirovanie, **4**, No.3, 40-52 (1992).

8 Anuchina N.N., Kucherenko Yu.A., Neuvazhaev V.E., Ogibina V.N., Shibarshov L.I. and Ya Ovlev V.G., Izv. AN SSSR, Ser. MZhG, **6**, 157-160 (1978).

9 Babakov A.V., Zhurn. Vychisl. Matematiki i Mat. Fiziki, **28**, 267-277 (1988).

10 Babakov A.V., Numerical modelling of problems of aerodynamics, VTs AN SSSR, Moscow (1986).

11 Babakov A.V., Belotserkovskii O.M. and Zyuzin A.P., Dokl. AN SSSR, **279**, No. 2, 315-318 (1984).

12 Bazhenov D.V. and Vazhenova L.A., in: Proc 2nd National Symposium on Physics of Acoustic-Hydrodynamic Phenomena in OptoacousticsPubl Nauka, Moscow (1982).

13 Vazhenov D.V., Bazhenova L.A. and Rimskii-Korsakov V.A., in: Proc. 5th National Conference on Theoretical and Applied Mechanics, Alma-Ata (1981).

14 Belen'kii S.Z. and Fradkin E.S., Tr. FIAN SSSR, **29**, 207-238 (1965).

15 Belotserkovskii O.M., Computing Mechanics: current problems and results, Publ Nauka, Moscow (1991).

16 Belotserkovskii O.M., Zhurn. Vychisl. Matematiki i Mat. Fiziki, **25** No.12, 1857-1882 (1984).

17 Belotserkovskii O.M., Numerical modelling in solid state mechanics, Publ Nauka, Moscow (1984), 2nd edition, Publ Fizmatlit, Moscow (1994).

18 Belotserkovskii O.M., Numerical experiment in turbulence, Publ Nauka, Moscow (1997), Belotserkovskii O.M. and Oparin A.M., as above, 2nd edition (suppl.), Publ Nauka, Moscow (2000).

19 Belotserkovskii O.M., Studies of turbulence, Publ Nauka, Moscow (1994).

20 Belotserkovskii O.M., Antonenko M.N., Konyukhov A.V., Kraginskii L.M. and Oparin A.M., in: Proc. 4th International Conference: Forest and Steppe Fires: formation, propagation, extinction and ecological consequences, Urkutsk (2002).

21 Belotserkovskii O.M. and Babakov A.V., Uspekhi Mekhaniki, Warsaw, **13**, No.3/4, 135-139 (1990).

22 Belotserkovskii O.M., Vinogradov A.V., Glazunov A.S., Zhuravlev Yu.I., et al, Dokl. AN SSSR, **261**, 1307-1310 (1981).

23 Belotserkovskii O.M., Gushchin A.V. and Kon'shin V.N., Zhurn. Vychisl. Matematiki i Mat. Fiziki, **27**, 594 (1987).

24 Belotserkovskii O.M. and Davydov Yu.M., Method of large particles in

gas dynamics, Publ Nauka, Moscow (1982).
25 Belotserkovskii O.M., Davydov Yu.M. and Dem'yanov A.Yu., Dokl. AN SSSR, **288**, 1071 (1986).
26 Belotserkovskii O.M., Demchenko V.V. and Oparin A.M., *ibid*, **354**, No.2, 190-193 (1986).
27 Belotserkovskii O.M., Demchenko V.V. and Oparin A.M., *ibid*, **334**, No.5, 581-583 (1994).
28 Belotserkovskii O.M., Erofeev A.I. and Yanitskii V.E., Uspekhi Mekhaniki. Warsaw, **5**, No.3/4, 11-40 (1982).
29 Belotserkovskii O.M., Ivanov S.A. and Yanitskii V.E., Zhurn. Vychisl. Matematiki i Mat. Fiziki, **38**, No.3, 489-503 (1988).
30 Belotserkovskii O.M. and Konyukhov A.V., *ibid*, **42**, No.2, 235-248 (2002).
31 Belotserkovskii O.M. and Nisht M.I., Detachment and , Publ Nauka, Moscow (1978).
32 Belotserkovskii O.M., Panarin A.I. and Shchennikov V.V., Zhurn. Vychisl. Matematiki i Mat. Fiziki, **24**, No.1, 65-74 (1984).
33 Belotserkovskii O.M. and Yanitskii V.E., ibid, **15**, No.5, Ch.1, 1195-1208; No.6, Ch.II, 1553-1567 (1975).
34 Batchelor J.K., Theory of uniform turbulence, Publ Inostr. Lit., Moscow (1955).
35 Bisikalo D.V., Boyarchuk A.A., Kuznetsov O.A., Popov Yu.P. and Chechetkin V.M., Astron. Zhurn., **72**, No.2, 190-202 (1995).
36 Bisikalo D.V., Boyarchuk A.A., Kuznetsov O.A., Khruzina T.S., Cherepashchuk A.M. and Chechetkin V.M., *ibid*, **75**, 40-53 (1998).
37 Bisikalo D.V., Boyarchuk A.A., Kuznetsov O.A. and Chechetkin V.M., *ibid*, **77**, No.1, 31-41 (2000).
38 Bisikalo D.V., Boyarchuk A.A., Kuznetsov O.A. and Chechetkin V.M., *ibid*, **74**, No.6, 889-897 (1997).
39 Bisikalo D.V., Boyarchuk A.A., Kuznetsov O.A. and Chechetkin V.M., *ibid*, **75**, 706-715 (1998).
40 Bisikalo D.V., Boyarchuk A.A., Kuznetsov O.A. and Chechetkin V.M., *ibid*, **74**, No.6, 880-888 (1997).
41 Bisikalo D.V., Boyarchuk A.A., Kuznetsov O.A. and Chechetkin V.M., *ibid*, **76**, No.12, 905-916 (1999).
42 Budyko M.I., Golitsyn G.S. and Izrael'Yu.A., Global climatic catastrophes, Publ Gidrometeoizdat, Moscow (1986).
43 Gerlakh N.N., Zueva N.M. and Solov'ev L.S., No.98, Preprint Institute of Applied Mechanics, AN SSSR (1975).
44 Godunov S.K., Zabrodin A.V., Ivanov M.Ya., et al, Numerical solution of multidimensional problems of gas dynamics, Publ Nauka, Moscow (1976).
45 Godunov S.K. and Ryaben'kii V.S., Difference systems: introduction into theory, Publ Nauka, Moscow (1977).
46 Gol'dshtik Yu.A. and Shtern V.N., Hydrodynamic stability and turbulence, Publ Nauka, Moscow (1977).
47 Gostintsev F.A., Kopylov N.P., Ryzhov A.M. and Khasanov I.R., Izv. RAN. Ser. MZhG., No.4, 47 (1990).
48 Gostintsev Yu.A., Kopylov N.P., Ryzhov A.M. and Khasanov I.R., Fizika Goreniya i Vzryva, No.6, 10 (1991).
49 Gostintsev Yu.A., Makhviladze G.M. and Novozhilov V.B., Izv. RAN. Ser. MZhG, No.1, 17 (1992).
50 Grishin A.M., Mathematical modelling of forest fires and new methods of

preventing them, Publ Novosibirsk, Nauka (1992).

51 Gruver M. and Zimmers E., SAPR and Automation of Production, Mir, Moscow (1987).

52 Davydov Y.A. and Panteleev M.S., Zhurn. Prikl. Matematiki i Teoret. Fiziki, **1**, 117 (1981).

53 Dem'yanov A.Yu. Inogamov N.A. and Oparin A.M., in: Proc. 9th School-Seminar: Current Problems of Aerohydrodynamics, Tuapse (2001).

54 Dem'yanov A.Yu., Inogamov N.A. and Oparin A.M., in: Proc International Conference: 6th Zababakhinsk Scientific Lextures, Sept 24-28, 2001, Snezhinsk (2002).

55 Dorodnitsyn A.A., Selected studies, in two volumes, Computing Centre, Russian Academy of Sciences, Moscow (1997).

56 Zhuravlev Yu.I., Probl. Kibernetiki, No.3, Nauka, Moscow (1978).

57 Zaets P.G., Onufriev A.T., Safarov N.A. and Safarov R.A., in: Applied mechanics and technical physics, Moscow (1991).

58 Zel'dovich Ya.B., Friction in liquids between rotating cylinders, Preprint, IPM AN SSSR, Moscow (1979).

59 Zueva N.M., Mikhailova M.S. and Solov'ev L.S., Convective instability of a gas sphere, Preprint No.65, IPM AN SSSR, Moscow (1977).

60 Ivanov M.F., Sultanov V.G., Oparin A.M. and Fortov V.E., Dokl. AN SSSR, **367**, No.5, 464-467 (1999).

61 Ivanov S.A. and Yanitskii V.E., An approach in statistical modelling of free turbulence, Preprint, VTs AN SSSR, Moscow (1992).

62 Ievlev V.M., Turbulent motion of high-temperature solids, Nauka, Moscow (1975).

63 Ievlev V.M., Dokl. AN SSSR, **208**, No.5, 1044-1047 (1973).

64 Inogamov N.A., Pis'ma v ZhTF, **4**, No.12, 743 (1978).

65 Inogamov N.A., Dem'yanov A.Yu. and Son E.E., Hydrodynamics of mixing, Izd. MFTI, Moscow (1999).

66 Inogamov N.A. and Oparin A.M., Zhurn. Eksperim. i Teoret. Fiziki, **116**, 908-939 (1999).

67 Inogamov N.A. and Oparin A.M., Pis'ma v ZhETF, **69**, No.10, 691-697 (1999).

68 Inogamov N.A., Oparin A.M., Dem'yanov A.Yu., Dembitskii L.N. and Khokhlov V.A., Zhurn. Eksperim. i Teoret. Fiziki, **119**, No.4, 822-852 (2001).

69 Belotserkovskii O.M. (ed), Informatics and medicine, Publ Nauka, Moscow (1997).

70 Kamchibekov M.D., Meshkov E.E., Nevmerzhitskii N.V. and Sotskov E.A., Turbulent mixing on a cylindrical gas–liquid interface, Preprint No.46-96, (1966).

71 Cantwell B.J., Organised motion in turbulent flows, Publ Mir, Moscow (1984).

72 Kolmogorov A.N., Izv. AN SSSR, Ser. Fiz., **6**, No.2, 56-58 (1942).

73 Konovalov N.A., Kryukov V.A., Mikhailov S.N. and Pogrebtsov A.A., Programmirovanie, No.1 (1995).

74 Konyukhov A.V., Meshcheryakov M.V. and Utyuzhnikov S.V., Teplofizika Vysokikh Temperatur., **32**, No.2, 236-241 (1994).

75 Konyukhov A.V., Meshcheryakov M.V. and Utyuzhnikov S.V., *ibid*, **33**, No.5, 726 (1995).

76 Konyukhov A.V., Meshcheryakov M.V. and Utyuzhnikov S.V., *ibid*, 904-911 (1999).

77 Konyukhov A.V., Meshcheryakov M.V., Utyuzhnikov S.V. and Chudov L.A.,

ibid. Izv. RAN. Ser. MZhG., No.3, 93 (1997).
78 Kopylov N.P., Ryzhov A.M. and Khasanov I.R., Fizika Goreniya i Vzryva, No.5, 51 (1985).
79 Kochin N.N., Kibel' I.A. and Roze N.V., Theoretical hydromechanics, Izd. Fiz.-Mat. Lit., Moscow (1963).
80 Cramer G., Mathematical methods of statistics, Publ Mir, Moscow (1975).
81 Kuznetsov O.M. and Popov S.G., Izv. AN SSSR. Ser. MZhG., No.2, 112-113 (1967).
82 Kuznetsov V.R. and Sabel'nikov V.A., Turbulence and combustion, Publ Nauka, Moscow (1986).
83 Kucherenko Yu.A., Neuvazhaev V.E. and Pylaev A.P., Dokl. AN SSSR, **334**, No.4, 445-448 (1994).
84 Landau L.D. and Livshits E.M., Hydrodynamics, Vol. 4, Publ Nauka, Moscow (1986).
85 Landau L.D. and Livshits E.M., Solid state mechanics, Publ Gostekhteorizdat, Moscow (1953).
86 Landau L.D. and Livshits E.M., Statistical physics, Publ Nauka, Moscow (1964-1995).
87 Magomedov K.M. and Kholodov A.S., Zhurn. Vyshisl. Matematiki i Mat. Fiziki, **9**, 373 (1969).
88 Magomedov K.M. and Kholodov A.S., Mesh-characteristic numerical methods, Publ Nauka, Moscow (1988).
89 Mardsen J., in: Strange attractors, Publ Mir, Moscow (1981).
90 Markelova L.P., Nemchinov I.V. and Shubadaeva L.P., Kvantovaya Elektronika, No.5, 1904-1906 (1987).
91 Marchuk G.I., Methods of computing mathematics, Publ Nauka, Moscow (1967).
92 Marshal V., Residual risks in chemical production, Publ Mir, Moscow (1989).
93 Merrit G., Raketnaya Tekhnika i Kosmonavtika, **12**, No.7, 73-85 (1974).
94 Methods of calculating turbulent flows, Publ Mir, Moscow (1984).
95 Meshkov E.E., Izv. AN SSSR, Ser. MZhG., No.5, 151-157 (1969).
96 Monin A.S. and Yaglom A.M., Statistical hydromechanics, Publ Nauka, Moscow, in two volumes, Vol.1, 639 (1965), Vol.2, 720 (1967), Vol.1, 694 (1992).
97 Muzafarov I.F. and Utyuzhnikov S.V., Teplofizika Vysokikh Temperatur, **33**, No.4, 594 (1995).
98 Neuvazhaev V.E., Prikl. Mekhanika i Tekhn. Fizika, No.6, 82 (1976).
99 Nikishin V.V., Tishkin V.F., Emitrenko N.V., Lebo I.G., Rozanov B.V. and Favorskii A.F., Preprint No.30, FIAN, Moscow (1997).
100 Nikois G. and Prigozhin P., Self-organisation in nonequilibrium systems, Publ Mir, Moscow (1979).
101 Onufriev A.T., Prikl. Mekhanika i Tekhn. Fizika, No.2, 62-72 (1970).
102 Oparin A.M., in: New in numerical modelling: algorithms, computing experimenyts, results, Publ Nauka, Moscow (2000).
103 Oparin A.M., Inogamov N.A. and Dem'yanov A.Yu., Pis'ma v ZhETF, **72**, No.10, 704-710 (2000).
104 Orseg S., in: Turbulence: principles and applications, Publ Mir, Moscow (1980).
105 Petrov A.A., Economics. Models. Computing experiments, Publ Nauka, Moscow (1996).
106 Prokhorov A.M., Anisimov S.I. and Pashinin P.P., Uspekhi. Fiz. Nauk, **119**,

No.3, 401-424 (1976).

107 Rozhdestvenskii B.L. and Yanenko N.N., Systems of quasilinear equations, Publ Nauka, Moscow (1978).

108 Rykov V.V., Numerical modelling of spatial flows of an incompressible viscous liquid, VTs AN SSSR, Moscow (1983).

109 Samarskii A.A., Theory of diffference schemes, Publ Nauka, Moscow (1967).

110 Samarskii A.A. and Nikolaev E.S., Methods of solving net equations, Publ Nauka, Moscow (1978).

111 Gol'dshtik M.A. (ed), Structural turbulence, Publ Izd-vo SO AN SSSR, Novosibirsk (1982).

112 Struminskii V.V., A new direction in examining turbulence problems. Turbulent flows, Publ Nauka, Moscow (1977).

113 Townsend A.A., Structure of the turbulent flow with cross shear, Inostr. Lit., Moscow (1959).

114 Tikhonov A.N. and Arsenin V.Ya., Methods of solving ill-posed problems, Publ Nauka, Moscow (1986).

115 Troshkin O.V. and Yanitskii V.E., Preprint, VTs AN SSSR, Moscow (1988).

116 Waterman D., A Handbook of expert systems, Publ Mir, Moscow (1989).

117 Ustyugov S.D. and Chechetkin V.M., Astron. Zhurn., **76**, No.11, 816-824 (1999).

118 Fedorenko R.P., Application of high-order accurate difference schemes in numerical solution of hyperbolic equations, Zhurn. Vychisl. Matematiki i Mat. Fiziki, **2**, No.6, 1122-1128 (1962).

119 Khintse I.O., Turbulence, Publ Fizmatgiz, Moscow (1963).

120 Hessard B., Kazarinov N. and Ven I., Theory and applications of bifurcation of cycle generation, Publ Mir, Moscow (1985).

121 Chapman D.R., Raketnaya Tekhnika i Kosmonavtika, **18**, No.2, 3-32 (1980).

122 Chechetkin V.M., Ustyugov S.D., Gorbunov A.A. and Polezhaev V.I., Pis'ma v AZh, **64**, No.12, 34-41 (1997).

123 Chzhen P., Separation flows, Publ Mir, Moscow (1973).

124 Shvets A.I. and Shvets I.T., Gas-dynamics of near wakes, Publ Naukova dumka, Kiev (1976).

125 Shets G., Turbulent flow: processes of entry fnad mixing, Publ Mir, Moscow (1984).

126 Slichting G., Boundary layer theory, Publ Nauka, Moscow (1969).

127 Schuman U., Grattsbach G. and Claizer L., in: Methods of calculating turbulent flows, Publ Mir, Moscow (1984).

128 Belotserkovskii O.M. (ed), Studies of turbulence, Publ Nauka, Moscow (1994).

129 Yanenko N.N., Methods of fine steps for solving multidimensional problems of mathematical physics, Publ Nauka, Novosibirsk (1967).

130 Yanilkin Yu.V., Vopr. Atomnoi Nauki Tekh., Ser. Mat. Modelirovanie Fiz. Protsessov., No.88, 4 (1999).

131 Yanitskii V.E., in: Studies of turbulence, Publ Nauka, Moscow (1994).

132 Yanitskii V.E., A statistical method of particles for solving some problems of the kinetic theory of gases and turbulence, Disseration, VTs AN SSSR, Moscow (1984).

133 Yanitskii V.E., Stochastic models of a perfect gas from a fibnite number of particles, Preprint VTs AN SSSR, Moscow (1988).

134 Abstracts of 7th Intern. Workshop on the Physics of Compressible Turbulent Mixing, St Peterburg, Russia (1999).

135 Alon U., Hecht J., Ofer D., and Shvarts D. Phys. Rev. Lett., **74**, 534 (1995).

136 Anisimov S.I., Zel'dovich Ya., Inogamov N.A. and Ivanov M.F., Shock waves, explosions and detonations. Progress in astronautics and aeronautics series, J.R. Bowen, J.C. Leyer and R.I. Soloukhin, (Summerfield, editor-in-chief). Vol. 87, AIAA, Washington (1983).

137 Antonenko M.N., Konyukhov A.V. Kraginskii L.M., Meshcheryakov M.V. and Utyuzhnikov S.V., Numerical modelling of intensive convective flows in atmosphere, induced by large-scale fire, Computational Fluid Dynamics J., **10**, No.3, 372-375 (2001).

138 Anuchina N.E., Es'kov N.S., Polinov A.V., Ilyutina O.S., Kozyrev O.M. and Volkov V.I., Pis'ma ZhTF, **4**, No.12, 743 (1978).

139 Armitage P.J. and Livio M., Accretion discs in interaction binaries: simulations of the stream-disc impact, Atrophys. J., **470**, 1024-1032 (1996).

140 Buchler J.R. and Kondrup H. (eds), Atrophysical turbulence and convection, Annals New York Acad. of Sci., **898** (2000).

141 Balabanov V.A., Konyukhov A.V., Meshkov M.A., et al, Physical and numerical modelling of convective flow over large area buoyancy source in atmosphere, Internat. Conf. "Dynamics of Ocean Atmosphere", Moscow (1995).

142 Belotserkovskii O.M., Turbulence and instabilities, Lewinston etc.: Edwin Mellen Press, USA (2000).

143 Bird C.A., Molecular gas dynamics, Clarendon Press, Oxford (1976).

144 Bradshaw P., The analogy between streamline curvature and buoyancy in turbulent shear flow, J. Fluid Mechanics, **36**, No.1, 177-191 (1969).

145 Bruenn S.W., Mezzacappa A. and Dineva T., Phys. Rep., **256**, 69 (1995).

146 Bisikalo D.V., Boyarchuk A.A., Chechetkin V.M., Kuznetsov O.A. and Molteni D., 3D numerical simulation of gaseous flows structure in semi-detached binaries, Mon. Not. R. Astron. Soc., J.R. Bowen, J.C. Leyer and R.I. Soloukhin, **300,** 39 (1998).

147 Bisnovatei-Kogan G.S. and Lovelace R.V.E., Advective discs and related problems including magnetic fields, New Astronomy Reviews, **45**, 663-742 (2001).

148 Canuto V.M., Turbulence and laminar structures: can they co-exist? Mon. Not. R. Astron. Soc. **317** 985-988 (2000).

149 Chandrasekhar C., Hydrodynamic and hydromagnetic stability, Clarendon Press, Oxford (1961).

150 Chandrasekhar C. and Lebovits N.P., Non-radial oscillation and convective instability of gaseous masses, Astrophys J., **138**, 185 (1963).

151 Chudow L.A., Konyukhov A.V., Meshcheryakov M.A., et al, Numerical modelling a large fire in the atmosphere. 2nd Internat. Seminar "Fire and explosion hazard of substances and venting of deflagrations". Moscow, Russia (1997), p.141.

152 Colgate S.A. and Buchler J.R., Coherent transport of angular moment. The Ranque-Hilsch tube as a paradigm, Annals of New York Acad. of Sci., **898**, 105-112 (2000).

153 Cowling T.G., The non-radial oscillations of politropic stars., Mon. Not. R. Astron. Soc., **101**, 367-374 (1942).

154 Deardorff J.W., The use of subgrid transport equations in 3D model of atmospheric turbulence, J. Fluid Eng., **95**, 429-438 (1973).

155 Dorodnicyn A.A., Review of methods for solving the Navier-Stokes equations, Lect. Notes Phys., **18**, No.1, 23-47 (1973).

156 Dotani T., Hayashida K., Inoue H., et al, Discovery of an unusual hard x-ray source in the region supernova in 1987A, Nature, **330**, 230-231 (1987).

157 Driscall R.J. and Kennedy K.A., A model for the turbulent energy spectrum, Phys. Fluids, **26**, No.5. 1228-1233 (1983).

158 Druden H.I., Recent advances in the mechanics of boundary layers flow, Adv. Appl. Mech., **1**, 1-40 (1948).

159 Favre A., Equations des gaz turbulents compressible, J. Mechanique, **4** No.3, 361 (1965).

160 Fersiger J.H., Large eddy numerical simulations of turbulent flows, AIAA J., **15**, No.9, 1261-1267 (1977).

161 Fox G., Jonson M., Lyzenga G., Otto S., Salmon J. and Walker D., Solving problems on concurrent processors, Vol. 1, Prentice-Hall, Englewood Cliffs, New York (1988).

162 Gardner G., Glimm J., McBryan O., Menikoff R., Sharp D.H. and Zhang Q., The dynamics of bubble growth for Rayleigh-Taylor instability, Phys. Fluids, **31**, 447-465 (1988).

163 Glaiser P., An approximate linearisation riemann solver for the Euler equations for real gases, J. Comput. Phys., **74**, 382-408 (1988).

164 Glimm J., Grove J.W., Li X.L., Shyue K.M. and Zhang Q., SIAM. J. Sci. Comp., 19, 703 (1998).

165 Glimm J., Li X.L., Menikoff R., Sharp D.H. and Zhang Q., A numerical study of bubble interactions in Rayleigh-Taylor instability for compressible fluids, Phys. Fluids, **2**, 2046-2054 (1990).

166 Hachisu K.A., Versatile method for obtaining structures of rapidly rotating stars. II. Three-dimensional self-consistent field method, Atrophys. J. Supl., **62**, 461-499 (1986).

167 Harlow F.H. and Welch J.E., Numerical study of large-amplitude free-surface motion, Phys. Fluids, **9**, 842 (1966).

168 Harten A., High resolution schemes for hyperbolic conservation laws, J. Comput. Phys., **49**, 151 (1983).

169 Harten A., On a class of high resolution total-variation stable finite-difference schemes, SIAM J. Numer. Analys., **21**, 1-23 (1984).

170 Harten A., Engquist B., Osher S. and Chakravarthy S., Uniformly high-order accurate essentially non-oscillatory schemes III, J. Comput. Phys., **71**, 231-303 (1987).

171 Harten A. and Osher S., Uniformly high-order accurate non-oscillatory schemes I, SIAM J. Numer. Analys. **24**, No.2, 279-309 (1987).

172 He X., Zhang R., Chen S. and Doolen G.D., On the three-dimensional Rayleigh-Taylor instability, Phys. Fluids, **11**, No.5, 1143-1152 (1999).

173 Herant M., et al, Inside the supernova. A powerful convective enquine, Astrophys. J., **435**, 339-361 (1994).

174 Hirh C., Numerical computation of internal and external flows, Vol.1/2, John Wiley & Sons, New York (1980).

175 Holmes R.L., Grove J.W. and Sharp D.H., Numerical investigation of Richtmeyer-Meshkov instability using front tracking, J. Fluid Mech., **301**, 51 (1995).

176 High Performance Fortran Forum, High-performance Fortran language specification, version 1.1, Rice University, Houston TX (1994).

177 Inogamov N.A., Astroph. Space Phys. Rev., **10**, II, 1 (1999).

178 Inogamov N.A., Tricottet M., Oparin A.M. and Bouquet S., Hydrodynamic instabilities in astrophysics, Euroconference "Hypersonic and aerotermic flows and shocks, and lasers", Observatoire de Paris, Site de Meudon, France (2001).

179 Inogamov N.A., Tricottet M., Oparin A.M. and Bouquet S., Three-dimensional morphology of vortex interfaces driven by Rayleigh-Taylor or Richtmeyer-Meshkov instability, http: xxx.archiv.org, arXiv:physics/0104084 (2001) (??? Physics Letters A).

180 Janka H.-Th. and Muller E., Neutrino heating, convection, and the mechanism of Type-II supernova explosions, Astronomy and Astrophysics, Vol.306 (1996).

181 Kalinov A. and Lastovetsky A., Heterogeneous distribution of computations while solving linear algebra problems on networks of heterogeneous computers, Proceedings of the 7th International Conference on High Performance Computing and Networking in Europe (HPCN Europe '99), Lecture notes in computer science 1593, Amsterdam, The Netherlands (1999), pp.191-200.

182 Keil W., Janka H.-Th. and Muller E., Ledoux convection in protoneutron stars - a clue to supernova nucleosynthesis, Astrophys. J. Lett., **473**, L111-L118 (1996).

183 Lanzafame G., Belvedre G. and Molteni D., 3D simulation of the SS CVG accretions disc in the quiescent phase, Monthly Notices Roy. Atron. Soc., **258**, 152-158 (1992).

184 Lastovetsky A., Arapov D., Kalinov A., and Ledovskih I., Parallel language and its programming system for heterogeneous networks, Concurrency: practice and experience, **12**, No.13, 1317-1343 (2000).

185 Lattimer J.M. and Swesty F.D., Nucl. Phys. A., **535**, 331 (1991).

186 Larne J. and Libby P., Statistical properties of the interface in the turbulent wake of a heated cylinder, Phys. Fluids, **19**, No.12, 1864-1875 (1976).

187 Leonard A., Energy cascade in large-eddy simulations of turbulent fluid flows, Adv. in Geophysics, **18A**, 237-248 (1974).

188 Li H., Finn F.M., Lovelace R.V.E. and Colgate S.A., Rossby wave instability of thin accretion disc, Ap. J. (2000).

189 Li X.L., A numerical study of three-dimensional bubble merger in the Rayleigh-Taylor instability, Phys. Fluids, **8**, 336-343 (1996).

190 Li X.L., Study of three-dimensional Rayleigh-Taylor instability, Phys. Fluids, **5**, 1904-1913 (1993).

191 Li X.L., Jin B.X. and Glimm J., Numerical study for the three-dimensional Rayleigh-Taylor instability through the TVD/AC scheme and parallel computation, J. Comp. Phys., **126**, 343-355 (1996).

192 Lin J.T. and Pao J.H., Wakes in Straitfield fluids, Ann. Review Fluid Mech., **11**, 317-338 (1979).

193 Linden P., Redondo J. and Youngs D., Molecular mixing in Rayleigh-Taylor instability, J. Fluid Mech., **265**, 97 (1994).

194 Lundgren T.S., Model equation for non-homogeneous turbulence, Phys. Fluid, **12**, No.3, 485-496 (1969).

195 Makita M., Miyawaki K. and Matsuda T., Two- and three-dimensional numerical simulations of accretion discs in close binaries systems, Monthly Notices Roy. Astron. Soc., **316**, 906-916 (2000).

196 Matsuda T., Sekino N., Shima E., Sawada K. and Spruit H., Mass transfer by tidally induced spiral shocks in an accretion disc, Astron. Astrophys., **235**, 211-218 (1990).

197 Marinak M.M., Glendinning S.G., Wallace R.J. and Remington B.A., Non-linear evolution of a three-dimensional multimode perturbation, Phys. Rev. Lett., **80**, 4426 (1998).

198 Message Passing Interface Forum, MPI: A message-passing interface standard, version 1.1 (June 1995).

199 Meyer F. and Meyer-Hofmeister E., On the elusive cause of cataclysmic variable outbursts, Astron. Astrophys. (Letters), **104**, L10 (1981).

200 Mezzacappa A., et al, An investigation of neutrino-driven convection and the core collapse supernova mechanism using multigroup neutrino transport, Astrophys. J., **495**, 911-930 (1998).

201 Molemaker M.J. and McWilliams J., Instability and equilibration of centrifugally stable stratified Taylor-Couette flow, Phys. Rev. Letts., **86**, No.23, 5270-5273 (2000).

202 Molteni D., Belvedere G. and Lanzfame G., Three-dimensional simulation of polytropic accretion discs, Monthly Notices Roy. Astron. Soc., **249**, 748-754 (1991).

203 Mottura L., Vigevano L. and Zaccanti M., An evaluation of Roe's scheme generalisations for equilibrium real gas flows, J. Comput. Phys., **138**, 354-399 (1997).

204 Nadyozhin D.K., Gravitational collapse of iron cores with masses 2 and 10 M, Astrophys. and Space Sci., **51**, 283-302 (1977).

205 NASA Press Release, N STScI-PR-97-03.

206 Naudasher E., Flow in the wake of self-propelled body and related sources of turbulence, J. Fluid Mech., **22**, No.4, 625-656 (1965).

207 Ofer D., Alon U., Shvarts D., McCroιy R.L. and Verdon C.P., Modal model for the non-linear multimide RTI, Phys. Plasmas, **3**, No.8, 3073-3090 (1996).

208 Oparin A.M., Numerical study of hydrodynamic instabilities, Computational Fluid Dynamics J., **10**, No.3: Special issue, 327-332 (2001).

209 Oparin A. and Abarzhi S., Three-dimensional bubbles in Rayleigh-Taylor instability, Phys. Fluids, **11**, No.11, 3306-3311 (1999).

210 Oparin A.M., Inogamov N.A. and Dem'yanov A.Yu., RT Turbulence: Dramatic dynamics of interpenetration (fast jets, sharp decelerations and accelerations), 8th International Workshop on the Physics of Compressible Turbulent Mixing, Pasadena, USA (2001).

211 Pekeris C.L., The propagation of pulse in the atmosphere, Prog. Roy. Soc., **171**, 434-449 (1939).

212 Penner J.A., Haselman L.C. and Edwards L.L., Smoke-plume distributions above large scale fires: Implications for simulations of "nuclear winter", J. Climate and Appl. Meteor., **25**, No.10, 434 (1986).

213 Pope S.B., The calculation of turbulent recirculating flows in general orthogonal coordinates, J. Comp. Phys., **26**, No.3, 197-217 (1978).

214 Porter D.H., Woodward P.R. and Jacobs M.L., Convection in Slab and Spherical Geometries, Annals of the New York Academy of Science, **898**, 1-20 (2000).

215 Jourdan G. and Houas L. (eds), Proceedings of 6th Intern. Workshop on the Physics of Compressible Turbulent Mixing, Marseille (1997).

216 Qian J., Variational approach to the closure problem of turbulence theory, Phys. Fluids, **26**, No.8, 2098-2104 (1983).

217 Read K.I., Experimental investigation of turbulent mixing by Rayleigh-Taylor instability, Physica D., **12**, No.1/3, 45-58 (1984).

218 Report on the 1998 HPF Users Group's annual meeting in Porto, IEEE Computational Science & Engineering, Vol. 5, No.3, Portugal (1998), pp.92-93.

219 Roe P.L., Approximate Riemann solvers, parameter vectors, and difference schemes, J. Comput. Phys., **43**, No.2, 357-372 (1981).

220 Roe P.L., Some contributions to the modelling of discontinuous flows, Lect. in the Appl. Math., **22**, 163-193 (1985).

221 Roe P.L., Characteristic-based schemes for the Euler equations, Ann. Rev. Fluid Mech., **18**, 337 (1986).

222 Roshko A., Experiments on the flow past a circular cylinder at very high Reynolds number, J. Fluid Mechanics, **10**, No.3, 345-356 (1961).

223 Sawada K., Matsuda T. and Hachisu I., Accretion shocks in a close binary system, Monthly Notices Roy. Anstron. Soc., **221**, 679-680 (1986).

224 Sawada K., Sekino N. and Shima E., Instability of accretion discs with spiral shocks, KEK Progress Report, **89**, 197-206 (1990).

225 Schneider M.B., Dimonte G. and Remington B., Phys. Rev. Lett., **80**, No.16, 1312 (1998).

226 Sharp D.H., An overview of Rayleigh-Taylor instability, Physica D., **12**, 3-18 (1984).

227 Shu C.W. and Osher S., Efficient implementation of essentially non-oscillatory shockcapturing schemes, J. Comput. Phys., **77**, 439-471 (1988).

228 Smagorinsky J., Manabe S. and Holloway J., Numerical results from a Ninelevel general circulation model of the atmosphere, Month. Weather Rev., **93**, 727-768 (1965).

229 Small R.D. and Heires K.E., Early cloud formation by large area fires, J. Appl. Meteorol., **27**, No.5, 654 (1988).

230 Snir M., Otto S., Huss-Lederman S., Walker D. and Dongarra J., MPI: The Complete Reference, MIT Press, New York (1996).

231 Structure and Mechanisms of Turbulence, I, II, Led. Notes Phys., Vol. 75, 76, Springer, Berlin etc.(1978).

232 Sunyaev R.A., Kaniovsky A.S., Efremov V.V., et al, Discovery of hard x-ray emission from supernova 1987A, Nature, Vol. 330, 227-229 (1987).

233 Taylor G., Proceeding Society L., Vol.157, 546 (1936).

234 Townsend A.A., The structure of turbulent shear flow, Emmanuel College, Cambridge (1956).

235 Trygvasson G. and Unverdi S.O., Computations of three-dimensional Rayleigh-Taylor instability, Phys. Fluids, **2**, 656 (1990).

236 Van Dyke M., Album of fluid motion, Pergamon Press, Standford, California, USA (1982).

237 Van Riper V.K., General relativisitic hydrodynamics and the adiabatic collapse of stellar cores, Astrophys. J., **232**, 558-571 (1979).

238 Wilson G.V., Practical parallel programming, MIT Press, New York (1995).

239 Wilson J.R. and Mayle R.W., Phys. Rep., **227**, 97 (1993).

240 Yabe T., Hoshino H. and Tsuchiya T., Two- and three-dimensional behaviour of Rayleigh-Taylor and Kelvin-Helmholtz instabilities, Phys. Rev. A., **44**, No.4, 2756-2758 (1991).

241 Yang J.Y. and Hsu C.A., High-resolution, non-oscillatory schemes for unsteady compressible flows, AIAA J., **30**, No.6, 1570-1575 (1992).

242 Youngs D.L., Numerical simulation of turbulent mixing by Rayleigh-Taylor instability, Physica D., **12**, 32-44 (1984).

243 Youngs D.L., Three-dimensional numerical simulation of turbulent mixing by Rayleigh-Taylor instability, Phys. Fluids, **3**, No.5, 1312-1320 (1991).

244 Zaets P.O., et al, Experimental study of the turbulent one-dimensional spectrum function in rotating pipe flow: Importing of the isotropic uniform turbulent model. Proc. of the V EPS Liquid State Conf., Moscow (1989), pp.33-36.

Index

L

M

N

O

P

R

S

T